Studies in Computational Intelligence

Volume 646

Series editor

Janusz Kacprzyk, Polish Academy of Sciences, Warsaw, Poland
e-mail: kacprzyk@ibspan.waw.pl

About this Series

The series "Studies in Computational Intelligence" (SCI) publishes new developments and advances in the various areas of computational intelligence—quickly and with a high quality. The intent is to cover the theory, applications, and design methods of computational intelligence, as embedded in the fields of engineering, computer science, physics and life sciences, as well as the methodologies behind them. The series contains monographs, lecture notes and edited volumes in computational intelligence spanning the areas of neural networks, connectionist systems, genetic algorithms, evolutionary computation, artificial intelligence, cellular automata, self-organizing systems, soft computing, fuzzy systems, and hybrid intelligent systems. Of particular value to both the contributors and the readership are the short publication timeframe and the worldwide distribution, which enable both wide and rapid dissemination of research output.

More information about this series at http://www.springer.com/series/7092

Kawa Nazemi

Adaptive Semantics Visualization

 Springer

Kawa Nazemi
Competence Center Information
   Visualization and Visual Analytics
Fraunhofer Institute for Computer Graphics
   Research (Fraunhofer IGD)
Darmstadt
Germany

ISSN 1860-949X   ISSN 1860-9503   (electronic)
Studies in Computational Intelligence
ISBN 978-3-319-80893-2   ISBN 978-3-319-30816-6   (eBook)
DOI 10.1007/978-3-319-30816-6

Printed on acid-free paper

This Springer imprint is published by Springer Nature
The registered company is Springer International Publishing AG Switzerland

The more I tried to make computers intelligent, the more I understood how intelligent humans are...

...to my loved ones, Mariam, Kian and Navid

Foreword

Every time I search online for certain textual information with my favorite search engine, I am quite happy that it knows my preferences. It probably knows more than I am aware of and uses this information to sort the search results in in order that is probably most appropriate to me. And to be honest, most of the time I am quite happy with that.

By now, searching text documents and presenting them is rather well established. The situation dramatically changes when we move to information in a generalized sense. Visualizing, browsing, and searching information pose additional problems when the preferences of the individual user are in focus. First of all, we now have to deal with visualizing sets of huge amounts of data of any kind. This is the typical area of information visualization. Second, objects of our searches usually have multiple meanings or relations. Here we enter the area of semantic visualization. Finally, presenting all results and allowing the user to browse and explore the data requires the possibility of continuous modification of the visualized results. This is being done in an adaptive visualization. The strength of this book lies in the combination of all the three areas. The presented model builds upon state of the art in information visualization, semantic visualization, and adaptive visualization. For each area an exhaustive review of literature is given. This gives a reader a good overview of the current status, but more importantly shows the need of the combination of them in order to be able to create satisfying adaptive semantics visualizations.

SemaVis is an example of such a system. Three different scenarios show the value of the approach. Instead of just searching for papers by author X, papers on Y, papers citing Z, one now has the full ability of combining all meta-information of several publishers and browse through relations papers and authors have. In this way, also new research trends can be identified. The approach is not limited to this more scientific field. It can be applied to any other area where we deal with huge semantic data. Another example is policy making, where the aim is to provide insight for nonexperts. So instead of new research trends, politicians can easily see the opinion of citizens—basically regarding any topic they can digitally react on via social media or more dedicated fora.

Is this "yet another approach"? That is a valid question that is answered in the last part of this book. The general "yet another approach" is often presented without validation, or with limited validation. To me, an additional strength of this book is the statistical evaluation of the concept. The model has been tested with more than enough participants in order to be able to put a firm and solid base under the experiments. Experiments are well designed and the evaluation is scientifically done. This is unfortunately far from trivial in many works, but taken care of appropriately in this book. They show that the model is significantly favorable above textual, nonadaptive, and partially adaptive visualizations, in terms of both attractiveness and effectiveness.

I strongly believe that this book will give the reader a good introduction into adaptive semantic visualization, its background, and usefulness.

January 2016   Prof. Dr. Arjan Kuiper

Preface

This book is a slightly revised version of my doctoral thesis supervised by Dieter Fellner, Chair of the Visual Computing Group of the Technische Universität Darmstadt and director of the Fraunhofer Institute for Computer Graphics Research (IGD) in Darmstadt, Rostock, Graz, and Singapore. The work is co-supervised by Stefan Wrobel, chair at the University of Bonn and director of Fraunhofer Institute for Intelligent Analysis and Information Systems (IAIS).

The original thesis is available through the digital libraries of the Eurographics Association and the Technical University Darmstadt. This version is revised in terms of the used figures in the state-of-art part. There are only figures used, which are either owned by me and my group or explicitly permitted by the owner. Further, a small part was added and revised in the literature review.

As the "Academia Europaea" awarded me with their Burgen scholarship, I was permitted to talk about several aspects of intelligent way of information processing by human and machines and the interplay of intelligence between human and machines.

I thank Janusz Kacprzyk for recognizing the value of my work and giving me the opportunity to republish it with Springer.

Darmstadt   Kawa Nazemi
December 2015

Acknowledgments

The research work presented in this thesis was carried out while working as researcher, project manager, and head of group at the Fraunhofer Institute for Computer Graphics Research (IGD) in a number of European, national, and industrial projects. Major parts of the research described in this thesis were conducted in the THESEUS Program, where I led the Core Technology Cluster for Innovative User Interfaces and Visualizations.

First and foremost, I am grateful to my supervisor Prof. Dr. Dieter Fellner for the invaluable support and inspiring guidance throughout the different stages of my work. Without his continuous support this work would not have been possible. Special thanks go also to Prof. Dr. Stefan Wrobel for acting as a co-referee. I am very grateful for the advice and support of Dr. Jörn Kohlhammer, who inspired me with his extensive knowledge in information visualization and visual analytics. PD. Dr. Arjan Kuijper supported me in organizing the thesis, publishing the intermediate results, and gave excellent advice for writing the thesis. I am very grateful that Arjan acted as my research coach and made the thesis possible.

My greatest thanks go to my former students that became colleagues and friends, Dirk Burkhardt, Matthias Breyer, Christian Stab, and Wilhelm Retz. They supported me in all situations with their excellent work, their theses, and their discussions. Without them, this thesis and SemaVis as a sustainable technology would not have been possible. I want to further thank all my colleagues from the department Information Visualization and Visual Analytics as well as my colleagues from the department 3D-Knowledge Worlds and Semantics Visualization at Fraunhofer IGD. Their continuous support and the inspiring working atmosphere unburdened the work on this thesis.

I thank all the students, who supported me during my work in the past years. In particular, I thank Reimond Retz, David Hoppe, and Maximillian Döpfmer. Further, I want to express my thanks to Dr. Oliver Christ and Prof. Dr. Constantin Rothkopf for supporting me in the evaluations and user studies in this thesis. My thanks also go to all the participants of the user studies.

I thank my friends and former colleagues, Dr. Michael Hellenschmidt and Prof. Dr. Eicke Godehardt, for supporting and motivating me. I also want to thank Dr. Ralf Schäfer for his continuous motivation during and after the THESEUS Program.

I am very glad that I had the opportunity to present my work on different conferences to get all the invaluable comments. In this context I would like to thank Cristina Conatti, Peter Brusilovsky, Paolo Buono, and Ben Steichen for their great comments and the rigorous discussions.

Finally, I thank my loved ones—my wife Mariam, my sons Kian and Navid, my brothers and parents for their unlimited support, love, and patience.

Contents

Abstract

Human access to the increasing amount of information and data plays an essential role in the professional-level and also in everyday life. While information visualization has developed new and remarkable ways for visualizing data and enabling the exploration process, adaptive systems focus on users' behavior to tailor information for supporting the information acquisition process. Recent research on adaptive visualization shows promising ways of synthesizing these two complementary approaches and make use of the surpluses of both disciplines. The emerged methods and systems aim to increase the performance, acceptance, and user experience of graphical data representations for a broad range of users. Although the evaluation results of the recently proposed systems are promising, some important aspects of information visualization are not considered in the adaptation process. The visual adaptation is commonly limited either to change visual parameters or to replace visualizations entirely. Further, no existing approach adapts the visualization based on data and user characteristics. Other limitations of existing approaches include the fact that the visualizations require training by experts in the field.

In this thesis, we introduce a novel model for adaptive visualization. In contrast to existing approaches, we have focused our investigation on the potentials of information visualization for adaptation. Our reference model for visual adaptation not only considers the entire transformation, from data to visual representation, but also enhances it to meet the requirements for visual adaptation. Our model adapts different visual layers that were identified based on various models and studies on human visual perception and information processing. In its adaptation process, our conceptual model considers the impact of both data and user on visualization adaptation. We investigate different approaches and models and their effects on system adaptation to gather implicit information about users and their behavior. These are then transformed and applied to affect the visual representation and model the human interaction behavior with visualizations and data to achieve a more appropriate visual adaptation. Our enhanced user model further makes use of the semantic hierarchy to enable a domain-independent adaptation.

    To face the problem of a system that requires to be trained by experts, we introduce the canonical user model that models the average usage behavior with the visualization environment. Our approach learns from the behavior of the average user to adapt the different visual layers and transformation steps. This approach is further enhanced with similarity and deviation analyses for individual users to determine a similar behavior at an individual level and identify differing behavior from the canonical model. Users with similar behavior get similar visualization and data recommendations, while behavioral anomalies lead to a lower level of adaptation. Our model includes a set of various visual layouts that can be used to compose a multi-visualization interface, a sort of "visualization cockpit." This model facilitates various visual layouts to provide different perspectives and enhances the ability to solve difficult and exploratory search challenges. Data from different data sources can be visualized and compared in a visual manner. These different visual perspectives on data can be chosen by users or can be automatically selected by the system.

    This thesis further introduces the implementation of our model that includes additional approaches for an efficient adaptation of visualizations as a proof of feasibility. We further conduct a comprehensive user study that aims to prove the benefits of our model and underscore limitations for future work. The user study with overall 53 participants focuses with its four conditions on our enhanced reference model to evaluate the adaptation effects of the different visual layers.

Chapter 1
Introduction

1.1 Motivation

Digital information resources are getting with every hour more complex, bigger, more decentralized, and more difficult to manage [1–3]. Users all over the world are putting digital information into digital resources, libraries publish primarily digital, and even 3D-objects are digitized and stored in decentralized data-sources. The term information overload is used for more than one decade to address the steadily increasing vast amount of data and information [2]. Beside the increasing amount of data, the structure of data and its complexity brought new challenges for research. Consequently information overload does not describe the digital information problem with all its facets. The variety, volume, and veracity of data emerged the new term "Big Data", to address beside the volume, the variety of data in distributed data-sources [4]. One of the main challenges of the vast and daily increasing amount of data is the human access to data. While analyzing, storing, and managing data can be processed commonly through the advances in new hardware technologies, the human access to data is dependent to a factor that is not easy to manage: the human factor.

The problem of human access to data was recognized and is investigated by a variety of disciplines and research areas. Information visualization and the related field of Visual Analytics investigates the human visual information processing [5–8] to provide an interactive picture of the data to amplify human's cognition and provide insights and knowledge [1, 9, 10]. For information visualization, the aspect of human perception and visual information processing is a matter of research. How can data be transformed to interactive graphical representations that amplify cognition, support the information acquisition process, and consequently the acquisition of knowledge? Visual Analytics investigates further the manipulation of data-analysis and transformation to provide unexpected patterns and thereby new insights [11, 12]. In both research fields the way from data-oriented visualization to a more human-centered information presentation plays a key role. Thereby two main aspects were proclaimed to enhance the interactive visual picture of data: *"The key challenge for*

© Springer International Publishing Switzerland 2016
K. Nazemi, *Adaptive Semantics Visualization*, Studies in Computational
Intelligence 646, DOI 10.1007/978-3-319-30816-6_1

visual analytics is to derive semantic content or meaning from images in real time"
[13, p. 112]. Thereby the inclusion of the semantics or context in information visualization [11] and Visual Analytics, respectively played already several years ago an
essential role. The role of semantic and the related acquisition of meaningful sentences and information [13], is one key challenge towards a more human-centered
visual representation of data. Moreover, the human as an implication and decision
factor for information visualization was placed in the foreground of research [1].
The increased involvement of user's intentions and preferences in the process of
information visualization got more important. Thereby the adaptation of information visualization systems by developing "*novel interaction algorithms incorporating
machine recognition of the actual user intent and appropriate adaptation of main
display parameters such as the level of detail, data selection, etc. by which the data
is presented*" [12, p. 162] was proposed as a main challenge of research. The proposed human-centered research challenges evoked new approaches and technologies
to ease the human access to data.

In contrast to information visualization with the main goals of amplifying cognition and providing a more exploratory way of information retrieval, semantic technologies aim at formalizing data as a "*conceptualization of knowledge*" [14]. As the
World Wide Web provides a crucial information resource, BERNERS- LEE proposed
the idea of a Web of Data that enables the access to the resources with sense of
"meanings" as *Semantic Web* [15, 16]. The main idea of Semantic Web is to formalize data and information in a machine-readable way [16]. The formalization aimed at
making the web "meaningful" based on a formalized notation of content followed by
a formalization of the underlying structure to provide a rule and meaning inference
for making the Web accessible for human and computer [17]. As the formal logical
representation of data as ontologies still exists, the more promising and disseminated way of knowledge formalization of *Linked-Data* occupied the Web [18–20].
With the broad dissemination of Linked-Data, Semantic Web has gained a lightly
differing character of a Web with interlinked and meaningful resources [21, 22].
Although, Linked-Data opened new ways for acquiring information and knowledge,
the related human-centered technologies are primarily aiming at providing answers
to questions that can be verbalized by human and require therefore prior knowledge
of a certain domain. Semantics visualizations are commonly designed for ontology
visualization and ontology engineering. The process of information acquisition in an
exploratory manner [23, 24] does not play any role for today's semantics visualization approaches. The semantics visualizations are focusing far more on overviewing
the data, rather than on navigating through the conceptualized information. They
remain on the abstract level of ontological concepts and do not provide a search or
information acquisition paradigm. Consequently, the proposed challenges to include
semantics for a more efficient way of visualizing information and amplifying users'
cognition was not yet responded by the research community.

As the research on semantics visualization did not brought sufficient solutions for
acquiring information through Linked-Data on Web, a new interdisciplinary research
area of adaptive information visualization emerged from the fields of adaptive systems and information visualization [25–27]. Adaptive systems provide a useful and

promising way to face in particular the variety of users [27, 28], context [29], and data [30] with adaptive methods that reduce human effort in complex information acquisition tasks. The main idea of adaptive systems can be summarized with helping users to achieve their intended tasks faster, easier, or with better results [31] through the support of system-use [32, 33] or information acquisition [34–36]. The general process of adaptation can be summarized by the acquisition of relevant information (influencing factors), the formal representation of this information, and the production of certain changes of the system behavior [37]. Adaptive systems dispose of a comprehensive pool of methods, systems and algorithms for recognizing and analyzing user related information. With these methods adaptive systems facilitate the handling with complex information and support users during their work process [29, 38]. Different existing systems e.g. intelligent help systems [39], personalization of web page navigation [32] or learning systems [34] are already using these methods and tailor the user interface to influencing factors, those information that influence the behavior, appearance, or view of a system.

Adaptive information visualization combines the areas of information visualization and adaptive systems to provide personalized and enhanced visualization. Recent research in adaptive visualizations showed significant advances in human information processing [40–42]. The adaptation techniques were in particular adopted to search and exploration tasks [3, 40]. The evaluation results of the implemented adaptive visualizations are promising, whereas the applied methods vary enormously [3, 42, 43]. Although this young research area has already provided interesting and promising approaches, a review on the last decade of developed systems and approaches in adaptive visualizations shows shortcomings and limitations. A first limitation refers the use of different influencing factors in adaptive visualizations. In information visualization two main aspects plays a key-role for a sophisticated design, the user with her visual abilities, prior knowledge, and aptitudes; and the main characteristics of data [8, 10]. The adaptation of existing systems is either affected by data [30, 44] or by user [26, 27]. A system or approach that adapts based on both influencing factors could not be found. The second limitation refers to the training of such self-learning adaptive visualizations. The systems and approaches that are adapting to users' characteristics have to be trained by visualization experts [43]. With each new visual layout the entire system have to be trained with commonly static behavioral patterns as repeated interaction sequences. To our best of knowledge there exists no method that replaces a system-training by experts. The third and in our opinion main limitation is that the transformation pipeline of data to visual representation is not considered in today's approaches. Although, there are many studies of visual perception, reference models for information visualization, and a huge treasure of methods, applications and their effects to human perception, the outcomes of these decades of work [5, 10, 12, 45] are not reflected in today's adaptive visualization approaches. Our review clearly signals that the emerging area of adaptive visualizations did not investigate the human interface adaptation in depth. The most systems are replacing visualization types based on some users' implicit or explicit demands. The focus of today's systems is more *to what* should be adapted

rather than *what can be adapted*. None of the today's systems adapts the entire range
of possible visual layers.

The young research field of adaptive visualization made impressive advances
and provided promising approaches. However, a coherent model that investigates
the potentials of information visualization with its various variables that influence our
perception and consequently the information acquisition is missing completely. The
transformation steps from data to interactive visual representation are not investigated
in the entire research field, even though these are the fundamentals of information
visualization.

1.2   Research Goals

In this thesis, we present a novel and coherent model for adaptive visualization for
information acquisition from distributed semantic data sources. In contrast to existing systems and approaches, we investigate in particular the potentials of information
visualization for adaptation. Our reference model for visual adaptation considers not
only the entire transformation pipeline [10] from data to visual representation. It
enhances far more the reference model to meet the requirements for adaptive visualizations. Our model provides an adaptation on different visual layers and enhances
the state of research. Each of the identified layers can be adapted automatically by
various influencing factors. The transformation steps from data to visualization are
enhanced to provide a fine granular adaptation of visual parameters. To identify the
visual layers that affect the human information processing, we investigate various
models and studies on visual perception. We further review the existing interaction
techniques, visualization methods, data types, and visualization tasks as foundations
for our model. In this context, the differentiation of visual layers and their effects on
human visual perception is of great interest for our research.

Our conceptual model adapts the visual representation of data not only to users'
characteristics. It considers in its adaptation process both influencing factors: data and
user. In this context, we investigate various existing classifications of data. Thereby,
our research focuses not only on semantic data. The entire data-types and categories
are considered as foundation for our reference model. For gathering implicit information about users and their behavior, we investigate different approaches and models
of interaction analysis and their effects on system adaptation. These are than transformed and applied to affect the visual representation based on combined models
that represent user and data. We introduce in this context an improved interaction
prediction algorithm that is used to load data on demand from Web repositories,
before the user selects a data entity. Further the prediction algorithm is used to
guide the attention of users to recommended content. The user model introduced
in this thesis combines the interaction behavior of users with the characteristics of
data and the content. The enhanced user model further makes use of the semantic
hierarchy to enable a domain-independent adaptation. We introduce in this context a
formal representation of users' behavior with data, visual layouts, and content. With a

subsumption approach on semantic concept level, the domain independent adaptation is achieved. Thereby the entire semantic structure is still part of the model to ensure a more detailed adaptation within a knowledge domain.

The conceptual design proposed in this thesis includes further the approach of a canonical user model that models the average usage behavior of users with the visualization environment. With this approach, an initial training of the visual environment by experts is not required. The visual environment learns from the average user behavior and adapts the entire visual transformation steps to the canonical user. This approach is further enhanced with similarity and deviation analysis of individual users. As the canonical user model represents the average usage behavior and provides a general adaptation to all users, users are able to login as individuals. Our enhanced approach measures the similarity of users and the deviation of the individual user to the canonical user. Based on these measurements a more personalized visual adaptation is possible. Users with similar behavior get visualization recommendations from similar user. Users with the interaction behavior that differs from the canonical user model, gets less adaptation based on the canonical user model and more through their individual as soon as it contains enough information. With our canonical user model and the related measurements, we provide an approach that addresses not only the initial training of systems by experts. It further provides a step towards solving the *new user* and *new context* problem.

Our model includes a set of various visual layouts that can be composed to a multi-visualization interface. The related visualization cockpit model enables the orchestration of visual layouts linked to semantic data-bases and interlinked with each other. In this context, we investigate models and approaches of *exploratory search* [23, 24, 46] and provide a conceptual design that supports the entire process of exploratory search based on semantic data. We identify different visual orchestration methods to enable solving analytical tasks by providing different perspectives on the same data, the same perspective on different data, or different perspectives on different data. Our visualization cockpit model enhances the traditional brushing and linking approach in information visualization by dislinking visual layouts from each other or from certain data-bases. The model further supports a simultaneous visualization of data from different data-sources or sub-sets of data from the same data-source. The visualization cockpit model enhances the adaptive behavior of our visualization environment by an automatic adding, dismissing or rearranging of visual layout on the so called visual interface. The adaptation and interlinking with data can be controlled by users too, whereas the appropriate visual layouts are recommended to support the users in an unobtrusive way.

Another research aspect of our work is the visualization of semantics. In this context, we investigate the different formalisms and data structures of semantics and the way how they are accessed by human. A comprehensive review on existing semantics visualizations enlightens limitations in existing systems. We clearly illustrate that existing semantics visualization does not support the mentioned paradigms of exploratory search. To face this challenge, we introduce first an approach that includes semantics resulted entities in semantic data-bases by iterative querying. The semantic structure and quantitative measures on the data builds our data model that is the foundation of the semantics visualization. We apply further our approach for non-semantic data-bases that returns just metadata as results and generate the semantics by our iterative querying approach. To visualize the relevance of the semantic neighbors of an entity, we introduce two algorithms that measure the contextual relevance of the semantic neighbors. The measured values are used to adapt the visual variables that guide the users' attention to certain data-entities.

We introduce in this thesis various models, approaches, and algorithms that enhances the idea of adaptive semantics visualization. We focus thereby on a replicable way of description and illustration of all our models. The theoretical approaches and models are the foundations of our work, but they need to be verified in terms of feasibility and added values. To prove the feasibility of our conceptual model, we introduce the so called SemaVis technology that implements the conceptual model of this thesis. We describe the technical interplay of the components based on a Model View Controller design pattern. In this context, we illustrate the main characteristics of SemaVis, a distributed system that can be used as single-client or client-server application. To demonstrate the implemented functionalities, we introduce three application scenarios with different data-bases, goals, and target audience. The main goal here is to demonstrate that SemaVis and consequently our conceptual model can be applied to different domains with its adaptive behavior. Beside the proof of feasibility, we conduct an empirical user study of the implemented system. The conditions in our study were chosen based on our reference model to validate the differentiation of the identified visual layers. Further two different task-types, were evaluated: simple and exploratory tasks according to the definition of exploratory tasks. We illustrate in our empirical user study that our conceptual model outperforms the tested conditions in terms of effectiveness, efficiency, cognitive effort, and satisfaction.

1.3   Contributions

This thesis investigates the adaptation of information visualization for distributed semantic Web data. The overall objective is to contribute with conceptual and technological advances for a more sophisticated and comprehensive adaptation of visualization based on user and data characteristics. The target audience is the research community of adaptive visualizations, whereas the researchers in the area of information visualization are addressed too. This section outlines the main contributions of this thesis in a comprehensible manner.

Overall Conceptual Model The comprehensive review on existing approaches for adaptive and semantics visualization illustrated clearly different gaps and limitations in both research areas. The conceptual model, as our main contribution, addresses the identified limitations and provides a novel model for adapting semantics visualizations based on user and data characteristics. Thereby the surpluses of existing models are used and combined with new approaches to provide a more reliable adaptation model. The conceptual model contains four main layers of influencing factors, knowledge model, process of adaptation, and visual adaptation. Each of these layers contains further components and models that enhance the existing approaches for visual adaptation.

Reference Model for Adaptive Visualization The transformation steps from data to visual representation are not investigated in today's adaptive visualization app roaches. We contribute here with a reference model for adaptive visualization that investigates all transformation steps for adaptation and enhances these with further relevant steps to provide a fine-granular adaptation. The reference model is based on a prominent and widespread model. Our enhancements for adaptation are based on models and study results from human visual perception. The reference model contains four adaptation layers, *Semantics*, *Visual Layout*, *Visual Variable*, and *Visual Interface* and includes the transformation steps of *data transformation*, *visual mapping*, *retinal variable mapping*, and *visual layout orchestration*. Beside the transformation steps, the four layers can be adapted by the conceptual model and the included adaptation processes. The main contribution here is the advanced reference model that can be applied to any kind of visual adaptation and enhances the state of the art with the various levels of adaptation based on human visual perception.

User Model Existing adaptive visualization approaches do not comprise data and users as influencing factor for the adaptation process. Further, system training by visualization experts is required to model the adaptation effects. We propose in this thesis a user model that **comprises both data and user** for the adaptation process. The user model includes thereby the combined interaction behavior with data and visual layouts. With the subsumption on concept-level, we further enhance existing approaches for user modeling to a **domain-independent** model. Trained user models in certain knowledge domains can be used for adaptation in other knowledge domains too. The introduced user model further makes use of the semantic hierarchy of data. Within a certain knowledge domain, the model provides **conceptual information** that leads to recommend data from the same semantic concepts. Beside the behavioral analysis of users an enhanced prediction algorithm is introduced that enables the guidance of users' attention to data or load not selected data, due to the prediction measures. One main concept of our user model is the appliance of a **canonical user model** that represents the average behavior of all users and leads to a general adaptation of the visualization environment without the necessity of an expert to train the system. A similarity algorithm measures the **similarity between users** and recommend in case of similar behavior to fill the gap of the user model with the information of other users. With our deviation analysis, the **differences between the canonical and individual** are measured. A differing behavior results in less adaptation based on the canonical user model.

Visualization Cockpit Model Existing semantics visualizations scarcely investigate the visualization of search results, whereas the process of exploratory search is not supported to our best of knowledge. We contribute in this thesis with our visualization cockpit model that aims at supporting the entire exploratory search process. Thereby visual layouts can be composed by the user or by the adaptive system in a juxtaposed manner on the visual interface to provide different perspectives on the same data, same or different perspectives on different sub-set of data, and even same or different perspectives on data from different data-sources. The main contribution is an enhanced brushing and linking metaphor that enables the placement, rearrangement, and displacement of visual layouts on screen. Each visual layout can be interlinked with another layout or with a data-base. The visualization cockpit model enables solving analytical and comparative tasks.

Proof of Feasibility: The SemaVis Technology We introduce in our thesis various model, approaches, and algorithms on a replicable but more theoretical level. It is therefore necessary to prove the feasibility of our conceptual model. To prove the feasibility, we introduce as one contribution the architectural design of the SemaVis technology that implements major parts of our conceptual model. With the implementation, we provide further three application scenarios, in which SemaVis were applied, digital library, Web-search, and policy modeling. We chose this way of introducing the technology to provide a comprehensible illustration of the system behavior. The main contribution here is not only the proof of feasibility but also the illustration of the adaptive system behavior.

Empirical User Study To validate our assumption that our conceptual model leads to significant advances in adaptive visualizations; we conducted a user study with 53 participants, four conditions, two interventions, and a total number of 40 tasks. For this purpose, we developed an evaluation-software that measured the task completion time, task correctness automatically and guided the users through the entire evaluation scenario. The study was conducted as a within-subject Latin-Square design. An 'a priori' power analysis was performed to measure the required sample size. The conditions were applied to our reference model for adaptation to investigate the effects of the different visual layers in adaptive visualizations. Beside a performance measure in terms of efficiency and effectiveness, two questionnaires were used to evaluate the perceived attractiveness, effort, cognitive load, and intuitiveness. For each measured valued a repeated measure ANOVA with pairwise t-Tests were applied to measure the significance of each condition. One of the main contributions of our thesis is the evaluation with the evaluation design and the results that give an insight on different performance and perceived satisfaction with our adaptive system.

Beside the introduced main contributions, we introduced various further novel approaches, algorithms, and models. An example is the iterative querying approach that enables the inclusion of semantics based on resulted semantic entities or the appliance of the iterative approach for metadata instead of semantic data. Further two algorithms were introduced that measure the contextual relevance of selected semantic instances to enable visualizing contextual relevance. We contributed further with novel visualization approaches that are integrated in our model as visual layouts and enable the adaptation on different levels. Listing all the contribution would go beyond the scope of this section. The thesis was partially published in various peer-reviewed journals, conference proceedings, book chapters, and work-shop proceedings. Interested audience finds a **record of publications** in Sect. A in the appendix of this thesis.

1.4 Dissertation Roadmap

This thesis is structured in three main parts of *Literature Review and State of the Art*, *Model for Adaptive Semantics Visualization*, and *Proof of the Conceptual Model* as illustrated in Fig. 1.1. The first part of the thesis starts with a chapter about informa-tion visualization (Chap. 2) as a canonical foundation. The main goal is to give an overview of the various disciplines, techniques, goals, and approaches that are cou-pled to interactive information visualization. In particular the investigation of human visual perception, visualization tasks, and data models are of great interest for our conceptual design. In this chapter, we introduce the reference model for information visualization, the differentiation of visual layers, and models of visual perception that are the foundations of our conceptual design. Further, we introduce in this chapter a high-level task classification based on existing classifications that will be used for our reference model.

Chapter 3 will give a short overview of the idea of semantic web and its technolo-gies. The main goal of this chapter is to give a comprehensive and comprehensible state of art and technology for semantics visualizations. For obtaining a clear pic-ture of the existing systems and approaches, we will first define the term *semantics*

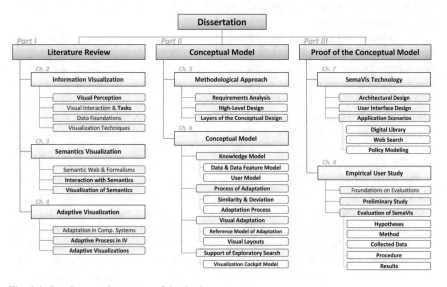

Fig. 1.1 Roadmap and structure of the thesis

visualization. With this definition of semantics *ontology visualizations* are a subset of semantics visualization and thereby part of our review. We will further introduce a classification of semantics visualization for providing a comprehensible picture of the existing systems. Our state of art review will cover the last decade, whereas the existing systems will be introduced based on our classification. The foundation of our review will be established exploratory search models. Our review will outline that none of the existing systems supports the exploratory search process, although semantics is predestinated to support this kind of search.

Chapter 4 will investigate in particular the adaptation process in information visualization. To provide a comprehensible way for conveying different adaptation processes, we will introduce three main aspects: *influencing factors* by means of *to what can visualizations be adapted*, *knowledge modeling* that refers to the way how the influencing factors can be formalized (represented) and which factors may play a role for the adaptation process, and *human interface adaptation* that refers to visualization and their capabilities for adaptation. The main goal of this chapter is to give a comprehensive and comprehensible state of art analysis for adaptive visualizations. For this purpose we will first define adaptive visualization based on the definition of adaptive systems and the definition of information visualization. Our review on the existing systems will cover again the last decade. The goal is to find systems or approaches that make use of all the defined adaptation criteria, but at least combine some of them to provide a real benefit out of the visual structures. Our review will clearly signal that the emerging area of adaptive visualizations did not investigate the human interface adaptation in depth, yet. The most systems are replacing visualization types and layouts respectively based on some users' implicit or explicit demands. The focus of today's systems is more *to what* should be adapted rather than *what can be adapted*. With this chapter, we conclude the review on existing systems and approaches.

The second part of our thesis introduces our conceptual model. Chapter 5 will summarize some of the main outcomes of our literature review and propose based on the identified potentials, requirements that should be fulfilled to provide scientific and technological advancements in adaptive visualizations. Therefore, we first identify the requirements that build the foundation on the conceptual work. Thereafter a high-level design of our conceptual model will be presented. The high-level design aims at giving a short and comprehensible overview of our main intentions and related contributions. This chapter can be seen as a roadmap of the detailed descriptions, algorithms and models of our conceptual model.

Chapter 6 introduces based on the high-level design a detailed and replicable illustration of our conceptual model. First the knowledge model with its three main components of data model, data feature model, and user model will be introduced. Thereby data model will describe the way semantic information is gathered from Web-sources and from non-semantic metadata. Here the approaches of iterative querying will be described that leads to a formal representation of data as data model. Data feature model will illustrate the retrieving of quantitative measure of the underlying data with the same iterative querying approach. In this context two weighting-algorithms that measure the relevance of semantic neighbors of selected instances will be introduced. Thereafter, the user model and the related concepts will be introduced that combines the interaction behavior of users with data and visual layouts. Based on a formal specification of users' interactions the approach for determining and weighting user behavior and predicting users' action will be described. In this context the formal description of the canonical user model and the group definition will be introduced followed by user similarity and deviation analysis. Then the general adaption process will be described that guides through the entire process of adaptation and illustrates when and how the measured values and models are applied. Thereafter we will introduce our layer based reference model of adaptation. Therefore, we first introduce an abstracted task model for information visualization and the different models of visual perception. The chapter concludes with the description of our visualization cockpit model and illustrates how this model can be applied to support the exploratory search with juxtaposed visual layouts.

The last part of this thesis will introduce the proof of our conceptual model. Therefore Chap. 7 we will first introduce the architectural design of our SemaVis technology that implements major parts of our conceptual model. The architectural design of SemaVis will be described based on the MVC design pattern. SemaVis as a visualization technology enables visualizing various data-types, adapting to various influencing factors, and provides the functionalities described in our conceptual model. SemaVis is implemented as client-server technology, whereas it can be used as a client application or compiled as desktop application with limited functionalities. The general architecture aims at providing the technical interplay of the introduced approaches, algorithms, and models. It gives an overview of the implementation strategy and enables a mapping to the already introduced high-level design. Beside the general architecture of SemaVis three exemplary application scenarios will be introduced. The main goal of the application scenarios is to provide a proof of feasibility and an insight of the adaptive behavior of the system.

Chapter 8 introduces the empirical study on our approach with an evaluation as a controlled experiment. We will start with a general introduction into the topic of evaluation with a theoretical overview of the underlying psychological methods. Thereafter a preliminary pilot study on evaluating only the effects of visual variables in context of information search will be introduced that was performed together with the psychological department of the Technische Universität Darmstadt. The main goal here was to find out, if the visual variables in terms of color and size have already an effect on search efficiency and enable us to identify appropriate questionnaires, limitations, and shortcomings. Although, the number of participants was limited to just 14, an effect to task completion and therewith to effectiveness could be observed. The small sample size led to big standard errors. Further limitations and shortcomings were identified to be eliminated in the main evaluation. The main evaluation was conducted based on the implemented SemaVis and the application scenario of digital libraries. We used a within-subject Latin-Square group design with four conditions. The conditions were aligned to our reference model to evaluate the effect of the adaptation of the different visual layers. Overall 53 persons participated in the evaluation in a time-period of two weeks in a laboratory of Fraunhofer IGD. To reduce human errors, we implemented an evaluation-software that collected data and guided the users through the evaluation. Overall four hypotheses and nine sub-hypotheses were deduced to measure beside performance in terms of efficiency and effectiveness, cognitive load and effort, and intuitiveness with two questionnaires. The results of the evaluation illustrate that all our assumptions are confirmed. The full-adaptive semantics visualization is more efficient, more effective, leads to less cognitive load and effort and to higher satisfaction and user experience. In all our hypotheses, the full adaptive SemaVis outperformed the non-adaptive visualization, the partially adaptive visualization, and the textual baseline.

References

1. D.A. Keim, F. Mansmann, J. Schneidewind, H. Ziegler, J. Thomas, *Visual Data Mining: Theory, Techniques and Tools for Visual Analytics*. Lecture Notes in Computer Science (LNCS) (Springer, 2008)
2. D. Bawden, L. Robinson, J. Inf. Sci. **35**(2), 180 (2009). doi:10.1177/0165551508095781. http://dx.doi.org/10.1177/0165551508095781
3. J.W. Ahn, Adaptive visualization for focused personalized information retrieval. Ph.D. thesis (School of Information Sciences, University of Pittsburgh, 2010). http://etd.library.pitt.edu/ETD/available/etd-08262010-150850/
4. A. Katal, M. Wazid, R. Goudar, in *2013 Sixth International Conference on Contemporary Computing (IC3)* (2013), pp. 404–409. doi:10.1109/IC3.2013.6612229
5. A.M. Treisman, G. Gelade, Cogn. Psychol. **12**(1), 97 (1980)
6. R.A. Rensink, Annu. Rev. Psychol. **53**, 245 (2002)
7. J.M. Wolfe, W. Gray (ed.), *Integrated Models of Cognitive Systems* (2007), pp. 99–119
8. C. Ware, *Information Visualization Perception for Design* (Morgan Kaufmann (Elsevier), 2013)
9. D. Keim, J. Kohlhammer, G. Ellis, F. Mansmann, *Matering the Information Age Solving Problems with Visual Analytics* (Eurographics Association, 2010)
10. S.K. Card, J.D. Mackinlay, B. Shneiderman, *Readings in Information Visualization: Using Vision to Think*, 1st edn. (Morgan Kaufmann, 1999). http://www.amazon.com/exec/obidos/redirect?tag=citeulike07-20&path=ASIN/1558605339
11. J. Thomas, in *2007 6th International Asia-Pacific Symposium on Visualization, 2007. APVIS'07* (2007), p. ix
12. D. Keim, G. Andrienko, J.D. Fekete, C. Görg, J. Kohlhammer, G. Melançon, in *Information Visualization*. Lecture Notes in Computer Science, vol. 4950, ed. by A. Kerren, J. Stasko, J.D. Fekete, C. North (Springer, Berlin/Heidelberg, 2008), pp. 154–175
13. J.J. Thomas, K.A. Cook, *Illuminating the Path: The Research and Development Agenda for Visual Analytics* (National Visualization and Analytics Ctr, 2005). http://www.worldcat.org/isbn/0769523234
14. T.R. Gruber, Knowl. Acquisition **5**(2), 199 (1993). http://dx.doi.org/10.1006/knac.1993.1008. http://www.sciencedirect.com/science/article/pii/S1042814383710083
15. T. Berners-Lee, Semantic web roadmap. W3C Online Publishing on http://www.w3.org/DesignIssues/Semantic.html (1998)
16. T. Berners-Lee, Weaving the Web: The Original Design and Ultimate Destiny of the World Wide Web (HarperBusiness, 2000)
17. T. Berners-Lee, J. Hendler, O. Lassila, Sci. Am. (2001). http://www.scientificamerican.com/1999/0599issue/0599bosak.html
18. C. Bizer, T. Heath, K. Idehen, T. Berners-Lee, in *Proceedings of the 17th International Conference on World Wide Web, WWW 2008*, Beijing, China, 21–25 Apr 2008 (2008), pp. 1265–1266
19. C. Bizer, T. Heath, T. Berners-Lee, Int. J. Semant. Web Inf. Syst. **5**(3), 1 (2009)

20. T. Heath, C. Bizer, *Linked Data—Evolving the Web into a Global Data Space*. Synthesis Lectures on the Semantic Web: Theory and Technology (Morgan & Claypool Publishers, 2011)

21. P.N. Mendes, M. Jakob, C. Bizer, in *Proceedings of the Eight International Conference on Language Resources and Evaluation (LREC'12)* (Istanbul, Turkey, 2012)

22. Google Press Center. The Knowledge Graph (2013). http://www.google.com/intl/en/insidesearch/features/search/knowledge.html. Accessed Aug 2013

23. B.S. Bloom, *Taxonomy of Educational Objectives* (David McKay Co., Inc., NY, New York, 1956)

24. G. Marchionini, Commun. ACM **49**(4), 41 (2006). doi:10.1145/1121949.1121979. http://doi.acm.org/10.1145/1121949.1121979

25. C. Conati, G. Carenini, M. Harati, D. Tocker, N. Fitzgerald, A. Flagg, in *AAAI Workshops* (2011). http://www.aaai.org/ocs/index.php/WS/AAAIW11/paper/view/3944

26. J.W. Ahn, P. Brusilovsky, Inf. Process. Manage. **49**(5), 1139 (2013). doi:10.1016/j.ipm.2013.01.007. http://www.sciencedirect.com/science/article/pii/S0306457313000137

27. G. Carenini, C. Conati, E. Hoque, B. Steichen, D. Toker, J. Enns, in *Proceedings of the SIGCHI Conference on Human Factors in Computing Systems, CHI'14* (ACM, New York, NY, USA, 2014), pp. 1835–1844. doi:10.1145/2556288.2557141. http://doi.acm.org/10.1145/2556288.2557141

28. B. Steichen, O. Schmid, C. Conati, G. Carenini, in *UMAP 2013 Extended Proceedings. First International Workshop on User-Adaptive Visualizations (WUAV 2013)* (2013)

29. M. Hartmann, Context-aware intelligent user interfaces for supporting system use. Ph.D. thesis (Technische Universität Darmstadt, 2010)

30. J. Mackinlay, P. Hanrahan, C. Stolte, IEEE Trans. Vis. Comput. Graph. **13**(6), 1137 (2007). doi:10.1109/TVCG.2007.70594. http://dx.doi.org/10.1109/TVCG.2007.70594

31. E. Ross, Intelligent user interfaces survey and research directions. Technical report (University of Bristol, Bristol UK, 2000)

32. P. Brusilovsky, in *The Adaptive Web*, ed. by P. Brusilovsky, A. Kobsa, W. Nejdl (Springer, Berlin, Heidelberg, 2007), pp. 263–290. http://dl.acm.org/citation.cfm?id=1768197.1768207

33. P. Brusilovsky, in *Proceedings of the Companion Publication of the 23rd International Conference on World Wide Web Companion, WWW Companion'14* (International World Wide Web Conferences Steering Committee, Republic and Canton of Geneva, Switzerland, 2014), pp. 1075–1076. doi:10.1145/2567948.2580052. http://dx.doi.org/10.1145/2567948.2580052

34. P. Brusilovsky, E. Millán, User models for adaptive hypermedia and adaptive educational systems, in *The Adaptive Web*, ed. by P. Brusilovsky, A. Kobsa, W. Nejdl (Springer, Berlin, Heidelberg, 2007), pp. 3–53. http://dl.acm.org/citation.cfm?id=1768197.1768199

35. A. Jameson, in *The Human-Computer Interaction Handbook: Fundamentals, Evolving Technologies and Emerging Applications*, 2nd edn., ed. by A. Sears, J.A. Jacko (CRC Press, Boca Raton, FL, 2008), pp. 433–458

36. J. Guerra, S. Sosnovsky, P. Brusilovsky, in *Scaling Up Learning for Sustained Impact*. Lecture Notes in Computer Science, vol. 8095, ed. by D. Herndez-Leo, T. Ley, R. Klamma, A. Harrer (Springer, Berlin, Heidelberg, 2013), pp. 125–138. http://dx.doi.org/10.1007/978-3-642-40814-4_11

37. A. Kobsa, J. Koenemann, W. Pohl, Knowl. Eng. Rev. **16**, 111 (2001)

38. D. Parra, P. Brusilovsky, C. Trattner, in *Proceedings of the 19th International Conference on Intelligent User Interfaces, IUI'14* (ACM, New York, NY, USA, 2014), pp. 235–240. doi:10.1145/2557500.2557542. http://doi.acm.org/10.1145/2557500.2557542

39. J. Noguez, L.E. Sucar, in *ENC'05: Proceedings of the Sixth Mexican International Conference on Computer Science* (IEEE Computer Society, Washington, DC, USA, 2005), pp. 2–9. doi:10.1109/ENC.2005.7

40. J.W. Ahn, P. Brusilovsky, Inf. Vis. **8**(3), 180 (2009). http://ivi.sagepub.com/cgi/content/short/8/3/167

41. D. Toker, C. Conati, G. Carenini, M. Haraty, in *User Modeling, Adaptation, and Personalization*. Lecture Notes in Computer Science, vol. 7379, ed. by J. Masthoff, B. Mobasher, M. Desmarais, R. Nkambou (Springer, Berlin, Heidelberg, 2012), pp. 274–285. doi:10.1007/978-3-642-31454-4_23. http://dx.doi.org/10.1007/978-3-642-31454-4_23

42. B. Steichen, G. Carenini, C. Conati, in *Proceedings of the 2013 International Conference on Intelligent User Interfaces, IUI'13* (ACM, New York, NY, USA, 2013), pp. 317–328. doi:10.1145/2449396.2449439. http://doi.acm.org/10.1145/2449396.2449439

43. D. Gotz, Z. When, J. Lu, P. Kissa, N. Cao, W.H. Qian, S.X. Liu, M.X. Zhou, in *Proceedings of the First International Workshop on Intelligent Visual Interfaces for Text Analysis, IVITA'10* (ACM, New York, NY, USA, 2010), pp. 1–4. doi:10.1145/2002353.2002355. http://doi.acm.org/10.1145/2002353.2002355

44. J. Mackinlay, ACM Trans. Graph. **5**, 110 (1986). http://doi.acm.org/10.1145/22949.2295

45. J. Bertin, *Semiology of Graphics* (University of Wisconsin Press, 1983)

46. R.W. White, R.A. Roth, *Exploratory Search: Beyond the Query-Response Paradigm*. Synthesis Lectures on Information Concepts, Retrieval, and Services, vol. 1, ed. by G. Marchionini (Morgan & Claypool Publishers, 2009). doi:10.2200/s00174ed1v01y200901icr003. http://dx.doi.org/10.2200/s00174ed1v01y200901icr003

Part I
Literature Review and State of the Art

Chapter 2
Information Visualization

This chapter introduces information visualization as a canonical foundation of this thesis. We will first try to differentiate information visualization from related areas. The goal is to have a common understanding of the term information visualization in context of this work. Thereafter we outline the interdisciplinary character of information visualization. For this we start with the human and introduce various models and research outcomes on visual perception. We will continue with our human centered view on information visualization and describe classifications for interaction with information visualization. Based on an appropriate classification for our purposes, we will describe the interaction with application examples. Thus interactive visualizations leads to solving tasks, the next chapter will introduce visual task classifications. We will find a common understanding on the way how tasks are classified in literature in contrast to interactions. Therefore an abstraction of the task classification will be performed. Based on this abstracted task classification, we will describe the task and classify them in order to have a more concrete understanding of visualization tasks. This will be important for our conceptual model. With this procedure we will have a view how human is involved in the visualization process and which tasks can be solved. Further it will be necessary to investigate the aspect of data in and for information visualization. We will continue with the same procedure and introduce classifications of data. Further we will slightly change an existing classification and introduce the data types based on this classification. The chapter will conclude with a section about technique and methods for visualizing information. This section will follow the same procedure and introduces first various existing classifications. Here again we will see that the proposed classifications are not appropriate for our purposes and will combine existing classifications to have a baseline for introducing the visualization techniques. The visualization techniques and methods will be introduced exemplary and do not claim to cover the state of the art. The main goal of this chapter that was partially published in [1, 2] is to have a common understanding about the terms, methods, and techniques of information visualization. Therefore we chose the view from human side, the tasks, and the data to describe information visualization. Figure 2.1 illustrates an abstract view on the structure of this chapter.

© Springer International Publishing Switzerland 2016
K. Nazemi, *Adaptive Semantics Visualization*, Studies in Computational
Intelligence 646, DOI 10.1007/978-3-319-30816-6_2

Fig. 2.1 Abstract view on
the structure of the chapter
information visualization

2.1 Terminological Distinction

The most common definition for information visualization in computational systems was brought by Card et al. [3]. They started with a more general definition of *visualization* in computational systems and defined visualization as *The use of computer-supported, interactive, visual representations of data to amplify cognition* [3, p. 6], whereas the *cognition* is further proposed as "acquisition or use of knowledge" [3, p. 6]. With this definition they worked out that the main goal of visualizations is to provide insights (*discovery, decision making, and explanation*) and not only pictures. Visualizations may represent different types of data. In case of visualizing physical data, Card et al. tends to the term *scientific visualization*. [3, p. 6] Based on the type of data to be visualized they define *information visualization* as:

> The use of computer-supported, interactive, visual representations of **abstract** data to amplify cognition. [3, p. 6]

The main difference in this definition is the term "abstract data", which is related to the fact that no obvious spatial mappings can be assigned to the data. Without a spatial abstraction, one challenge is the problem of rendering the data into an effective visual [3, p. 7]. To face the mapping problem of raw data to visual forms Card et al. proposed a reference model for visualization [3, p. 17], using outcomes of previous works on non-computational visualization of abstract data [4]. The proposed reference model for visualization counts today as the most influential reference model for information visualization. It provides a data transformation process from raw data to views involving the human in the interaction processing. The reference model is an excellent groundwork to understand, define and distinct information visualization. Figure 2.2 illustrates the reference model with its transformation steps.

The series of transformations begins with *raw data* and ends after three transformation steps with the human, who gains insights from the visual presentations. Vice versa the human is enabled to operate and thereby manipulate and adjust the transformation steps (user interaction on different level). The first step of transformation is *data transformation*, with the diverse raw data formats to relations or sets

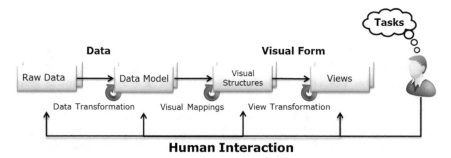

Fig. 2.2 Reference model for visualization (adapted from [3, p. 17] *with kind permission of B. Shneiderman*)

of relations (*data table*) that are structured and easier to visualize [3]. Card et al. define these relations mathematically as a set of tuples (see Eq. 2.1).

$$\{< Value_{ix}, Value_{iy}, ... >, < Value_{jx}, Value_{jy}, ... >, ...\} \qquad (2.1)$$

A *Data Table* combines relations with their describing metadata [3]. A data table is represented by rows, which contains variables as set of values in the tables and cases as set values for each variable. In context of data tables they introduce a categorization of the data variables and their possible sequences. They propose that there are three basic types of variables, *nominal*, *ordinal*, and *quantitative*. Nominal variables are unordered sets (are only $=$ or $\neq$ to other values), ordinal variables are ordered sets (obeys a $<$ relation), and quantitative variables are numeric ranges (can do arithmetic on them) [3, pp. 17–23].

The next step in the transformation process of the reference model is the mapping of the data tables to *Visual Structures*. Here the work of Bertin [4] builds the foundation of visual variables and structures to provide an effective mapping [3, pp. 23–31]. The reference model proposes that two main factors are important to provide an effective mapping to visual structures. The mapping should preserve the data with their type of variables and emphasize the important information to be perceived well by the human. The visualization should enable the human to interpret faster, distinct graphical entities, or make to fewer errors [3, p. 23]. In today's evaluation methods the two main factors for measuring the efficiency of visualizations are *task completion time* (faster interpretation) and *task completion correctness* (fewer errors). The visual structures of the reference model are enhancements of Bertin's work on graphical semiology [3, 4]. While Bertin subdivided the visual variables into retinal variables and layout, the reference model does not propose such a differentiation [3, p. 26]. It enhances the model of Bertin and consists of spatial substrates, marks, and graphical properties. Although the authors propose that some visual encodings are more appropriate for *uncontrolled processing* (or *preattentive*) (see Sect. 2.2.1) in tasks like search or pattern detection and others for *controlled processing* (see Sect. 2.2.2) [3, p. 25] the reference model itself does not propagate this separation. It focuses more on a general transformation of data tables and their sequential characteristics to visual structures. Visual structures may appear as *Spatial Substrates*, *Marks*, *Connection and Enclosures*, *Retinal Properties*, and *Temporal Encodings*, whereas the transformation encloses the entire spectrum of visual structures.

The final step of the reference model completes the loop between human and visualizations (visual forms) [3, p. 31]. It transforms static graphical presentation by incorporating humans' interaction to create different views of visual structures and provide an interactive visual environment. Card et al. lists three main view manipulations: (1) Location probes use location to reveal additional information from data tables, (2) Viewpoint controls magnify or change the viewpoint, e.g. by zooming or panning, and (3) Distortion provides a modification of the visual structure by creating a context plus focus view [3, p. 31]. The view manipulation techniques will be investigated in more detail in Sect. 2.3.2. The introduced reference model describes in a comprehensible way the transformation processes from raw data to visual structures, the view manipulations, and human operations on different levels back to the transformation steps. These steps focus on the how abstract data can be visualized interactively with computational systems and provide a well-established explanation of information visualization.

In recent years, the research field of *Visual Analytics* evolved from Information Visualization and other areas to emphasize the knowledge generation aspect. Visual Analytics were often used synonymous to information visualization, although both terms gained established definitions. The early and most influential definition of Visual Analytics was proposed by Thomas and Cook [5]:

> *Visual analytics is the science of analytical reasoning facilitated by interactive visual interfaces.* [5, p. 4]

Their definition emphasizes the "overwhelming amounts of disparate, conflicting, and dynamic information" [5, p. 2] in particular for security related analysis tasks. One of the main focuses of Visual Analytics is to "detect the expected and discover the unexpected" [5, p. 4] from massive and ambiguous data. They outlined that the main areas of the interdisciplinary field of Visual Analytics are:

- *Analytical reasoning techniques*: for obtaining insights and support analytical tasks such as decision making.
- *Visual representations and interaction techniques*: for enabling users to explore and understand large amounts of data, and interact with them with their visual perception abilities.
- *Data representations and transformations*: to convert all types of data, even conflicting and dynamic, to support visualization and analysis.
- *Production, presentation and dissemination*: to provide a reporting ability for a broader audience and communicate the analysis results. [5, p. 4]

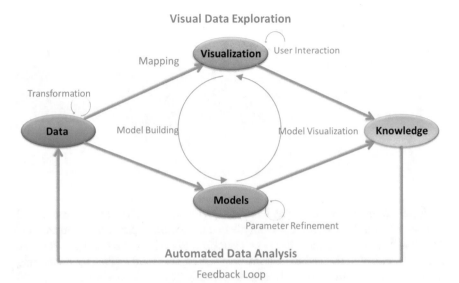

Fig. 2.3 The visual analytics process (adapted from [11, p. 10])

The definition of Visual Analytics gained a series of revisions to precise the abstract formulation [6–11]. Keim et al. commented that the definition of such an interdisciplinary field is not easy [11]. A more precise definition is:

> *Visual analytics combines automated analysis techniques with interactive visualizations for an effective understanding, reasoning and decision making on the basis of very large and complex datasets.* [11, p. 7]

This definition stated more precisely the interdisciplinary nature of Visual Analytics by introducing and outlining the combined use of analysis techniques and interactive information visualizations. In addition, it emphasizes the challenge of data amount, thus this confines Visual Analytics to "very large" data-sets. The main characteristics of solving analytical tasks with interactive information visualizations still remain. This definition of Visual Analytics is illustrated by a model for the Visual Analytics process. Figure 2.3 illustrates the process that targets on providing a tight coupling of visual and automatic analysis methods through human interaction to enable human to gain insights and knowledge [11, p. 10].

The visual analytics process models the different stages represented by oval forms and their transitions with arrows. The process starts with the data that may need to be preprocessed and transformed to an adequate way (indicated with the transformation arrow). After the transformation stage the "analyst" may choose to visualize the data or to use automatic analysis methods [11]. Keim et al. does not use the term "user" in their process. It may indicate that the Visual Analytic model is a dedicated design for "analysts" with the necessary of previous knowledge about the processes or tasks (analysis). If the automatic analysis is chosen, techniques from data mining are applied to generate models from the underlying data. These models can further be

Fig. 2.4 The computer's and
human' role in visualization
in context of policy
modeling (from [12, p. 85])

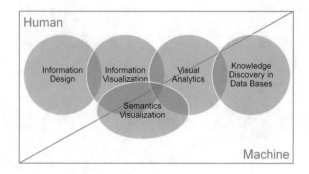

evaluated, refined, or specified by interacting with data [11]. Visualizations are used
to interact with the models and manipulate and refine the parameters. Further the
selection of alternating models can be visualized to evaluate the findings out of the
generated model. If the analyst decide to visually explore the data first, the underlying
model has to be confirmed based on this hypothesis. The visual representations reveal
insights, which can further be refined by interactions on the visualizations [11]. The
entire Visual Analytics process tightly couples the visualization and automatic data
modeling (data analysis) methods. It provides an interactive process to make use
of both, the interactive visual representations and data modeling approaches for
acquiring knowledge and insights, which build the last stage of the process [11]. The
role of human and the possibilities to interact in the stages of the visual analytics
process remains as they are proposed in the reference model for visualization [3].
The main difference is the interactively combined techniques for visualizing and
analyzing data.

Kohlhammer et al. proposed a differentiation of visualizations in context of policy
modeling [12]. Their differentiation proposes a classification based on the role of
human and machine in the data processing pipeline. Thus Visual Analytics make use
of more automatic processing and modeling techniques than information visualization, the model distinguishes visual analytics based on the role of the involvement
of automatic (computer-based) methods. Figure 2.4 illustrates the differentiation and
introduces further the field of *Semantics Visualization*, which will be described more
detailed in Chap. 3.

In this work we use the definition of information visualization as defined by Card
et al. [3]. Thus Visual Analytics makes use of information visualization for the visual
stages [11], we use the term information visualization for the visualization aspects
of visual analytics too. When describing Visual Analytics systems, our focus will be
the way how information is visualized and human are interacting with and perceiving
the visualizations. Amplifying cognition and acquiring knowledge [3, 11] with the
use of human's visual perception is essential for this work, whereas the automated
data processing and data analytics methods are not in scope of this work.

2.2   Visual Perception and Processing

Visualization is strongly related to the way how human perceive and process visual information. Physiological and psychological studies showed that vision processing consist of two main stages of attention, preattentive and attentive processing. Understanding these stages is essential for the identification of those visual attributes and variables that should be considered for the visual representation of information. This section introduces the terms preattentive and attentive visual processing and summarizes some of the most common theories. It further builds the foundation for the adaptation of the visual attributes. Physiological aspects of human image and vision perception will not be discussed in this section. For further readings in physiological aspects of vision perception the work of Hubel [13] is recommended.

2.2.1   Preattentive Processing

The process how human perceives visualizations were investigated in research for several years [14]. A fundamental result was the discovery of a limited set of visual properties, which is rapidly detected by the low-level visual system [14]. The so called preattentive features are detected by human in less than 250 milliseconds, which suggests that certain information can be processed in parallel [13–15]. A unique visual property allows identifying an object preattentively. This unique visual property might be length, width, size, curvature, number, terminators, intersection, closure, hue, intensity, flicker, direction of motion, binocular luster, stereoscopic depth, 3D depth cues, and lighting direction. All this variables are associated with the four primitive variables luminance and brightness, color, shape and texture, [14] which provide processing of visual information prior to selection [16]. This visual stimulus is called 'pop-out effect', an uncontrolled movement of eyes to visual features. Ward et al. name four tasks, which uses the pop-out effect in psychological experiments for performing tasks [14]:

- **Target detection**: The task is to detect presence or absence of a target with unique visual features within a field of distractors.
- **Boundary detection**: Users have to detect a texture boundary between two groups of objects, where each group has common visual features.
- **Region tracking**: Users track one element as it moves in time and space.
- **Counting and estimation**: The task is to count or estimate the number of objects with different visual attributes [14].

Treisman's **Feature Integration Theory** [17] has become one of the most influential theories in the area of preattentive visual information processing. It gives insights into the preattentive detection of boundary and targets in fields of distractors. To evidence the preattentive perception of visual features, she designed a set of tests. Therefore a target with a unique visual attribute (*target detection*) or a group

Fig. 2.5 Asymmetric and
symmetric preattentive
visual features (adapted from
[14, 20])

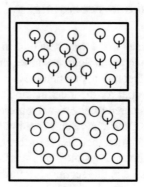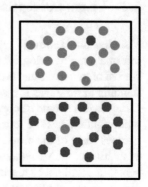

of target elements with unique visual features (*boundary detection*) was placed in a
field of distractors. The subjects had to communicate as fast as possible, if the target
is absent or present, while the amount of distractors was increased. Treisman and
other researchers tested the accuracy and the time of the responses. They assumed,
if the visual information would be processed serially, the subjects would need more
time, when the amount of the distractors increases. And if the amount of the distractors plays no role for the measured time and accuracy, the visual task was processed
in parallel, according to that preattentively [17]. In a further test (*accuracy model*)
a screen with a target or a group of targets was shown to the subjects just for 200
to 250 msec. In this time frame the subjects had no time to focus attentively to a
certain object. So if they give the accurate answer to the presence or absence of the
visual targets, the task was solved preattentively. Treisman and others used this test
to identify a list of preattentive visual features [17–19]. Further they detected that
some of the visual features are asymmetric, while others are symmetric. A circle with
a line (as a visual feature) in a sea of circles can be processed preattentively, while
a circle without a line in a sea with circle with lines is not preattentively processed
[15, 17, 19]. Figure 2.5 illustrates the difference between symmetric and asymmetric
visual features.

Treisman and Souther explained the phenomenon of preattentive visual processing
using a model of low-level human vision made up of a feature map and a master map
of locations [18]. They proposed to use a manageable set of features, consisting of the
main visual attributes. The feature map therefore consisted of the visual variables,
color, size, orientation, luminance and contrast. Whereas each of the features had
their own map and for the color the four primaries red, green, blue and yellows [18]
and the three primaries red, yellow, blue [19]. The feature map was expanded [20],
in which the features luminance and contrast were replaced by *stereo distance*.

The *master map of locations* in their *theoretical framework* [19] is a medium in
which the attention operates. This map "specifies *where* in the display things are, but
not *what* they are" [19, p. 17]. With a unique visual feature or unique visual features
compared to the distractors, a localization of the target or boundary is enabled with
the master map of locations. The more an object differs from the distractors, the

better it can be processed. A green square, for example, in a sea of red circles can be better recognized preattentively than a red square. This phenomenon shows that there exist differences between the preattentive processing of visual features. For that reason Treisman expanded her model in later works [19, 21], not only proposing a strict dichotomy of features being processed serial or parallel. These are more two ends of a spectrum [19–21].

Treisman proposed in her theory two main stages of visual perception, the preattentive and the focused attention stage. The preattentive stage is strongly related to one unique visual feature that stimulates a 'pop-out effect'. In this stage neither the target is localized, nor is it identified. One main finding of the *Feature Integration Theory* was that the localization of a target object is processed serially on the master map of locations. She evidenced that the presence and absence of a target object with unique visual features can be processed preattentively, but the identification and localization of the object on the master map of locations requires focused attention [17]. She evidenced her theory with the *illusionary conjunction*, where subjects identified not existing target objects in a sea of distractors with more than one unique visual feature.

The strict bisection of serial and parallel low-level visual processing based on the conjunction of visual attributes is not advocated by all researchers. Quinlan and Humphrey for instance propose that the search time for visually detecting objects depends on two other factors. Firstly on the number of items of information required to identify the target and secondly on how easily a target can be distinguished from the distractors, whereas unique visual features play no role on their '*Similarity Model*' [14, 22]. The model introduces the criteria *target to non-target similarity (T-N similarity)* and *non-target to non-target similarity (N-N similarity)*. Visual search time is based on *T-N similarity*, which defines the similarity between target and distractors and *N-N similarity*, the similarity between the distractors. The proposed model assumes that as *T-N similarity* increases, the search time decreases. Further as *N-N similarity* decreases, the search time increases and the search efficiency decreases. *T-N* and *N-N similarity* are related and comprehend each other. If *T-N* decreases and *N-N* similarity decreases too, a preattentive perception of the target will not be registered. If both similarities increase, the effect of a preattentive perception will get lower [14, 22]. The *Similarity Theory* preaches that the more an object distinguishes from the distractors and the more the distractors are similar, the better and faster it can be perceived, regardless of any unique visual attributes.

A more recent model of a two stage paradigm of preattentive and attentive visual perception was proposed by Wolfe in his **Guided Search** model [23–26]. In his first attempt his Guided Search model had a preattentive and an attentive stage, based on Treisman's Feature Integration Theory and explaining more the Similarity Theory. He further proposed that the information from the first stage could be used to guide the attention to the attentive stage [23]. An object with unique visual variables would lead in the preattentive stage the focus of the subject to the visual object, this attention is further present in the second attentive stage. The future versions of the Guided Search model, including the recent version *Guided Search 4.0* [26], proposed more smooth transition between the two strictly bisected stages of attention. One main finding of Wolfe was that the preattentive visual activation is not only stimulus-driven (bottom-up), like proposed in the Feature Integration Theory, but also user-driven (top-down). The Guided Search model argues the differentiation with its *feature maps*. Stimuli are assumed to be in parallel across the entire visual field. "At some point, independent parallel representations are generated for a limited set of basic visual features" [24, p. 204]. These sets of limited visual features are feature maps. The feature maps or independent maps for each visual attribute, e.g. color, size and orientation. Each of these maps may contain further maps, e.g. the color map may contain a map for green, red etc. Wolfe listed in his second model a set of visual features containing *orientation, color, motion, size, stereoscopic depth, other depth cues, binocular lusture, vernier offset, curvature, terminators* and *intersection*. [24] In case of localizing a target object, the feature maps are activated. And this activation can be either bottom-up or top-down [24].

The bottom-up activation is stimulus driven and thereby not depended to the subject's knowledge or preferences in a visual task. This activation is based on the differences between a target object and the neighboring distractors. The neighbor-hood of the target can be bounded in a 5×5 array around the identified object. Guided Search assumes that the bottom-up activation is calculated separately for each feature in the feature map. The bottom-up activation guides attention to a distinctive item in a field of distractors, if the visual features of the object are unusual. In contrast to the bottom-up activation, the top-down activation is user-driven and depends strongly to the task, knowledge and preferences of the user. [26] For instance, if a red circle is placed in a field of distractors of mixed color circles, the bottom-up activation will not be registered. But if the user is instructed to search for a red circle within the field of the heterogeneous distractors, the knowledge of the task will guide him to the red circle. This user-driven activation can be registered in a similar time-frame to the stimulus-driven activation [24].

Wolfe proposes that the strict dichotomy of parallel and serial visual processing does not hold. [24, 26] The Guided Search model assumes that the information from the first preattentive stage is forwarded to the second stage. The direct attention is guided through the preattentive processing, whereas the region of the target object is in the attentive processing further the region of interest. Wolfe evidences his model with triple conjunction of color, size and form (orientation) [23, p. 430]. The fact that three visual attributes are forwarded to the serial process leads to faster search process and reaction time.

The active involvement of users and the consideration of their pre-knowledge, preferences and tasks play an important role in the Guided Search model. If a user has an imagination of the searched target object, the reacting time decreases. The involvement of users' pre-knowledge played more and more a key-role in further works of computational modeling of visual attention in both stages. The challenge for the implementation of a comprehensive computational model of visual attention is the consideration of both activation types, [27] and consequently the involvement of users' pre-knowledge and visual tasks.

2.2.2 Attentive Processing

Human are able to detect certain visual features in parallel and thereby preattentively. The preattentive processing of information depends on visual features of targets and distractors. The 'pop-out effect' in this stage guides the attention of human to certain visual features, whereas this guidance is in most cases uncontrolled (see top-down and bottom-up activation in Sect. 2.2.1). The 'pop-out effect' does not include the localization or the target detection.

The attentive processing of visual information (or *postattentive vision* [14], *directed attention* [26]) begins, when we stop attending to the out-popped target (assuming there exists one) and look at something else [14]. Although, the strict dichotomy of parallel and serial processing is still disputed, Ware has proposed a three stage model, subdividing the attentive processing of visual information into a serial stage of *Pattern Recognition* and a further stage of *Sequential Goal-Directed Processing*, beside a preattentive stage [28].

Ware's model of perceptual processing is a simplification of several methods and models. The first of his three staged model is the preattentive stage, based on the proposed models of Treisman and Wolfe. Here information is processed in *parallel to extract low-level properties of the visual scene* [28, p. 20]. Similar to the described models, the parallel information processing cannot be consciously controlled by the user, is rapid and extracts basic visual features. The visual features that are investigated in this model are *orientation, color, texture*, and *movement patterns*. Based on the original Feature Integration Theory the parallel activation is bottom-up. Instead of using termini like stimulus-driven or feature-driven bottom-up activation, Ware introduces a *data-driven model of processing*. At the second *Pattern Recognition* stage of his model, rapid but active processes divide the visual field into regions and simple patterns. In this serial stage, regions and localizations can be identified, e.g. regions of similar or same colors. The flexibility of this stage can be influenced by both, the bottom-up activation from the previous parallel stage and the top-down activation. The top-down activation is driven by visual queries in this model. The visual queries are analog to Treisman's *feature maps*. Ware characterizes the second pattern perception stage including slow serial processing, with more emphasis on arbitrary aspects of symbols and the fluent combination of the bottom-up and top-down feature processing [28, p. 22].

The last stage of the three-stage model, the *Sequential Goal-Directed Processing*, is the highest level of perception involving active attention. The use of external visualizations let us "construct a sequence of visual queries that are answered through visual search strategies" [28, p. 22]. At this stage only few objects are in focus of attention, which are constructed by the subject from available patterns to solve a given visual query task. One main aspect in this stage is the use of the term *construct* that leads to the assumption that knowledge from the long-term memory (pre-knowledge) is associated to the visual patterns and new knowledge is constructed by human. In the context of knowledge construction it is necessary to introduce two terms that are essential for gathering knowledge through visualization, namely *recall* and *recognition*. Ware proposes that *recall* "consists of the activation of particular pathway" [29, p. 388] of associations stored in the long-term memory. Recall makes use of visual or verbal-propositional information to activate the traces of the long-term memory. It is necessary to describe (verbal or visual) some patterns and traces of our memory without the use of an indicator. Ware constitutes that *recognition* is superior to *recall*, thus in *recognition* a visual memory trace is reawakened [29, p. 388].

This phenomenon is one main reason, why visual system should consider in their design the knowledge of users. With the use of *recognition* instead of *recall* the efficiency of the problem-solving process in visualizations can be improved.

The aspect of post-attentional processing in terms of dynamic generation of visual representation was investigated by Rensink in his **Triadic Architecture** [30]. He argues based on the *Coherence Theory* [30, p. 19] that focused attention is needed to see changes at the time they occur and only one object in a scene (screen) can be given a coherent representation [30]. Moreover, the representation is limited in the amount of displayed information. So it is necessary to shift the attention to the appropriate objects at the right time. He discards the assumption that all visual processing pass a single attention locus (*attento-centric*) and proposes a triadic architecture with independent information processing systems. The first system, the *low-level vision*, makes use of the preattentive features to shift the attention to the location of interest. This level creates a high-detailed, volatile structure [30, p. 34]. In this system of early processing the resultant structures (*proto-objects*) may be sophisticated, the spatial coherence is limited and simply replaceable by new stimuli [31, p. 262].

The second system, the *Object (attentional)*, investigates the spatial arrangement (*Layout*) in the scene and activates a focused attention [31]. This provides a non-volatile representation of the locations of various structures on limited-capacity attentional system. This is used when attention is already directed. The third system, the *setting (nonattentional)* facilitates the perception via *gist* (meaning) and *layout*. Rensink proposed that "the most abstract aspect of a scene is its meaning" (gist) [30, p. 36]. It is a result from the context of an object and is used to refer to the properties of the long-term memory to *recognize* an image. The most important aspect in this context is the unification of *Layout* in terms of spatial arrangement of objects. Rensink proposes that one important aspect of the scene structure is *Layout*, "without regards to visual properties or semantic identity" [30, p. 36]. *Layout* is used to support the problem solving process as knowledge about the relationships of coherent-objects is needed [29]. The associated collection of representations is *scene schema*. Rensink proposes that gist and layout involve short-term or working representations, whereas the scene schema is long-term structures [30].

2.3   Visual Interaction

Today's information visualization systems do not just offer a static picture. Most of the existing visualizations provide different interaction techniques that allow solving the given visualization task through graphical interaction. The provided interaction method is one of the key-features of the visualization system and the issue of interacting with visualization was already investigated by various researchers. Several classifications, concepts and techniques were introduced to affirm the importance of interaction in visualizations. This section gives an overview about some of the classifications interactive visualizations. The section does not claim to be complete and aims to give an overview of the idea of *interaction* in information visualization systems.

2.3.1 Classifications of Visual Interactions

An abstract classification of interaction in visualizations was brought by Ware, who proposes a classification of interlocking feedback loops of *data manipulation, exploration and navigation*, and *problem-solving* [28]. At the lowest level, the *data manipulation loop*, objects are selected and moved using the basic skill of eye-hand coordination. In this loop the system and human reaction delay is an important factor for efficient interaction with visualizations. Ware introduces several measurements criteria and rules for measuring reaction time, e.g. reaction time (*Hick-Hyman law* [32]), selection time (*Fitts' law* [33]), and path tracing [34]. The selection and reaction time in the lowest level of interaction is constraint to the users' knowledge in interacting with the systems. Over time, people become more skilled in operating low-level interactions with visualizations. The informal learning of interaction with systems is introduced by Ware with the simple expression, known as *power law of practice*. This law describes the users' learning curve as:

$$\log(T_n) = C - \alpha \log(n) \tag{2.2}$$

where $C = \log(T_1)$ is the first performance of the user with a system, T_n is the time required to perform the nth trail, and α is a constant that represent the learning curve.

At the intermediate loop of *exploration and manipulation*, the way through large visual data space is found. In this interaction level the known similarities are "recognized" to find the way through the data. The differentiation between *knowledge types*, e.g. declarative, procedural, and topological knowledge [35] plays an essential role to find the path to the targeted knowledge and build a *cognitive spatial map* [28]. The highest level of the model the *problem-solving loop* provides the ability to form hypotheses. The augmented visualization process provides refinements and reformulations until a possible solution is identified. The iterative character of this level can further be enhanced by replacing and revising visualizations.

A classification of visual interaction methods is the *Visual Information Seeking Mantra* proposed by Shneiderman [36]. His mantra is not explicitly declared as a classification for interaction methods. It is far more a starting point for designing advanced graphical user interfaces and the foundation of the *Task by Data Type Taxonomy* of information visualization [36]. Shneiderman proposes in his mantra *overview first, zoom and filter, then details on demand.* The interactions on visual environments are ordered sequentially and have an iterative character. This classification is according to Ware's model on the highest *problem solving loop* of visual interactions. The mantra further correlates seven data types to seven tasks on the highest level of abstraction.

Cockburn et al. enhanced the interaction aspects of the *Visual Information Seeking Mantra* to survey and categorize visualizations [37]. They defined "overview plus context" as *Spatial Separation* between focused information entities and contextual information. The "zooming" interaction was reduced to the temporal separation of entities, whereas "focus plus context" minimizes the seam within the contextual information. Further proposed "cue-based" techniques selectively highlight information within the information context [37].

Keim enhanced and refined the *Visual Information Seeking Mantra* too and introduced the following interaction classification of information visualization [36, 38, 39]: *1. Data-to-Visualization Mapping, 2. Overview, 3. Zoom, 4. Filter, 5. Details on Demand, 6. Relate, 7. History, 8. Extract,* and *9. Linking & Brushing.* In this model the *interaction techniques* (Mapping, Projection, Filtering Link&Brush and Zooming) [38, p. 81] were categorized to *distortion techniques* and *data visualization techniques.* In this enhanced model *distortion* is categorized as an interaction technique. Further the *standard* interaction technique is introduced to conclude the whole spectrum of possible interaction techniques.

Hearst lists the following main techniques for interacting and navigating with information visualization within abstract data space: *brushing and linking, panning and zooming, focus-plus-context, magic lenses* and *animation* to retain the context [40, p. 260]. Further the combination of interactions as for example *overview plus detail* are proclaimed for solving tasks with interaction methods. The techniques are seen as foundations for the design and implementation of visualization techniques. In contrast to Keim's classification, Hearst proclaims a more *context-oriented* interaction. *Zooming* is mentioned in combination with *panning*, where the panning-action enables to view the overview-context before zooming into the visual area of interest. A similar procedure is proclaimed for all the interaction techniques, classified by Hearst. In the classification of Ware the identified interaction techniques can be positioned vertical to the whole spectrum of the interaction-loops. The main target of the interaction in this classification is problem-solving, whereas the context-orientation is addicted to the exploration and manipulation loop as well as to the data manipulation loop. Data entities, relations and attributes are visualized dynamically through the visualization techniques that access the data directly.

Ward and Yang introduced in [41] a classification of interaction in visualizations that distinguishes between *interaction operators* and *interaction operands.* They proposed that there is a significant difference if an interaction is *operated* to different objects or spaces. These objects or spaces are the operands of the interaction procedure. "To determine the result of an interactive operation, one needs to know within what space the interaction is to take place" [41, p. 2]. In their first attempt they classified three *interaction operations* and thereby *interaction operators.* With *navigation, selection* and *distortion* a significant percentage of the interaction operations in visualization systems was identifies [41, p. 1]. The interaction operands were classified in section of spaces upon which an interactive operator is applied. Their proposed framework contained following spaces as *operands*: *screen-space (pixels)*, *data value-space*, *data structure-space*, *attribute-space*, *object-space (3D surfaces)* and *visualization structure-space*.

The *screen-space* consists of actions directly on the screen with no impacts on the data. This contains transformation on screen-level, such as *panning*, *zooming*, *rotation* or *pixel-level* operation, e.g. *transformation*, *sampling* or *replication*. Interactions on the *data-value space* involve the data values for view specification. On this space *panning and zooming* or other interaction operations change the data values being displayed. An interaction operation on this space is similar to a database query for specifying data values. Interaction and navigation operations on the *data structure-space* involve view transformation along the structure of the data. Operation on this space allows identifying regions of interest in the data structure, e.g. selection of data in a cluster hierarchy [42]. Operations on the *attribute-space* are similar to that on *data value-space*; they involve a view transformation based on the attributes of the graphical objects. Whereas interactions on the *object-space* are defined as direct manipulation of graphical object, primarily 3D-objects, which can be turned transformed etc. Interaction operations on the *visualization structure-space* involve the view transformation of the visualized structure. The data are not manipulated on this level, but the user is able to rearrange the visual structure.

The classification of Ward and Yang is part of their unified framework for interactions on visualizations. This framework further proposed the parameterization of the operands [41, pp. 6–8] to define an extensive assortment of interaction operators. Ward et al. extended their framework in their recent work [14, pp. 315–354]. In this extended classification they enhanced the interaction operators by *filtering*, *reconfiguring, encoding, connection* and *abstraction/elaboration*. Further the *distortion* operation is not considered as a class of interaction operations. The interaction operands and the parameterization remain in the new version of their framework and classification.

This section gave an overview of some classifications of interaction methods in visualization systems. We presented heterogeneous classifications that investigated visual interaction in different levels of abstractions. The introduced classifications were rarely published as interaction classifications; they are rather evolved from design guidelines for visualizations or from classifications of visualizations. Nevertheless we could conclude that interactions in visual environments are investigated in various abstraction levels. Interactions transform the view on the visualized data by manipulating the data or the visual representation. The manipulation on both, data and visualization can be further classified as the model of Ware showed.

2.3.2   Visual Interaction Techniques

This section introduces the most common interaction techniques based on the classification of Hearst [40]. This classification describes the interaction techniques at a lower level of abstraction similar to Keim's classification. It further considers the context of users' interactions and is therefore adequate to explain the interaction techniques in the context of this thesis, which will further describe semantics visualizations. We added the interaction techniques *semantics zoom*, *dynamic queries* and *direct manipulation* to the model of Hearst for covering a wider range of possible interaction techniques.

Brushing and Linking
In multiple-visualization user interfaces, different visual representations of the same data give a view on various perspectives to the data. To not lose the visual context, "brushing and linking" provides a highlighting or selection of visual objects between different views [40]. The highlighting may occur in various ways, e.g. by changing the color or size of the *brushed* objects. The main target is to provide a visual differentiation to the displayed objects and distractors. The work on preattentive perception described in Sect. 2.2.1 provides important visual features to perform this interaction and provide visual features to distinguish brushed objects in linked visualizations.

Panning and Zooming
Panning and zooming provides the change of the viewpoint to the visualized data [40]. Card et al. use the term "panning and zooming" in their listing of interaction techniques as an equivalent to *camera movement and zooms* [3]. Panning and zooming targets to refine the visual area of interest by moving the screen or the view on the screen (pan) and zooming into the area of interest. Furnas and Bederson introduced an analytical framework by *space-scale diagrams* for a direct visualization and analysis of important scale relates issues. [43] They represented the panning and zooming interaction as space-scale diagrams by trajectories [43].

Focus plus Context
Zooming leads to the problem of getting more details about the zoomed part and losing the surrounding information. The higher the zoom-factor is, the more details can be shown about particular items, but the overall structure and the information context get lost with the increasing zoom-factor. To face this problem the interaction metaphor *focus plus context* technique offers more details in the zoomed part but keeps the context in a lower level of detail [3, p. 307].

One of the earliest techniques for focus plus context is the *fisheye view* [44, 45]. The model of "Degree of Interest" [46] was the pioneering foundation for the work on the fisheye views. In contrast to zoom, which is a transformation on the view level, the focus plus context interaction is categorized as transformation on data-level [3].

Beside the distortion techniques (fisheye views), Card et al. list *filtering*, *selective aggregation*, *micro-macro readings* and *highlighting* as selective information reduction methods for keeping the contextual area [3].

Semantic Zooming
Traditional zooming techniques operate on the visual level of a graphical data representation and manipulate the view. According to Ware's model [28] the zooming interaction occurs on the *problem solving loop*, e.g. by changing the size of a zoomed object. In contrast to the ordinary zoom, semantic zoom uncovers detailed information to encompass the context and meaning of a zoomed target [47]. Semantic zoom is for example recently used in ontology visualization for reducing the complexity of large ontologies [48].

Animation
Interactions in information visualization manipulates the visual representation of the underlying data on different levels [28], e.g. by data transformations, visual mappings, and view transformations. In these types of interactions users acts directly with visualizations and change the view. In contrast to the manipulating interaction techniques, animation does not provide the user with manipulation functionalities. The animation is more an implication of the users' interaction [49]. The literature in visual perception suggests the use of animation for the improvement of interaction and understanding [50, 51]. Attracting attention, perceiving in peripheral vision and comprehending the visual changes are the most argued reasons for the implementation of animation as consequences of interactions [50–52]. Further the continuous changes of visual parameters in visualization can be easily followed and understood [52].

Overview plus Detail
Keeping the information context, while gathering detailed information about a specific area of interest, is the main goal of *overview plus detail*. It displays the information with different levels of details in two or more linked visualizations. Card et al. differentiate between time- and space-multiplexed overview plus detail [3]. Time-multiplexed overview plus detail is conceptually similar to panning and zooming or just zooming, thus the interaction is processed serially. Here the main attribute of overview plus detail can be named as the fact, that the time-dependent serial interaction steps provide two main views, an overview of the information and a detailed view of the area or objects of interest. In contrast to that, space-multiplexed overview plus detail conveys both information detail-levels at the same time, in two separated areas of the display (views). This is most common way, how overview plus detail is used in visualization systems [3]. In contrast to panning and zooming, semantic zoom or focus plus context, the detailed and overview information are visualized at the same time in mostly separated display or display areas (space-multiplexed).

Dynamic Queries

The access of information in a human-understandable way is one main goal of information visualization. Today's information databases contain huge amount of data. The visualization of the whole data-set on a single visualization is often overcharging the human perception and information acquisition abilities. Dynamic queries can help to access the required information interactively, which can satisfy users' heterogeneous needs of information acquisition [53, 54]. Dynamic queries provide a well-known and successful approach for exploring [55] and visually seeking vast amount of data [53]. Visual interactive query formulations and refinements enable the reduction of the visualized information to a comprehensible and relevant scale [55].

Direct Manipulation

All the introduced visual interaction methods manipulate the visualization, either on a data-transformation level or on the level of visual transformation. Direct manipulation provides a direct interaction with the user interface or visualization without the need of commands. It bridges the gap between human and machine with a more intuitive graphical metaphor of interaction and avoids the barrier to translate ideas into commands [56]. Golbeck introduces the idea of direct manipulation with the example of deleting items through the trash [57]. The physical selection of an item and putting it into the trash is more intuitive an obvious interaction as a command-line expression for the same action Shneiderman defined various criteria such as visibility of objects and actions of interest; rapid, reversible, incremental actions; and replacement of command-language syntax for direct manipulation [56, 58]. In information visualization, where the data are represented with graphical metaphors or representatives, direct manipulation is essential for a natural interaction. Each interaction, which substitutes a command line expression, can be defined as direct manipulation.

2.4   Visualization Tasks

Interaction with visualizations enables the dialog between user and the visual representation of the underlying data. The interactive manipulation of the data, the visual structure or the visual representations provides the ability to solve various tasks and uncover insights. The term "task" in the context of information visualization is often used ambiguously. Often, interactions and tasks are not distinguished for visualization design, whereas the knowledge about the task to be solved with the visualization is of great importance for its design and thereby for the adaptation. This section starts with the introduction of taxonomies and classifications of tasks in visualization systems. The classifications will enlighten the heterogeneous view on visualization tasks and enable getting an overview of the differences. The classifications will enable to

Table 2.1 Task classification by Shneiderman [36, p. 337]

Task	Description
Overview	Gain an overview of the entire collection
Zoom	Zoom in on items of interest
Details-on-demand	Select an item or group and get details when needed
Relate	View relationships among items
History	Keep a history of actions to support undo, replay, and progressive refinement
Extract	Allow extraction of sub-collections and of the query parameters

investigate high-level tasks in more detail. These high-level tasks will be introduced in the second part of this section and conclude the section.

2.4.1 Classifications of Visual Tasks

One classification of tasks in visualization is the already mentioned *Task by Data Type Taxonomy* of Shneiderman (see Sect. 2.3) [36]. With the assumption that users are viewing collections of data with multiple attributes, he proposes that a basic search task is the selection of items that satisfies the search intents. This classification enhances Shneiderman's *Visual Information Seeking Mantra* with the tasks *relate*, *history*, and *extract*. Table 2.1 illustrates the seven tasks.

The overall tasks in this classification can be abstracted to the high-level tasks *exploration* and *search* and leads to finding (relevant) information.

Buja et al. proposed in their early work [59] a classification concept that investigates the interaction with visualizations (*view manipulation*) and the tasks that are supported by these interactions. They supposed that the purpose of the view manipulations is to support the search for structures in data [59]. For this search they identified three fundamental tasks for data exploration, namely *finding gestalt*, *posing queries*, and *making comparisons*. Finding certain patterns of interest, e.g. clusters, discreteness or discontinuities, are classified in the task *finding gestalt*. *Posing queries* is the next step after gestalt features of interest were found and further information are desired to get an comprehensible view on the chosen parts of the data. For the task *making comparisons* they distinguish between two types of comparisons. First the comparison of variables or projections and second the comparison of subsets of data. The comparison of variables enables the "view from different sides" [45, 59], which illustrates the data from different perspectives, whereas the data subset comparison provide a "view of different slices" and thereby of different subset of data [59].

Further they proposed that the identified tasks are optimally related to three manipulation views. For *gestalt finding* they identified the *focusing individual views*. Here

Table 2.2 Task classification by Buja et al. [59, p. 80]

Task	Manipulation view	Interaction
Finding gestalt	*Focusing individual views*	Choice of projection, aspect ratio, zoom, pan, order, scale, scale-aspect ratio, animation, and 3-D rotation
Posing queries	*Linking multiple views*	Brushing as conditioning/sectioning, database query
Making comparisons	*Arranging many views*	Arranging scatter plot matrix and conditional plot

focusing provide any operation of that manipulates the subset of data or view. The choice of projection, for viewing or the choice of aspect, ratio, and zoom are examples of focusing. For *posing queries* they identified *linking multiple views*. The linking contains view manipulation as brushing or query issuing by highlighting. *Making comparisons* is related to *arranging many views*. They propose that the arrangement of large numbers of related plots for simultaneous comparison is a powerful informal technique [59].

With this tasks and manipulation views they further propose a set of low-level interaction techniques that are related to each high level task. Table 2.2 provides an overview of the proposed task, manipulation views and interactions that are related to each other.

Another approach, which correlates low-level interactions with visualization tasks, was proposed by Chuah and Roth [60]. They summarized their "basic visualization interactions" as a set of low-level-interactions with the attributes input, output, and operation and abstracted them to three basic visualization tasks [60, p. 31]. At the lowest level they propose "Data Operations", which contains interactions affecting the elements within visualizations, e.g. add, delete or derive attributes. The higher level considers "Set Operations", which refers to operations on sets, which may have group characteristics. The highest level investigates "Graphical Operations", which are divided into encode-data, set-graphical-value, and manipulate-objects. While the classes encode-data and set-graphical-value change graphical attributes or the mapping between graphical objects and data, the class manipulate objects operates on graphical objects as a unit of manipulation [60, p. 33–36]. The investigated tasks in this classification focus on comparison and finding patterns as graphs or in data. The high-level task of this classification can be abstracted as "analysis". The aspect of analysis was investigated in various works. One early example is the classification of Wehrend and Lewis [61]. They proposed a taxonomy with ten analytical tasks: *location, identity, distinguish, categorize, cluster, distribution, rank, compare within entities, associate*, and *correlate*. Zhou and Feiner [62] proposed an approach by considering not only the interaction and manipulation abilities of visualizations. They

Table 2.3 Visual task classification by Keller and Keller (adapted from [14, p. 380])

Task	Description
Identify	Recognition of objects based on the presented characteristics
Locate	Identification of the position of an object
Distinguish	Determination the difference of objects
Categorize	Classification of objects into distinct types
Cluster	Grouping of objects based on similarities
Rank	Ordering objects by intended relevance
Compare	Examination of similarities and differences of objects
Associate	Drawing relationships between two or more objects
Correlate	Finding causal or reciprocal relationships between objects

investigated the human perception and the intended task of the visual presentation method in their classification to provide a more user centered task-classification.

Based on various existing classifications, they characterized visual tasks along two dimensions. In the dimension *Visual Accomplishments* the focus lies on the intention of the visual presentation [62, p. 394]. They assumed that a presentation intends either to convey the presenter's message or to help user solving a perceptual task. Based on this assumption, visual tasks are distinguished at the highest level between tasks that *inform* users by *elaborating* or *summarizing* and those, which *enables* users to perform a visual *exploration* or *computation*. Their second dimension *Visual Implications* considers research outcomes of the human visual perception. Based on these outcomes they summarize three types of visual perception and cognition principles: (1) the *visual organization* principle investigates how people organize and perceive a visual presentation, (2) the *visual signaling* principle investigates the manner how people interpret visual cues and infer meanings and (3) the *visual transformation* principle explains how people perceive visual cues and switch their attention. This incorporates the outcomes of the preattentive visual perception too (see Sect. 2.2.1). Zhou and Feiner use these principles to infer visual tasks and assign them to the first dimension of *Visual Accomplishments*.

A more user-centered approach for classifying task was proposed by Keller and Keller [63]. Their classification considers the goals and intentions of the users and suggest based on these certain visual representations [14, p. 164 and p. 380]. They classify the user-intended tasks in nine task categories (see Table 2.3). The main characteristic of their classification is that only analytical aspects play a role for users interacting with visualizations. Previous general tasks like exploration or search does not play any role.

Table 2.4 Visual task categorization by Yi et al. (adapted from [64, p. 1226])

Category	Description
Select	Mark something as interesting to enable the following of the object
Explore	Show something else e.g., different subsets of data
Reconfigure	Provide a different view or arrangement of the underlying data
Encode	Provide a different fundamental view by selecting another visualization technique
Abstract/elaborate	Provide a different level of detail on the data e.g., by details-on-demand techniques (see Sect. 2.3.2)
Filter	Provide a view with certain (predefined) criteria
Connect	Provide a visual connection (e.g. by brushing) between the same objects on different views

A comprehensive classification of users tasks based on user intentions and the interaction role in information visualization was provided by Yi et al. in [64]. Their classification attempts to abstract the most used interaction possibilities with users' intentions to provide categories of interaction. They classify the user tasks based on the role of interaction in information visualization in seven categories (see Table 2.4).

Although the identified categories are abstract views on the interaction roles, the level of abstraction differs enormously. The category *select* for example, can be defined as simple and low-level interaction. Here a user marks an object of interest to be able to follow this object in changed views [64]. In contrast to *select* the category *explore* provide a real abstraction of interaction to a user task. Here the user is able to view on various subset of data to see different characteristics and perform a various number of low-level task e.g., comparing subsets or identifying relevant objects.

Pike et al. extends the proposed approach of Yi et al. [64] by differentiating between low-level and high-level interactions intending to meet high- and low-level user tasks and goals and propose a mutual feedback between user goals and tasks and the affordance of interactive visualizations [65]. They define seven categories of high-level tasks, which can be achieved by a number of low-level tasks and interactions respectively. Further they relate the representation and interaction intents of interactive visualizations, similar to the proposed classification of Zhou and Feiner [62] to low-level representation and interaction techniques. The proposed approach relates the classifications of user goals and tasks with the abilities and goals of interactive visualization in a "mutual feedback" [65]. The relationship of the proposed techniques and the user's goals and tasks is the "analytical discourse", which investigates the low-level interaction and user goals to form a feedback between them [65, p. 265].

The classification of Pike et al. considers the interaction value and user's goal and tasks from both perspectives, information visualization and Visual Analytics and gives a good overview of the high-level tasks intended by users and provided by interactive visualizations. Nevertheless, the differentiation of high- and low-level tasks is not clearly defined. A "compare" task could be a part and therefore a low-level task of "assess" or "analyze", while important tasks like "decision making" [66, 67] are not considered at all.

Fluit et al. proposed in [68] a very simple classification of visualization tasks in the special domain of ontology visualizations in the categories *Analysis*, *Query*, and *Navigation*. Therefore they define the *Analysis* task for getting a global view on data, the *Query* task for finding a narrow set of items, and the *Navigation* task for graphically navigating through the data [68]. In their revised work [69] the last category *Navigation* was replaced by *Exploration*. They propose that *Analysis* can be performed within a single domain with various perspectives, in various sets of data, and by monitoring the changes of data over time. The category *Query* is divided into the processes of query formulation, initiation of actions, and review of results. The task category *Exploration* is defined as finding information that are loosely of interest for the users [69]. Here a further subdivision is not proposed.

2.4.2   High-Level Visual Tasks

The previous section could work out that visual tasks are defined and classified in various levels of granularity. The described approaches are mostly using similar tasks or interactions to describe the problem solving process in visual interfaces. This section targets on the identification of high-level visual tasks as foundation for the visualization design and the adaptation process. We define, in this work, high-level tasks as tasks that are a summarization of visual tasks and provide a higher level of abstraction.

The described classifications consider in various ways the aspect of *search*. Shneiderman's *Visual Information Seeking Mantra* proposes a top-down information seeking approach [36]. Buja et al. propose searching information as the main task, which can be solved by manipulating the view [59]. Zhouh and Feiner investigate the way from information to user [62]. They elaborated the enabling and informing users. Informing users by elaboration and summarization, premises the information searching task. They further propose that *search* is a sub-task of *exploration* in their *enabling* category [62]. Fluit et al. propose in their simplified task classification *search* as one of the three high-level tasks and call it *query* [68, 69]. As the most of the presented classifications consider *search* as an important and fundamental task in visualizations, *search* is considered in this thesis as one of the high-level visual tasks. Beside *search* the high-level visual task *explore* plays an essential role. The task classifications show that exploring information plays a key-role for each classification. Shneiderman's model proposes a top-down seeking model with the characteristics of exploration [36]. From overview to detailed information can be assigned as an exploration task [36]. Zhou and Feiner explicitly name the task *explore* is a higher-level task of search and verify [62]. Yi et al. have their own categorization for *explore*, although their classification is not considering the high-level tasks [64]. The classification of Pike et al. [65] assigns the task *explore* as the high-level task on the user-goal and tasks level. In particular the classification of Keller and Keller [63] proposes a different view on solving visual task. They propose that the main visual interactions are solving more analytical task (see Table 2.3. Their task classification can be abstracted to *analyze*. Zhouh and Feiner differentiate in their model different aspects of *analyze*. In particular, the task category enable and verify leads to the higher level task *analyze*, whereas some aspects of inform and summarize are addressing the analysis task [62]. Pike et al. identified the task *analyze* already as a high-level task in their model [65]. Further they assigned the task "compare" as a high-level task too, whereas other works (e.g., [63]) assigns compare as a sub-task of the analytical visual problem solving process. As the analytical tasks plays an important role in all presented classification, *analyze* should be assigned as a high-level visual task.

On balance, the classification of Fluit et al. seems to be a well elaborated high-level task definition, whereas the definition of each task cannot be intermeatable with other definitions. In this work we investigate *search*, *explore* and *analyze* as high-level tasks. Further we assign the identified tasks as sub-tasks of the identified three high-level tasks. This is to ensure that the major tasks are categorized to the high-level tasks. This classification and task definition will be applied in this work. In Fig. 2.6 the classification and the assignment of the lower-level tasks are illustrated as applied in this work.

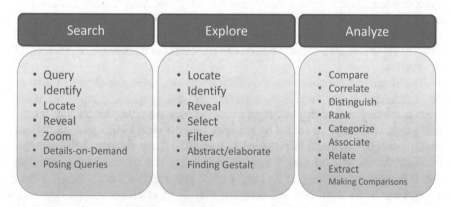

Search	Explore	Analyze
• Query	• Locate	• Compare
• Identify	• Identify	• Correlate
• Locate	• Reveal	• Distinguish
• Reveal	• Select	• Rank
• Zoom	• Filter	• Categorize
• Details-on-Demand	• Abstract/elaborate	• Associate
• Posing Queries	• Finding Gestalt	• Relate
		• Extract
		• Making Comparisons

Fig. 2.6 High-level tasks with their sub-tasks

2.5 Data Foundations

A fundamental component in information visualization is data. As the reference model for visualization [3] and the visual analytics process [11] already illustrated (see Sect. 2.1), the process of visualizing information starts with the underlying data. It is essential to process the data in order find the adequate way for visualization. Therefore different aspects of data play a role in visualization. This section introduces some of the most common aspects of the data that should be considered in visualizations. First a number of common classifications on data will be introduced. Afterward different types of data will be described based on an established classification.

2.5.1 Classifications of Data

The starting point of the visual transformation is the data, thus the classification of data is essential for visualizations [3, 11, 70, 71]. In this section various classifications of data will be presented. Some of these classifications were already mentioned in context of interaction and tasks. Further, most of the visualization techniques are described based on the data classes and the way of their categorization [14]. Altogether the classifications of data can be abstracted to three main ways of categorizing data: by data values (level of measurement), by the transformation steps of data, and the data dimensions.

Card and Mackinlay introduced [71] and enhanced [3] a classification based on the value of data. This considers the level if measurement of data values and their ability to order. As already introduced (Sect. 2.1) they propose that data values can be:

- nominal: without any value that can be ordered
- ordinal: possess a value that can be ordered by relations between the values
- quantitative: numerical values and provide thereby a natural order [3, 71].

Ward et al. define the ability of numerical order of data value as "ordinal" [14]. They define that ordinal data values can be binary (e.g. 0 and 1), discrete, or continuous [14, p. 46]. Both, discrete and continuous data types may have numerical values. Further they introduce the mathematical concept of *scale* [14]. They define the data type *scale* as values with ordering relation with distance metric, with which the distances between the values can be computed, and with the existence of absolute zero for the definition of a fixed lowest value [14, pp. 46–47].

Chi introduced a taxonomy for visualizations [72] by using their *Data State Reference Model* [73]. Although, the taxonomy was proposed for classifying visualizations, the aspect that the data transformation and data types are the baseline, is interesting for data classification. The Data State Reference Model [72, 73] proposes three types for data transformation, four data stages and four types of operations within the model [72, pp. 1–2]. The model starts with the *value* (raw data) and

Table 2.5 Data type classification by Shneiderman [36]

Data type by Shneiderman	
1-dimensional	Linear data types
2-dimensional	Planar or map data
3-dimensional	Real world objects
Temporal	1-dimensional data with start and finish time
Multi-dimensional	Data in relational and statistical data-bases with n attributes
Tree	Data with a link to (one) parent
Networks	Data items linked to an arbitrary number of other items

generates some form of *analytical abstraction* (*data transformation*). The analytical abstraction contains information about the data (meta-data). In the transformation operations *visualization transformation* and *visual mapping transformation*, appropriate visualizations are chosen (visualization abstraction) and the visual *picture* is generated [72]. The main aspect of data categorization is the differentiation between the raw data (value) and the meta-data (analytical abstraction), that contain structured information about the raw data [72, 73].

Another way of classifying data is by their dimensionality [3, 14, 36, 74]. This classification that was already introduced in context of classifying tasks and interactions (Sects. 2.3 and 2.4), is the most common way to differentiate data and their mapping to visualizations, tasks, and interactions [36, 74, 75]. This classification was proposed by Shneiderman in context of tasks to be solved with visualizations and his *Visual Information Seeking Mantra* [36]. The classification subdivides data based on their dimension in seven categories (illustrated in Table 2.5). The purpose of the classification was not to cover all types of data. Shneiderman proposed that there may exist further data types, e.g. $2\frac{1}{2}$-, 4-dimensions or multitrees. His categorization "reflects an abstraction of reality" [36, p. 339] and various visualizations may use combinations of them [36].

Keim et al. proposed based on Shneiderman's classification a "data type" classification [74], which defines the number of data variables as the dimensionality of data [74, p. 4]. In their classification One-dimensional data have one dense dimension, but they count temporal data in this category. They propose that each point of time may have further variables assigned (and can be multi-dimensional). Further they consolidated Shneiderman's 3- and multi-dimensional data into the category "Multi-dimensional data" and the categories Tree & Networks into the category of "Hierarchies & Graphs". Further they put two new categories into their classification "Text & Hypertext" and "Algorithms & Software" [74]. Table 2.6 illustrates their classification.

Table 2.6   Data type classification by Keim et al. [74]

Data types by Keim et al.	
One-dimensional	Data with one dense dimension, e.g. temporal data
Two dimensional	Data with two dense dimensions, e.g. X-Y-plots or geographical data
Multi-dimensional	Data with more than three dimension, also called multivariate data (We assume that three-dimensional data are investigated here too), e.g. relational databases
Text & Hypertext	Data with unknown dimensions and number. In particular the interlinked (hyperlinked) data (text, multimedia content in the World Wide Web)
Hierarchies & Graphs	Data with relationships to other information entities. The relation can ordered, arbitrary or hierarchical, e.g. e-mail relationships of persons, hyperlink relations on web
Algorithm & Software	Written representation (program code) of complex algorithms

The introduced classification investigated different aspects of data from different viewpoints. The ability to order data values plays an important role for classifying data in [3, 71]. The stages and transformation of data, in particular the differentiation between data and metadata, plays a role in [72, 73]. The main differentiation aspects for Shneiderman [36] and Keim et al. [74] were the dimensions of the data variables, which further may have an order too.

2.5.2   Data Types

The introduced classifications showed that the way how data types are distinguished is tightly coupled to visualization design. We assume that all data variables and values can be distinguished based on their level of measurement according to [36]. Let us assume for example that we have data-set with two variables, time and books. Let us further assume that there is no more information about the variable book, so that we are not able to categorize (order) it. We can now categorize this set of data as two-dimensional (or one-dimensional according to [74]) data with one quantitative (or ordinal according to [14]) and one nominal variable. Based on this information a fitting visualization could be chosen or designed. According to our example all variables can be categorized according to [36], therefore we introduce the data types based on their dimensionality according to Shneiderman [36] and Keim et al. [74]. It is important to have a common understanding of dimension in context of this

Table 2.7 Data types in context of this work (adapted from [36, 72, 74]

Data types in context of this work	
One-dimensional	Data with only one variable, e.g. list of words or temporal data without any associated variables
Two-dimensional	Data with two exactly two variables, e.g. X-Y-plots of time and books
Multi-dimensional	Data with more than two variables, e.g. relational databases
Hierarchies & Graphs	Data with relationships to data. The relation can be ordered, arbitrary or hierarchical
Metadata	Structured data with associations (links, identifiers) to other unstructured data, e.g. markup descriptions of textual webpages

work. Further some data types are not of interest, e.g. "Algorithms & Software", and other can be included to another data type. The aspect of metadata is of interest in context of this work, therefore we enhance the categorization with metadata. The data types will be described according to a slightly different categorization as illustrated in Table 2.7.

One-Dimensional Data
The variable of these data can be ordinal, nominal or quantitative. The way how the values of the variable are ordered (sequential manner as proposed in [36]) is an important factor for the visualization. One example for one dimensional data can be a data-set of countries with the variable "country name". As defined by Shneiderman [36], this variable or data value is nominal and does not own a "natural" order. It may be obvious from our experience with such data, that the nominal variable in this case is ordered alphabetically to provide an appropriate order for searching the relevant information in a more efficient way. This data-set can be visualized as list using the alphabetical order. One-dimensional data can be represented as a table with one column (variable) and a set of rows.

Two-Dimensional Data
Two-dimensional data have exactly two variables that are associated with each other. The values can be represented in a data table as two columns. The two variables may be ordinal, nominal or quantitative. An example could be a data-set consisting of countries with the variables "country name" and "population". As the nominal variable "country name" may be ordered alphabetically, the variable "population" owns a natural quantitative order. Two-dimensional data can be visually represented by X-Y-plots, whereas many chart visualizations are designed to illustrate these kinds of data. In the mentioned example a pie-chart may be an adequate way of illustration; based on the fact that one dimension has a quantitative order and the other nominal. Another example for two-dimensional data may be a data-set of "events" with the variables "quantity" and "time". The variable quantity represents the amount of

events associated with time as a second variable. The main difference between the data-sets is their order.

Multi-dimensional

Data-sets with three and more variables are investigated in this thesis as multi-dimensional data. Shneiderman differentiates in his classification between three and multi-dimensional data [36], while Keim et al. do not classify three-dimensional data at all [74]. The reason is quite simple, thus visualizing the third dimension on a two-dimensional screen is easy. The third dimension or variable can be presented by an adequate visual variable, e.g. color or size. The choice of the visual variable depends on the ability to order the data variable [3, 76]. If the data variable or dimension is for example quantitative, the size of the icons may be the right visual variable to illustrate this issue, in case of a nominal variable the differentiation may be performed by color. Data-sets with more than three dimensions are called in the literature multi-dimensional or multivariate data. There are many data-sets that consist of more than three variables [74]. With each variable or dimension the complexity of visualizing the data increases. Each of these variables may be nominal, quantitative or ordinal. Their visualization can overwhelm and confuse even experts, if hundreds of dimensions are visualized at the same time. Shneiderman proposes the use of buttons, if the cardinality is small, further he introduces a slider to control two-dimensional *scatter-grams* [36, 77]. Different visualization approaches focus on multi-dimensional data, the interaction with the visualizations and the control of their dimensions [78–81].

Hierarchies & Graphs

Data entities in data-sets may have different relations to each other and provide thereby a hierarchical or network structure. Keim et al. differentiates between arbitrary, ordered and hierarchical relationships of data entities [74, p. 4]. The relationships of the entities may have different structures. Shneiderman lists as examples, *acyclic, lattices, rooted, unrooted, directed, and undirected* as examples [36, p. 339], but proposes to investigate them all as network data. Keim et al. include the hierarchical structure as the same data type and do not differentiate anymore between hierarchies and networks [74]. Example of these relations may be the inter-linking of web-pages, the correspondence of emails or relations in relational data bases. [36, 74] There are various visualization techniques that face in particular the problem of large graphs and huge amount of entities [82–88].

Metadata

Metadata can be simply defined as data or information about data. Chi introduced in his Data State Reference Model a transformation step that produces some form of *analytical abstraction* from the underlying data [72]. The analytical abstraction is one way to generate a data model that provides information about the underlying data. Thus metadata depends in general on the underlying data; their dimensionality can depend to the dimensions of the data too. It is further possible that metadata have less or more dimensions as the raw data. It depends what the purpose of the metadata is. In general, metadata is represented in a structured form using an annotation (markup)

language. The developments in World Wide Web and mobile devices fostered the
use and application of metadata in the recent past. There exist a number of common
languages for describing metadata, e.g. the Extensible Markup Language (XML) or
the Hypertext Markup Language (HTML).

2.6 Methods and Techniques in Information Visualization

There exist many classifications of visualization techniques that use various criteria for categorizing visualizations and provide different views [89]. This chapter
already introduced some of the most common classifications. These are using data,
interactions, tasks, or the stages of data processing for classifying the visual representation. The following part of this section will focus on a classification of the
visual representations and convey correlations to the introduced classifications.

2.6.1 Classifications of Visualization Techniques

The most common classifications or taxonomies for information visualization use
one or more of the introduced criterion, namely data, interactions, tasks, or the
stages of data processing for classifying the visual representations. Although, most
of the visualization aspects were introduced, the step from data to a visual mapping
[3] is also essential to understand information visualization as foundation of this
work. This section will outline the aspect of visualization techniques from introduced
classifications and will further enhance with those classification that use other as the
introduced criterion.

Card and Mackinlay classified visualizations based on data value types [71]. Their
classification uses the ability to order data values (see Sect. 2.5.1) as criterion for
categorizing visualizations. With the use of graphical properties [4] and the mapping
to these properties they classified visualizations based on data value types (ordered,
nominal, and quantitative) into *scientific visualization, GIS, multi-dimensional plots,
multi-dimensional tables, information landscapes and spaces, node and link, trees,*
and *text transformation.*[71] In their revised work [3], as they defined *information
visualization*, they do not classify scientific visualization as a class or category of
information visualization [3]. This definition was applied and is today valid too, that
the scientific visualization builds an own class of visualizations [5, 14, 90].

Data as criterion for classifying information visualization played an important
role and are still today an important factor for classifications. The most common
classification beside the classification based on data value order is the *type by data
taxonomy* [36, p. 337] (see Sects. 2.4.1 and 2.5.1). This classification investigates
both, the data types as described in Table 2.5 and correlates them to the visualization tasks introduced in Table 2.1 [36, 76]. The correlation of tasks and data types
leads according to Shneiderman to solve visualization problems. He proposes to

Table 2.8   Visualization techniques (adapted from [74]

Visualization techniques according to Keim [74]	
Standard 2D/3D display	X-Y and X-Y-Z plots for standard visualization, e.g. barcharts, linecharts, or piecharts
Geometrically-transformed display	Techniques with exploratory statistics to find interesting transformation and patterns in multi-dimensional data-sets
Iconic display	Techniques that map attributes of data values from multi-dimensional data-sets to the features of icons, which may appear as *little faces, needle icons, star icons, color icons* etc.
Dense pixel display	Techniques that divide the screen into multiple subwindows based on the amount of the dimensions in the data-sets. Each data value of the dimensions is mapped to one pixel of the screen
Stacked display	Tailored techniques to present data partitioned in hierarchical manner. Therefore a coordinate system is embedded to another one and this may have further embedded coordinate systems. Each of the coordinates visualizes two attributes and provides with their stacked character the visualization of multi-dimensional data-set

arrange the data types in this taxonomy on the left side, which describes the task-domain. The information objects are correlating to users intentions for solving the tasks. [36] With this matrix a taxonomy is proposed that focuses on data types but investigates the problem-solving process (visual information seeking mantra) by the categorized tasks. [36, 76] The correlations of tasks (interactions) to data and this kind of mapping were enhanced by Keim et al. [74]. They used a slightly different data type classification (see Table 2.6), the interaction techniques (see Sect. 2.3.2) and provided a correlation to visualization techniques. They defined five visualization techniques that correspond to the display mode. They proposed that the visualization classes are "basic visualization principles" [74, p. 5] and can be combined to provide efficient visualizations. Table 2.8 illustrates the visualization classes identified by Keim et al. [74].

The classification of Keim et al. focuses on the visualization of multi-dimensional data. The identified data types include graphs and hierarchies, graph-layout algorithms are not mentioned at all. This type of data is correlated to geometrically-transformed data, which may contain graph-layouts, but this type of visualization is not mentioned [74].

Table 2.9 Visualization techniques of Keim in contrast with each other

Visualization techniques according to Keim [38, 74, 75]

[38, 75]	[74]	Description
Graph-based	–	Structured visualization of graphs and networks, e.g. basic graphs
Hierarchical	–	Subdividing the k-dimensional space and presenting the subspaces as hierarchies, e.g. treemap
Pixel-oriented	Dense pixel display	Techniques that map data values to the features of screen pixels
Icon-based	Iconic display	Techniques that map attributes of data values to the features of icons
Geometric	Geometrically-transformed	Techniques with exploratory statistics to find interesting transformation and patterns in multi-dimensional data-sets
–	Standard 2D/3D display	X-Y and X-Y-Z plots for standard visualization, e.g. barcharts, linecharts, or piecharts
Hybrid	–	Arbitrary combinations of the introduced techniques

Previous works of Keim [38, 75, 91] proposed a similar classification of visualizations, whereas a dedicated class for Graph-Layout was identified. This classification used more the interaction techniques rather than the data dimensions for identifying the categories of visualization techniques [38, p. T6-6]. Further this classification proposes more the combination of the different visualization classes and classes for 3D-, Dynamic- and Hierarchical techniques. Table 2.9 illustrates this classification and compares it to the visualization techniques in Table 2.6. The descriptions of the visualization techniques are used based on the work in [38, 75, 91] and may slightly be different. Just in case of the standard visualizations 2D and 3D Displays no description were given, so this is according to [74].

A high-level taxonomy to categorize visualization (both: scientific and information visualization) was proposed by Tory and Möller [90]. They investigated the visual space as whole and included factors like user models to get ideas about the "object of study". The involvement of a user model that affect the understanding of what data represent was novel in contrast to the work that was previously focused just on data, interactions and tasks [90]. The classification of visualization provides a high-level classification on data level. The authors differentiate between "discrete" and "continuous" character of "design models", "the conceptualization of a system that the designer has in mind" [90, p. 3] based on Norman's definition [92]. The differentiation leads to the choice of various display attributes, e.g. color or transparency [90]. In the next step of their classification they introduce low-level taxonomies based on the dependent and independent variables (data values) in visualizations. The continuous model visualization correlates the type of the dependent variables to the number of the independent variables. Dependent variables can occur as scalar (color gradients, isolines), vector (glyphs, particle traces), and tensor (ellipsoid-shaped glyphs) [90, p. 5]. The discrete model visualization first differentiates between data structure and data value. Value is defined by the dimension of the underlying data. Tory and Möller differentiate between 2D, 3D, and nD [90]. The structure may occur as node-link, hierarchies, and space-filling mosaics [90].

However, other criteria have been proposed to classify visualizations: e.g. by space, by changes of the data over time or their transformation steps, by number of visual attributes, by tasks and interactions, by the several aspects of data, or by human factors [37, 67, 71, 72, 89]. Further some special criteria were investigated for the classification of visualizations. Grimstead et al. for example investigated visualizations in context of collaboration [93]. Some visual classifications appeared in context of the application domain. Gelernter for example investigated visualizations in context of digital libraries and proposed a categorization in hierarchical lists, concept maps, tree maps, and self-organizing maps [94]. The main criterion for classification still remained the data based classification. Ward et al. presented a taxonomy of visualizations based on the data structure and subdivide each class again by the data types (data value) or visual attributes [14]. This more recent example amplifies the assumption that the visual transformation of the data type, structure and value is an adequate way to categorize visualization types. The classification of Ward et al. is illustrated in an abstract way in Table 2.10.

Table 2.10 Visualization classification by Ward et al. (adapted from [14])

Spatial data	Geospatial data	Multivariate data	Trees, networks and graphs	Text and documents
One-dimensional data	Visualization of point data	Point-based technique	Hierarchical structures	Single document visualizations
Two-dimensional data	Visualization of line data	Line-based techniques	Arbitrary graphs and networks	Document collection visualizations
Three-dimensional data	Visualization of area data	Region-based technique		Extended text visualizations
Dynamic data		Combination of techniques		
Combining technique				

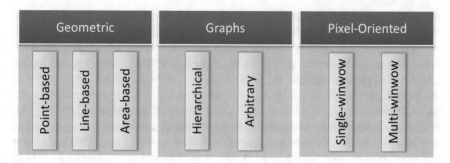

Fig. 2.7 Visualization classification used in this work (adapted and combined from [14, 74, 75]

2.6.2 Visualization Techniques

The classifications of the visualization techniques could outline that many criteria can be used for categorizing visualization. The main criteria still remain the data structure and data value [14, 38]. This section will give an overview about the visualization techniques that are classified above. For introducing the visualizations, a combination of the most relevant classifications will be used. The baseline of the classification used in this work, is the visual categorization by Keim et al. [38, 74, 75, 91]. Thus the different classifications use various terms for similar classes of visualization (see Table 2.9), the introduced techniques will be introduced by recently used terms in the research literature. Further some categories are either obvious (e.g. hybrid), not of interest in context of this work, or not anymore in context of information visualization. These classes will not be considered. Based on these factors a slightly different classification will be introduced that considers the research goals of our thesis. Figure 2.7 illustrates the visualization classification as a combined result of the works of Keim et al. and Ward et al. The introduced visualization techniques serve as examples and do not claim to present the state-of-the-art in information visualization.

Geometric Visualizations
Geometric visualization techniques [38] (Geometric Projection Techniques [91], Geometrically transformed Techniques [74]) transform data values of one- to multi-dimensional data into graphically transformed object. Every transformation that goes beyond the mapping of data values to pixel or a transformation to graph-layout can be count as geometric visualization technique. Keim et al. propose that this transformation aims at finding interesting patterns, in particular in multi-dimensional data-sets [74, p. 5]. In context of this thesis simple transformation in one and two dimensional data-sets are counted as geometric visualization techniques. Thus the "standard 2D /3D" visualization techniques are investigated as part of this visualization technique. The geometric transformation can be performed in various ways. In two dimensional data, e.g. the transformation is a mapping to geometric objects. In multi-dimensional data-sets the use of exploratory statistics, e.g. Principal Concept Analysis (PCA) [95]

or Factor analysis and multidimensional scaling, enables such geometric projections [74]. The outcome of the transformation is geometric objects that may appear as different object types. After the review of recent and traditional visualization techniques that use a geometric transformation, the classification of Ward et al. for multivariate data [14, p. 237] is partially applied to outline the main "geometric objects". Therefore we classify the geometric visualization techniques into their output geometry that commonly appears as "points", "lines", and "areas".

Point-Based Visualizations

The geometric transformation to point-based techniques includes the visual projection of data values to graphical representations as points, marks or other "aesthetic entities" [14, p. 237]. According to given attributes the visual representations of each record are placed on the screen to derive a visual representation of the whole data set. Depending on the selected attributes and the chosen layout method, point-based techniques are suitable to compare certain data characteristics, identify outliers or irregularities in the data, recognize relationships among data entities, and identify unexpected or previously unknown clusters and patterns. Usually each data record is projected from its n-dimensional space to a k-dimensional space (usually two- or three-dimensional) and the visual representation of the record is represented at the k-dimensional point on the screen.

One of the most common and established approaches for point-based visualizations are the use of Cartesian coordinates for positioning points [96]. The most common way perform such a positioning are scatterplot (two- or three-dimensional scatterplots [97]). Each axis of the coordinate system represents one dimension of the underlying data. The type of each dimension can either be ordinal or nominal. With the use of linear interpolation or other interpolation and projection methods [98] the data values are mapped to the dimensions of the plot. The representations of each data record are drawn at the location in the coordinate system that corresponds to the attribute values for the dimension in the record. So, different records can be compared according to the chosen dimensions. Another important feature of scatterplots is that the amount of the visualized dimensions is not limited to the number of axes of the coordinate system. Current approaches [97–99] make use of visual properties of the depicted marks to include further dimensions into the representation. One common method for including further dimensions in a scatterplot is mapping of data values to visual variables, e.g. color, size or shape.

The projection of points to Cartesian coordinates is just one way for the geometric transformation of data values as points on the screen. Further examples for point-projections are *Barycentric mappings* that consider coordinates as weighted sums of the anchor positions [100] or projections to circles, constructing multiple dimensions by a flattening process [101]. A recent example of Dinkla et al. visualizes network data with strong structural characteristics as point-based matrices (Compressed Adjacency Matrices) [102]. These matrices allow an easy detection of sub networks with specific structures and motifs [14, 102].

Line-Based Visualizations

Line-based visualizations project data values as lines on the screen. Therefore the vertical axis represents traditionally the value of a data variable and the horizontal axis the order. In many cases this order is a temporal (quantitative) variable that visualizes a continuous value. Line-based visualizations are the most common way to graphically represent the continuous value, e.g. in stock markets and financial sectors. Due to the high distribution and the high familiarity, these visualization techniques are effective for users to analyze and explore data. Each line represents one data value in correlation to two dimensions. With this characteristic the basic form of line-based visualizations are univariate. In contrast to point-based visualization techniques, line-based techniques provide also visual patterns for slopes, curvature, crossing, and further line patterns [14].

The most common way of line-based visualizations is line-graph. Similar to the traditional scatterplots, line-graphs visualizes data values on the axes of Cartesian coordinates. The line-graph is originally a univariate data visualization that can be easily extended for visualizing multivariate data [14]. According to Ward et al. there are four main strategies for providing visualizations of multivariate data with line-graphs: (a) superimposition (b) stacking, (c) ordered superimposition, and (d) ordered stacked [14, p. 246]. The idea behind each of these strategies is to represent each dimension with an own line in the same coordinate system using one of the four strategies:

- *Superimposing*: Superimposing multiple lines in one coordinate system allows directly comparing different dimensions and records in a single visualization. Crossings and shared trends can be easily recognized with this strategy. However if the dimensionality increases this approach becomes unclear.
- *Stacking*: Stacking lines on top of each other avoid crossings of lines. The idea is to start with the first line and use it as base line for the following. The quantitative value of each record is reflected by the distance of the line at a specific point to the line beneath it. So it is difficult to recognize the actual value for each record but this strategy is well suited to explore aggregated values of multiple records.
- *Ordered superimposition*: The ordering or sorting of records based on a specific dimension is another strategy for visualizing multivariate data in a line-graph. The ordering of the records has direct impact to the expressiveness of the visualization. Adequate orderings alleviate the recognition of patterns and relations among the data set.
- *Ordered stacking*: Analogue to the superimposition strategy, the ordering can also be applied to stacked line-graphs. The ordering of records according to a certain dimension may reveal interesting patterns and is a direct factor for the expressiveness of the visualization technique [14, p. 264]

Another well-known and established method of the geometric transformation to line-based visualizations is "parallel coordinates". This projection technique was introduced by Inselberg for studying high-dimensional geometry [103] and found its way through many applications and enhancements to multi-dimensional data. In contrast to line-graphs, parallel coordinates do not make use of orthogonal axes. They order the axes, which may represent the data dimensions parallel to each other. Spaced vertical and horizontal lines represent the ordering or value of the data. Traditionally each data record is plotted as a polyline across the parallel arranged axes. The polyline crosses each axis at the position proportional to its value to represent the characteristics of the data record. Parallel coordinates can be used to identify clusters in the data by means of similar curve shapes of the visualized records and to identify correlations and outliers [14, 104, 105]. Furthermore parallel coordinates allow two basic ordering methods that can be controlled by the user in most of existing implementations: (1) order of axes and (2) ordering of values [14]. The order of axis determines which dimensions are arranged next to each other and the ordering of values the position of values according to each axis. Analogue to line charts the ordering of axes as well as the ordering of the values in each dimension influences the expressiveness of the visualization. If an unfavorable ordering is chosen relevant patterns in the underlying data may remain hidden from the user.

Since the development of parallel coordinates many approaches extended the idea to provide more efficient visual pattern recognition. Fua and Ward introduced a hierarchical approach for visualizing aggregated information of large data-sets [106]. Miller and Wegemann introduced the concept of line densities to replace the raw data with a density plot that reveals clusters in multivariate data [107]. Yang proposed an interactive hierarchical dimension ordering, spacing and filtering approach that is based on dimension hierarchies derived from similarities among dimensions and introduced an approach for visualizing association rules in parallel coordinates [108]. Peng et al. presented an approach for reducing visual clutter by using dimension reordering strategies [109]. Novotny and Hauser proposed an approach for visualizing context at several levels of abstraction in parallel coordinates for representing outliers and trends [110] [14, p. 248]. Yuan et al. enhanced the line-based approach of parallel coordinates with point based techniques of scatterplots to enhance the usability in large multi-dimensional data [104]. Zhou et al. proposed a splatting approach to reduce clutter in parallel coordinates and reveal patterns [105]. Pilhöfer et al. presented ordering and optimization algorithms based on Bertin's classification [4] to reveal cluster with visual variables [111].

Line-graphs and parallel coordinates were example of line-based visualization, using orthogonal or parallel axes to visualize data values and dimensions, their correlations and visual patterns. The idea of visualizing lines across axes or parallel to axes was applied on different further approaches. Ward et al. outline in particular the radial axes techniques that project a given range of values to a circular scale [14]. They propose that each record is plotted as a line offset from the circular base for representing the data set. This technique is especially useful for analyzing periodic or cyclic data.

Area-Based Visualizations
Area-based (region-based, space-based, space-filling) visualization techniques make use of filled polygons or spaces on the screen to project data values and dimensions on the screen [14]. Usually instantiations of area-based techniques incorporate

different properties of the given data into the visual design of the polygons to convey additional values and data characteristics to offer the possibility of comparing different features of data. For instance the size, the shape or the color of the visual representation of a data record can be utilized for visually representing additional dimensions of the data set. Due to the ability of the human perception which enables an effective differentiation of the length or the size of presented polygons, area-based visualizations are successfully applied for representing and analyzing information encoded in the data [14].

The most common representative of area-based visualization technique is the bar chart that was successfully integrated in many different applications for comparing and analyzing data. Similar to scatterplots or line-graphs, bar charts are based on a Cartesian coordinates that include commonly two or three dimensions. Usually the vertical axis represents the range of the values, whereas the horizontal axis represents an ordering of the given records [14]. Each data record is represented as a bar and the length visualizes the data value. According to Ward et al. there are two different strategies to present each data record in a bar chart, stacked or clustered [14]. In a stacked bar chart the values of each dimension and record are stacked together in an aggregated bar. In a clustered bar chart the value of each record for each dimension is represented in a bar and positioned next to each other [14]. Other approaches of bar chart visualizations utilize a three dimensional coordinate system to separate the different dimensions into a new coordinate.

The geometric transformation of data in area-based visualization techniques do not need to be arranged to certain axes. Thus the polygons and spaces provide various visual variables for projection, the polygons themselves may contain information represented by appropriate visual variables. For instance the data in tables that already owns relationships are predestinated to be visualized as polygons [14]. Multivariate data are often stored in tables (see also Sect. 2.1) where each row represents a data record containing the values for each of the dimensions represented by the columns of the table. Tabular displays like heatmaps [112, 113] or table lens visualizations [114] directly exploit this tabular structure to visualize data. Heatmaps utilize a color gradient to map each value of the given table to a specific color and fill the corresponding cell of the table with the derived color [14]. For instance such an approach visualizes higher values with a more intensive color for visually separating these values. The approach of visualizing multivariate data with heatmaps is especially useful for the task of identifying outliers in the data by means of strongly deviating colors.

The Table Lens approach introduced by Rao and Card [114] represents each data record in a row of the table and encodes the values for each dimension and record respectively with a visual representation. For instance a numerical data entry is represented as a horizontal bar and the length represents the value. Usually implementations of Table Lens include different ordering functions that allow users to order the data records in the table with respect to a selected dimension. Another commonly integrated feature of table lens is the ability to expand specific rows by selection and to inspect the data records in its textual form. Table Lens visualizations are especially useful for inspecting a large amount of data and to get an overview of its characteristic as well as analyzing the data distribution of specific dimensions.

Beside tabular and multidimensional data, various data projections were developed with area-based visualizations. Shneiderman proposed a visualization approach that transforms hierarchical data structures in nested rectangular areas by splitting the screen into vertical and horizontal rectangles [115]. This *Treemap* visualization is a classic representative of area-based visualization techniques that is not constrained to any data structure. It is obvious that hierarchies can be conveyed with this visualization in an effective way, but further enhancements of the Treemap proofed its applicability for various data [116–122]. Brunetti for instance applied the Treemap visualization for semantic data [122]. Cox applied a circular Treemap to visualize the customer price index of New York [118] and used color of the spaces to visualize the temporal changes.

The introduced techniques for area-based visualizations should be considered as examples that may illustrate the effectiveness of these visualization techniques. There exist many further approaches that make use of the transformation of data values to polygons and their visual variables. Further examples could be the *Docuburst* introduced by Collins et al., an approach for a radial space-filling tree visualization to explore textual documents [123], Zhou et al. introduces an approach that make use of a splatting framework to reduce the clutters and transforms lines to spaces [105], and we have proposed a superimposed stacked-graph using polygons for visualizing trends and the occurrence of related documents for detecting latent trends over time [124]. Shin et al. proposed a hierarchical tree visualization for mobile devices by integrating focus plus context [125]. Their Tablorer system integrates space-filling and indented lists to visualize hierarchies.

Many applications and methods used combination of the above introduced techniques. A famous example is the *Table Lens* proposed by Rao and Card [114]. Table Lens make use of all three mapping methods based on the value type of the data, e.g. quantitative variables are presented by bars [114]. This approach was applied and enhanced by further applications (e.g. [126, 127]).

Graph Visualizations

Many data sets provide a kind of relationship between data entities. These relationships may be stored in tables, as metadata in structured documents, or appear as a result of data processing. The result of the data entity relationships can be called graphs. A graph consists of data entities that represents as a set of nodes and their connections to each other, called edges or links [74]. Diestel describes a graph as a pair $G = (V, E)$ of sets, where $E \subseteq [V^2]$, $V \cap E = 0$ and the elements of V are vertices or nodes and the elements of E are edges or links [128, p. 2]. The resulting relationships may occur in different ways and can consequently be visualized in various ways. Keim et al. proposes for visualizing graphs ordered, hierarchical, and arbitrary network relation visualizations [74, p. 4]. von Landesberger differentiates graph visualization techniques between node-link, matrix, and combined techniques [129]. This thesis differentiates between those relationships that describe a parent to child relation (hierarchical) and relations that have not any hierarchical correlations

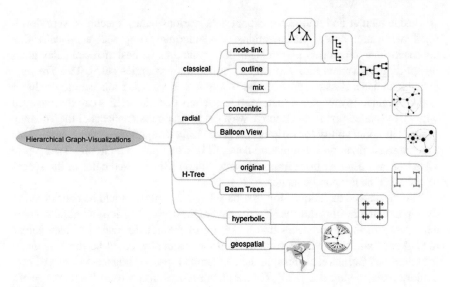

Fig. 2.8 A taxonomy for visualizing hierarchies with graph-layouts (adapted from [130, p. 579] and based on [84, p. 4])

(arbitrary). For differentiating the graph-layouts Herman et al. proposed a survey of different graph-layout algorithms and their application scenarios [84].

Hierarchical Graph-Visualizations
Hierarchical visualizations (or Tree visualizations [14]) aim at interactively visualize the data relationships that can be described as parent to child relation. Graph-visualizations are predestinated for visualizing these kinds of relations. Although the previously introduced Treemaps and further space-filling visual techniques are used to interactively present the hierarchical structure, one main way of visualizing hierarchical relations still remains the use graphs and their node-link diagrams in different ways. We elaborated the literature on graph-visualization and worked out a taxonomy of graph-layouts for visualizing hierarchies [130, 131]. Figure 2.8 illustrates an adapted version of our taxonomy. Although the list of graph-layouts for visualizing hierarchies is much longer, the illustration gives an overview of the how graph-layout could be used to present hierarchies.

One prominent way of using node-link diagrams for visualizing graph-based hierarchies are *dendrograms* [132]. Dendrograms are binary three structures that are characterized by the fact that all nodes of a hierarchy level are in the same line. This attribute improves the visual arrangement of the hierarchical structure. The simple graph-structure of dendrograms allows complex information presentation, whereas huge numbers of entities may overcharge users. Chen et al. investigated this aspect and proposed an overview plus context approach for dendrograms that separates a dynamically-linked overview and detail-view dendrogram to allow more complex information visualization [132].

D'Ambros and Lanza proposed an approach of visualizing hierarchies of discrete "time figures" [133]. Their approach investigated the problem of bug-finding and

-reporting in software systems and visualized it in a hierarchical graph-diagram. Therefore they used a heatmap similar approach for coloring time periods in rectangles, which are then visualized in a hierarchical node-link diagram. Holten and van Wijk introduced a visualization approach for comparing different hierarchies [134]. They proposed a visual clutter reduction method (Hierarchical Edge Bundles) that provides an easy interaction with the complex hierarchical structures. Dinkla et al. proposed a Visual Analytic system that supports comparisons of hierarchies by including node-link diagrams for the hierarchy representation of the weights of the related instances [135]. With a further linked-visualization they provided a detailed view on hierarchy structure, weights and metadata with a user-customizable analysis algorithm for ordering the weights as heatmap rectangles and find interesting nodes [135].

Arbitrary Graph-Visualizations

Hierarchical graph-visualizations are one specific type of graph-visualizations. They premise that at least a parent-child dependency exists. Even, if these hierarchies are just one subgroup of graph-visualizations, the placement algorithms could achieve high complexity as we described in the previous pages. Compared to hierarchical graph-visualizations, the placement of nodes in arbitrary graph layouts that fulfill certain optimization constraints is more complex brandes03. Arbitrary graph-visualizations illustrate information that does not contain a known class or structure [14]. There exist many ways to visualize graph structures, e.g. as node-link diagrams or matrices [136]. Ward et al. subdivides the use of node-link diagrams for graph-visualization into planar and force-directed graph drawing [14]. Force-directed [137, 138] graph drawing makes use of mass-spring-simulations, the so called "spring-embedders", and model the optimality criteria as an energy function [137]. Each pair of nodes is connected with two forces, one caused by the spring between them and the other a repulsion that keep nodes from getting too close to each other [14, p. 278]. Planar graph drawing makes the assumption that the underlying graph is planar and contains for instance no crossing edges. The research work on graph-drawing contains various methods for drawing graphs and visualizing information and their relationships as visual patterns. Each of the classes proposed by Ward et al. may rise in various ways and provide slightly different views on data and information. We worked out a kind of taxonomy for arbitrary graph-visualization [130]. Figure 2.9 illustrates an adapted version of our taxonomy. Similar to the hierarchical approaches graph-drawing methods, the list of graph-layouts for visualizing non-structured networks is much longer, the illustration aims to give just an overview of some algorithms.

There exist many approaches for visualizing graphs and these were applied in various ways to visualize information. One of the main problems of arbitrary graph-visualization is the complexity of huge amounts of nodes and edges. In particular, in arbitrary graph visualizations, where the structure and classes are unknown [14], the graph-visualization may become difficult for human to understand. Therefore many approaches faced this problem from different point of views. Abello et al. for instance proposed an approach for navigating in large graphs by displaying an overview of the

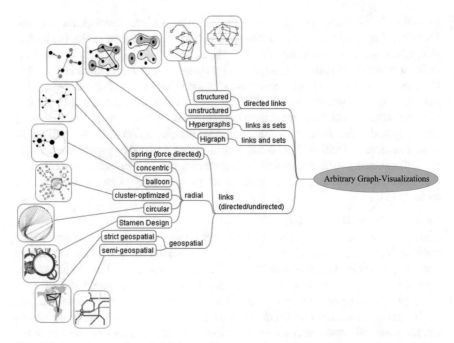

Fig. 2.9 A taxonomy for visualizing arbitrary graphs with graph-layouts (adapted from [130, p. 579] and [84, p. 4])

graph and provide with this overview a navigation support [86]. This linked overview further enables a filter functionality to collapse or expand sub-graphs in their detailed graph visualization. van Ham and Perer [87] faced the same problem from an opposite viewpoint. They proposed that the procedure of getting first an overview and then detailed information (see Sect. 2.4.2 for *visual information seeking mantra*), may not be appropriate for all visualization tasks or groups. They applied the "Degree of Interest" concept of Furnas [45] to graph-visualizations and proposed an interaction model that starts with a user interest-based *search* on initial nodes [87, pp. 954–955]. The second step of their interaction model is *show context* that provides a sub-graph of the focused node. Their model concludes with *expand on demand step*, where users can decide to expand this sub-graph and get more contextual information or an overview. They applied their model to a massive data-set of legal document citations in order to provide a comprehensible view on complex court decisions.

May et al. used the degree-of-interest approach of van Ham and Perer and enhanced it with a multiple focal node selection based on an enhanced focus plus context metaphor [88]. Their approach uses a symbolic (arrows) representation to point along the shortest-path to regions of interest in the graph that might be worth exploring. Further they added landmarks as graphical cues to give information on the context of the visible sub-graph.

Graph visualizations were applied in various domains, e.g. social networks [139, 140] for heterogeneous tasks, e.g. threat detection [141] and with various methods [142, 143]. This thesis does not claim to give a comprehensive view on graph visualizations or their algorithmic optimization methods. There exist a huge number of literature that investigate the aspect of graph visualization, graph drawing [82, 85, 144] and statistics [96] in depth. Further various surveys [84, 136, 145, 146] exist that give an excellent overview on these graph visualization techniques.

Pixel-Oriented Visualizations

Pixel-oriented visualization techniques project each data value of a data-set to one pixel of the screen and present them related to the dimensions [74, 147, 148]. This visualization approach is appropriate for massive data, thus the screen provides a huge amount of pixels and thereby huge visualization capabilities for an excellent overview of massive data. The visualization of data values as pixel has limitations too. One pixel may have the visual attributes of color, including brightness, hue, and saturation or make use of the Grey-scale and its values [148]. Further variables, e.g. shape or texture cannot be used on the pixel level. Massive data can be visualized with the pixel-oriented visualizations with two main limitations: Firstly the amount of visual variables, in most cases color, is limited to a certain range [149]. The second limitation is about arranging the pixels related to the data set. The visualization approach can be seen as a function that projects values from high-dimensional space to a two-dimensional screen [148].

To face the problem of the limitation of visual variables in pixel-oriented visualizations, Oelke et al. has investigated various visual variables for their "boosting" effect [150]. For their work they investigated the visual properties of Ware [28, 29] with respect to their applicability in pixel-oriented visualizations [150]. Whereas the restriction of one data value per pixel could not be applied to many of the proposed visual variables. Thus the variable *halo* for instance needs the space surrounding the pixel, which visualizes the data value [150, p. 9]. They introduced a differentiation of pixel-oriented visualization based on *image-driven* and *data-driven* boosting, which was further subdivided into *parse* and *dense* data sets. They evaluated eight visual variables based on the mentioned classification.

Keim proposed a differentiation of pixel-oriented visualization in *Query-Independent* and *Query-Dependent* visualization techniques [147, p. 2]. While query-independent visualizations visualizes data values by mapping them directly to color, query-dependent visualizations consider the user intention, as her query on the data set. Therefore the distances of attribute values to the query are mapped to colors. The mapping of color in query-independent visualization can be performed by naturally orders of data values [148] or by any other attribute of interest, e.g. by the quantity of data value appearance [80]. Keim introduced the question of using subwindows in various ways [148], e.g. as circle segment techniques or rectangular techniques. The use of subwindows in pixel-oriented visualization is common, but there exist various techniques that make use of a single window for visualizing the colored pixel for conveying information. In this thesis we investigate some examples and separate the pixel-oriented visualization techniques from the human point of view in *single-window visualizations* and *multi-window visualizations*. For perceiving visual information it is in our opinion important how information is visualized.

Single-Window Pixel-Oriented Visualizations
According to Keim, single-window pixel-oriented visualizations make use of the screen to project data values as colored pixel [148]. The order of the pixel depends on the intention of the visualization, if the data value contains a natural order, e.g. quantitative values. The mapping is assigned to this order. May introduced an approach for a single-window pixel-oriented visualization by applying the expressions of the *Karnaugh map* [151] to the visual appearance [152]. His approach enables the visualization of multi-dimensional data on a two-dimensional single-screen display and provides the recognition and detection of visual patterns in the multi-dimensional data-set [152, p. 227]. Another example for such a single-window pixel-oriented visualization was brought by Stein et al. [153]. They used the single-window pixel-oriented approach for visualizing social networks.

Multi-window Pixel-Oriented Visualizations
Pixel-oriented visualizations make maximum use of the screen. Thus each data value can be mapped to one single pixel; the most common approaches are dividing the screen into subwindows [148]. The multi-window approach provides more possibilities to detect and recognize clusters, visual patterns, and correlations [148]. The most common approach for subdividing the screen into multiple windows is rectangular subwindows. Andrienko et al. introduced such an approach for visualizing spatio-temporal data in a pixel-oriented multi-window visualization by using self-organizing maps [154]. Although the rectangular subwindows are the most common way of visualization in this context, there exist further approaches for sub-windowing the visual area. Keim et al. proposed a multi-resolution approach for pixel-oriented visualization in circle segments [155]. Their approach focused on improving the scalability of pixel-oriented visualization by introducing a multi-resolution pixel-oriented visual exploration approach for large data-sets. Therefore they combined clustering techniques with pixel-oriented projections to preserve local clusters by using circle segmentation, an enhanced type of *CircleView*.

2.7 Summary and Findings

This chapter introduced information visualization as canonical foundation of this thesis. The terminological distinction aimed at clarifying the term information visualization in contrast to visualization, scientific visualization and the recently rising term of Visual Analytics. We defined information visualization in context of this work as interactive visualization of abstract data that includes the visualization of data models and provide an interactive character with interlinking to data and their operands to amplify cognition and provide insights and knowledge. In context of this work the human with his ability to perceive and process visual information is mainly focused. We investigated human perception and human visual processing to give an insight how human perceives visualizations. We could outline that beside heterogeneous classification of human visual processing, the so called parallel and sequential (or serial) processing plays an important role, in particular for choosing the proper visual variables. Different research outcomes of studies in cognitive sciences were introduced that prove at least a continuous differentiation of visual variables and the way how and when they are perceived. Further the results of the studies can be used to improve in particular the visual appearance of abstract data.

With the interactive character of information visualization, we could depict that information visualization is more than only pictures. We introduced various classifications of interaction in information visualization and selected one classification to describe interaction on data and visual level. Thus the human interaction leads to solving tasks with information visualization, different classifications of visualization tasks were introduced. We could illustrate that a clear bisection between tasks and interactions is not possible with the existing classifications. We introduced a high-level tasks classification and tried to categorize the different existing classifications into the abstract model, which was partially published in [2] and assigned categorized the existing types of tasks into the abstract model. With visual perception, tasks and interactions, we covered the human-interaction with visual information systems. The data level completed the process of data transformation to an interactive visual representation. In this context various classifications of data provided different views on data, their value, and their dimensions or variables. For describing the data types the three introduced classifications were merged into a slightly different classification. The goal was to give a common understanding of the terms that was often used ambiguously in context of data and visualizing data. We described the most common appearances of data in context information visualization based on our classification. The outcomes of the data structures, values and variables were used to introduce information visualization methods and techniques. Therefore various classifications were introduced that give different views on visualization techniques. We chose one common classification and changed it slightly to introduce an overview of possible visualizations. Based on the introduced classification we introduced exemplary visualization techniques and methods.

The main goal of this chapter was to give an overview of the various disciplines, techniques, goals, and approaches that are coupled to interactive information visualization: cognitive scientists investigates the perception of visual illustrations, algorithmic methods optimize layouting, data models and visualization techniques, the area of human computer interaction investigates the behavior of human and appropriate reactions of computer systems, and further research areas, e.g. Visual Analytics work on the optimization for coupling these methods. Further this chapter worked out that information visualization as an interdisciplinary research area makes use of ambiguous terms and classifies their approaches in various ways. Therefore it is essential to have a common understanding of the terms at least in context of this work.

The next two chapters will focus on specific domains of information visualization and its applications. First we will introduce a general view on semantic technologies and data. The goal is to give a more detailed view on the state of the art in semantics visualization. Thereafter we introduce the general idea of adaptive systems and survey the existing approaches on adaptive information visualizations.

References

1. K. Nazemi, M. Steiger, D. Burkhardt, J. Kohlhammer, in *Handbook of Research on Advanced ICT Integration for Governance and Policy Modeling*, ed. by P. Sonntagbauer, K. Nazemi, S. Sonntagbauer, G. Prister, D. Burkhardt (Business Science Reference (IGI Global, Hershey PA, USA, 2014)
2. K. Nazemi, J. Kohlhammer, in *1st International Workshop on User-Adaptive Visualization (WUAV 2013)*. *Extended Proceedings of UMAP 2013, CEUR Workshop Proceedings*, vol. 997 (2013). ISSN:1613-0073
3. S.K. Card, J.D. Mackinlay, B. Shneiderman, *Readings in Information Visualization: Using Vision to Think*, 1st edn. (Morgan Kaufmann, San Francisco, 1999). http://www.amazon.com/exec/obidos/redirect?tag=citeulike07-20&path=ASIN/1558605339
4. J. Bertin, *Semiology of Graphics* (University of Wisconsin Press, Madison, 1983)
5. J.J. Thomas, K.A. Cook, *Illuminating the Path: The Research and Development Agenda for Visual Analytics* (National Visualization and Analytics Ctr, 2005). http://www.worldcat.org/isbn/0769523234
6. D. Keim, G. Andrienko, J.D. Fekete, C. Görg, J. Kohlhammer, G. Melançon, in *Information Visualization*. Lecture Notes in Computer Science, vol. 4950, ed. by A. Kerren, J. Stasko, J.D. Fekete, C. North (Springer, Berlin/Heidelberg, 2008), pp. 154–175
7. D.A. Keim, F. Mansmann, J. Schneidewind, H. Ziegler, J. Thomas, *Visual Data Mining: Theory, Techniques and Tools for Visual Analytics*, Lecture Notes in Computer Science (LNCS) (Springer, Berlin, 2008)
8. J. Thomas, J. Kielman, Challenges for visual analytics. *Information Visualization* (2009). doi:10.1057/ivs.2009.26. http://www.cs.uml.edu/~grinstei/InfoVisJournal-2009-2011/Information%20Visualization-2009-Thomas-309-14.pdf
9. J. Thomas, in *2007 6th International Asia-Pacific Symposium on Visualization*. APVIS'07 (2007), p. ix
10. J. Thomoas, J. Kielman, Inf. Vis. J. **11**, 309 (2009). Special Issue: Foundations and Frontiers of Visual Analytics
11. D. Keim, J. Kohlhammer, G. Ellis, F. Mansmann, *Matering the Information Age Solving Problems with Visual Analytics* (Eurographics Association, Goslar, 2010)
12. J. Kohlhammer, K. Nazemi, T. Ruppert, D. Burkhardt, IEEE Comput. Graph. Appl. **32**(5), 84 (2012). doi:10.1109/MCG.2012.107. http://bibcd.igd.fraunhofer.de/bibcd/INI_Science/papers/2012/12p079.pdf
13. D.H. Hubel, *Scientific American Library* (W. H. Freeman, San Francisco, 1988)
14. M. Ward, G. Grinstein, D. Keim, *Interactive Data Visualizations Foundations, Techniques, and Applications* (A K Peters, Ltd., Natick, 2010)
15. A.M. Treisman, Comput. Vis. Graph. Image Process. **31**(2), 156 (1985)
16. J. Wolfe, A.M. Treisman, What shall we do with the preattentive processing stage use it or lose it? Poster presented at the Third Annual Meeting of the Vision Sciences Society, Sarasota, FL (2003)
17. A.M. Treisman, G. Gelade, Cogn. Psychol. **12**(1), 97 (1980)
18. A.M. Treisman, J. Souther, J. Exp. Psychol.: Hum. Percept. Perform. **12**, 107 (1986)
19. A.M. Treisman, S. Gormican, Psychol. Rev. **95**(1), 15 (1988)
20. A.M. Treisman, Sci. Am. **255**, 106 (1986)
21. A.M. Treisman, J. Exp. Psychol.: Hum. Percept. Perform. **17**(3), 652 (1991)
22. P. Quinlan, G. Humphreys, Attention Percept. Psychophys. **41**, 455 (1987)
23. J.M. Wolfe, J. Exp. Psychol. Hum. Percept. Perform. **15**(3), 419 (1989)
24. J.M. Wolfe, Psychon. Bull. Rev. **1**(2), 202 (1994)
25. J.M. Wolfe, G.G. Gancarz, Lakshminarayanan, V. (ed.), *Basic and Clinical Applications of Vision Science* (Springer, Netherlands, 1999), pp. 189–192
26. J.M. Wolfe, W. Gray (ed.), *Integrated Models of Cognitive Systems*, pp. 99–119 (2007)
27. L. Itti, C. Koch, Nat. Rev. Neurosci. **2**, 1 (2001)

28. C. Ware, *Information Visualization Perception for Design* (Morgan Kaufmann Publishers, San Francisco, 2004)
29. C. Ware, *Information Visualization Perception for Design* (Morgan Kaufmann (Elsevier), Amsterdam, 2013)
30. R.A. Rensink, Vis. Cogn. **7**, 17 (2000)
31. R.A. Rensink, Annu. Rev. Psychol. **53**, 245 (2002)
32. R. Hyman, J. Exp. Psychol. **45**, 423 (1953)
33. P. Fitts, J. Exp. Psychol. **47**, 381 (1954)
34. J. Accot, S. Zhai, in *Proceedings of CHI 1997* (ACM, 1997), pp. 295–302
35. A.W. Siegel, S.H. White, H.W. Reese (ed.), *Advances in Child Development and Behavior*, vol. 10 (1975), p. 9
36. B. Shneiderman, in *VL* (1996), pp. 336–343
37. A. Cockburn, A. Karlson, B.B. Bederson, ACM Comput. Surv. **41**(1), 1 (2008)
38. D.A. Keim, Visual techniques for exploring databases, in *Invited Tutorial, International Conference on Knowledge Discovery in Databases (KDD'97)* (1997)
39. M. Leissler, A generic framework for the development of 3d information visualisation applications. Ph.D. thesis, Technische Universität Darmstadt (2004)
40. M.A. Hearst, User interfaces and visualization, in *Modern Information Retrieval*, ed. by R.A. Beaza-Yates, B. Ribeiro-Neto (Addison Wesley Longman, ACM Press, 1999), pp. 257–324
41. M.O. Ward, J. Yang, in *Joint Eurographics/IEEE TCVG Symposium on Visualization*, Konstanz, Germany, ed. by P. Brunet, D. Fellner (2004), pp. 137–145
42. E.A. Rundensteiner, Y. Fua, M.O. Ward, IEEE Trans. Vis. Comput. Graph. **6**, 150 (2000)
43. G.W. Furnas, B.B. Bederson, in *Proceedings of the SIGCHI Conference on Human Factors in Computing Systems, CHI'95* (ACM Press/Addison-Wesley Publishing Co., New York, NY, USA, 1995), pp. 234–241. doi:10.1145/223904.223934, http://dx.doi.org/10.1145/223904.223934
44. R. Spence, M. Apperley, Behav. Inf. Technol. **1**(1), 43 (1982). doi:10.1080/01449298208914435. http://www.tandfonline.com/doi/abs/10.1080/01449298208914435
45. G.W. Furnas, in *Proceedings of the SIGCHI Conference on Human Factors in Computing Systems, CHI'86* (ACM, New York, NY, USA, 1986), pp. 16–23. doi:10.1145/22627.22342
46. G.W. Furnas, The fisheye view: A new look at structured files. Technical report, Murray Hill, NJ (AT&T Bell Laboratories) (1981)
47. K. Boulos, Int. J. Health Geograph. **2**, 1 (2003)
48. J. Garcia, R. Theron, F. Garcia, in *Highlights in Practical Applications of Agents and Multiagent Systems*. Advances in Intelligent and Soft Computing, vol. 89, ed. by J.B. Pérez, J.M. Corchado, M.N.M. García, V. Julián, P. Mathieu, J.C. Bago, A. Ortega, A. Fernández-Caballero (Springer, Berlin/Heidelberg, 2011), pp. 85–92. http://dx.doi.org/10.1007/978-3-642-19917-2_11
49. L. Bartram, in *1997 IEEE International Conference on Systems, Man, and Cybernetics, 1997. Computational Cybernetics and Simulation*, vol. 2 (1997), pp. 1686–1692. doi:10.1109/ICSMC.1997.638254
50. S. Palmer, *Vision Science: Photons to Phenomenology* (MIT Press, Cambridge, 1999)
51. G. Robertson, M. Czerwinski, D. Fisher, B. Lee, Rev. Hum. Factors Ergon. **5**(1), 41 (2009). doi:10.1518/155723409X448017, http://rev.sagepub.com/content/5/1/41.abstract
52. G.G. Robertson, J.D. Mackinlay, in *Proceedings of the 6th Annual ACM Symposium on User Interface Software and Technology, UIST'93* (ACM, New York, NY, USA, 1993), pp. 101–108. doi:10.1145/168642.168652, http://doi.acm.org/10.1145/168642.168652
53. B. Shneiderman, IEEE Softw. **11**(6), 70 (1994). doi:10.1109/52.329404, http://dx.doi.org/10.1109/52.329404
54. C. Ahlberg, B. Shneiderman, in *Proceedings of the SIGCHI Conference on Human Factors in Computing Systems: Celebrating Interdependence, CHI'94* (ACM, New York, NY, USA, 1994), pp. 313–317. doi:10.1145/191666.191775, http://doi.acm.org/10.1145/191666.191775

55. C. Ahlberg, C. Williamson, B. Shneiderman, in *Proceedings of the SIGCHI Conference on Human Factors in Computing Systems, CHI'92* (ACM, New York, NY, USA, 1992), pp. 619–626. doi:10.1145/142750.143054, http://doi.acm.org/10.1145/142750.143054

56. B. Shneiderman, C. Plaisant, *Designing the User Interface*, 4th edn. (Pearson Education, Toronto, 2005)

57. J. Golbeck, Direct manipulation (Univerity of Maryland, 2002). http://www.cs.umd.edu/class/fall2002/cmsc838s/tichi/dirman.html. Accessed June 2012

58. B. Shneiderman, Direct manipulation: a step beyond programming languages, in *Human-Computer Interaction*, ed. by R.M. Baecker (Morgan Kaufmann Publishers Inc., San Francisco, CA, USA, 1987), pp. 461–467. http://dl.acm.org/citation.cfm?id=58076.58115

59. A. Buja, D. Cook, D.F. Swayne, J. Comput. Graph. Stat. **5**, 78 (1996)

60. M.C. Chuah, S.F. Roth, in *INFOVIS'96: Proceedings of the 1996 IEEE Symposium on Information Visualization (INFOVIS'96)* (IEEE Computer Society, Washington, DC, USA, 1996), p. 29

61. S. Wehrend, C. Lewis, in *VIS'90: Proceedings of the 1st Conference on Visualization'90* (IEEE Computer Society Press, Los Alamitos, CA, USA, 1990), pp. 139–143

62. M.X. Zhou, S.K. Feiner, in *CHI'98: Proceedings of the SIGCHI Conference on Human Factors in Computing Systems* (ACM Press/Addison-Wesley Publishing Co., New York, NY, USA, 1998), pp. 392–399

63. P.R. Keller, M.M. Keller, *Visual Cues: Practical Data Visualization* (IEEE Computer Society Press, Los Alamitos, CA, 1994)

64. J.S. Yi, Y. ah Kang, J. Stasko, J. Jacko, IEEE Trans. Vis. Comput. Graph. **13**(6), 1224 (2007). doi:10.1109/TVCG.2007.70515

65. W.A. Pike, J. Stasko, R. Chang, T.A. O'Connell, Inf. Vis. **8**(4), 263 (2009). doi:10.1057/ivs.2009.22, http://dx.doi.org/10.1057/ivs.2009.22

66. J. Kohlhammer, Knowledge representation for decision-centered visualization. Ph.D. thesis, Technische Universität Darmstadt (2005)

67. R. Amar, J. Stasko, in *IEEE Symposium on Information Visualization. INFOVIS 2004* (2004), pp. 143–150. doi:10.1109/INFVIS.2004.10

68. C. Fluit, M. Sabou, F. van Harmelen, *Handbook on Ontologies*. International Handbooks on Information Systems, ed. by S. Staab, R. Studer (Springer, Berlin, 2004), pp. 415–434

69. C. Fluit, M. Sabou, F. van Harmelen, in *Visualizing the Semantic Web: XML-Based Internet and Information Visualization*, 2nd edn., ed. by V. Geroimenko, C. Chen (Springer-Verlag, London, 2006)

70. D.A. Keim, Commun. ACM **44**, 38 (2001). doi:http://doi.acm.org/10.1145/381641.381656, http://doi.acm.org/10.1145/381641.381656

71. S.K. Card, J.D. Mackinlay, in *INFOVIS* (1997), pp. 92–99

72. E.H. Chi, in *INFOVIS'00: Proceedings of the IEEE Symposium on Information Vizualization 2000* (IEEE Computer Society, Washington, DC, USA, 2000), p. 69

73. E.H. Chi, J. Riedl, in *INFOVIS'98: Proceedings of the 1998 IEEE Symposium on Information Visualization* (IEEE Computer Society, Washington, DC, USA, 1998), pp. 63–70

74. D.A. Keim, C. Panse, M. Sips, in *Exploring Geovisualization*, ed. by J. Dykes, A. MacEachren, M.J. Kraak (Elsevier, Oxford, 2003)

75. D.A. Keim, in *SIGMOD Conference* (1996), p. 543

76. B. Shneiderman, Olive: On-Line Library of Information Visualization Environments (1999). http://otal.umd.edu/Olive/, Accessed July 2013

77. C. Ahlberg, B. Shneiderman, in *Proceedings of ACM CHI94 Conference on Human Factors in Computing Systems* (1994), pp. 365–371

78. A. Inselberg, B. Dimsdale, in *Proceedings of the 1st Conference on Visualization'90, VIS'90* (IEEE Computer Society Press, Los Alamitos, CA, USA, 1990), pp. 361–378. http://portal.acm.org/citation.cfm?id=949531.949588

79. M. Gahegan, Comput. Environ. Urban Syst. **22**(1), 43 (1998). doi:10.1016/S0198-9715(98)00018-0, http://www.sciencedirect.com/science/article/pii/S0198971598000180

80. T. May, J. Kohlhammer, Comput. Graph. Forum **27**(3), 911 (2008). http://bibcd.igd. fraunhofer.de/bibcd/INI_Science/papers/2008/08p049.pdf
81. T. May, J. Davey, in *IEEE Conference on Visual Analytics Science and Technology 2010. Proceedings. IEEE Computer Society Visualization and Graphics Technical Committee (VGTC)* (IEEE Press, New York, 2010), pp. 239–240. http://bibcd.igd.fraunhofer.de/bibcd/ INI_Science/posters-etc/10dp008.pdfbibcd.igd.fraunhofer.de/bibcd/INI_Science/posters-etc/10dp008a.pdf
82. P. Mutzel, C. Gutwenger, R. Brockenauer, S. Fialko, G.W. Klau, M. Krüger, T. Ziegler, S. Näher, D. Alberts, D. Ambras, G. Koch, M. Jünger, C. Buchheim, S. Leipert, in *Whitesides* [118], pp. 456–457
83. S. Whitesides (ed.), *Graph Drawing, 6th International Symposium, GD'98*, Montréal, Canada, August 1998. Lecture Notes in Computer Science, vol. 1547 (Springer, Berlin, 1998)
84. I. Herman, G. Melançon, M. Marshall, IEEE Trans. Vis. Comput. Graph. **6**(1), 24 (2000)
85. C. Gutwenger, M. Jünger, G.W. Klau, S. Leipert, P. Mutzel, R. Weiskircher, in *Graph Drawing* (2001), pp. 473–474
86. J. Abello, F. van Ham, N. Krishnan, IEEE Trans. Vis. Comput. Graph. **12**(5), 669 (2006). doi:10.1109/TVCG.2006.120
87. F. van Ham, A. Perer, IEEE Trans. Vis. Comput. Graph. **15**, 953 (2009)
88. T. May, M. Steiger, J. Davey, J. Kohlhammer, Comput. Graph. Forum **31**(3), 985 (2012). doi:10.1111/j.1467-8659.2012.03091.x, http://bibcd.igd.fraunhofer.de/bibcd/INI_ Science/papers/2012/12p037.pdf
89. Q. Chengzhi, Z. Chenghu, P. Tao, in *Asia GIS Conference* (2003)
90. M. Tory, T. Möller, in *Proceedings of the IEEE Symposium on Information Visualization, INFOVIS'04* (IEEE Computer Society, Washington, DC, USA, 2004), pp. 151–158. doi:10. 1109/INFOVIS.2004.59, http://dx.doi.org/10.1109/INFOVIS.2004.59
91. D.A. Keim, H.P. Kriege, IEEE Trans. Knowl. Data Eng. **8**, 923 (1996)
92. D.A. Norman, *The Design of Everyday Things* (Doubleday, New York, 2002). First published in 1988
93. I.J. Grimstead, D.W. Walker, N.J. Avis, in *DS-RT'05: Proceedings of the 9th IEEE International Symposium on Distributed Simulation and Real-Time Applications* (IEEE Computer Society, Washington, DC, USA, 2005), pp. 61–69
94. J. Gelernter, Knowl. Organ. **34**(3), 128 (2007)
95. I.T. Jolliffe, *Principal Component Analysis*, 2nd edn. (Springer, New York, 2002)
96. S.B. Jarrell, *Basic Statistics* (Brown (William C.) Co., USA, 1994)
97. H. Sanftmann, D. Weiskopf, IEEE Trans. Vis. Comput. Graph. **18**(11), 1969 (2012). doi:10. 1109/TVCG.2012.35
98. S. Bachthaler, D. Weiskopf, in *Proceedings of the 11th Eurographics/IEEE—VGTC Conference on Visualization, EuroVis'09* (Eurographics Association, Aire-la-Ville, Switzerland, 2009), pp. 743–750. doi:10.1111/j.1467-8659.2009.01478.x, http://dx.doi.org/10.1111/ j.1467-8659.2009.01478.x
99. F. Jourdan, A. Paris, P.Y. Koenig, G. MelanÃSon, in *Pixelization Paradigm*. Lecture Notes in Computer Science, vol. 4370, ed. by P. LÃvy, B. Grand, F. Poulet, M. Soto, L. Darago, L. Toubiana, J.F. Vibert (Springer, Berlin, 2007), pp. 202–215. doi:10.1007/978-3-540-71027_ 18, http://dx.doi.org/10.1007/978-3-540-71027-1_18
100. F. Piekniewski, L. Rybicki, in *Artificial Intelligence and Soft Computing–ICAISC 2004*. Lecture Notes in Computer Science, vol. 3070, ed. by L. Rutkowski, J. Siekmann, R. Tadeusiewicz, L. Zadeh (Springer, Berlin, 2004), pp. 247–252. doi:10.1007/978-3-540-24844-6_33, http:// dx.doi.org/10.1007/978-3-540-24844-6_33
101. J. Sharko, G. Grinstein, K. Marx, IEEE Trans. Vis. Comput. Graph. **14**(6), 1444 (2008). doi:10.1109/TVCG.2008.173
102. K. Dinkla, M. Westenberg, J. van Wijk, IEEE Trans. Vis. Comput. Graph. **18**(12), 2457 (2012). doi:10.1109/TVCG.2012.208
103. A. Inselberg, Vis. Comput. **1**(2), 69 (1985). doi:10.1007/BF01898350, http://dx.doi.org/10. 1007/BF01898350

104. X. Yuan, P. Guo, H. Xiao, H. Zhou, H. Qu, IEEE Trans. Vis. Comput. Graph. **15**(6), 1001 (2009). doi:10.1109/TVCG.2009.179

105. H. Zhou, W. Cui, H. Qu, Y. Wu, X. Yuan, W. Zhuo, in *Proceedings of the 11th Eurographics/IEEE—VGTC Conference on Visualization, EuroVis'09* (Eurographics Association, Aire-la-Ville, Switzerland, 2009), pp. 759–766. doi:10.1111/j.1467-8659.2009.01476. x, http://dx.doi.org/10.1111/j.1467-8659.2009.01476.x

106. Y.H. Fua, M. Ward, E. Rundensteiner, in *Proceedings of the Conference on Visualization'99: Celebrating Ten Years, VIS'99* (IEEE Computer Society Press, Los Alamitos, CA, USA, 1999), pp. 43–50. http://dl.acm.org/citation.cfm?id=319351.319355

107. J.J. Miller, E.J. Wegman, Construction of line densities for parallel coordinate plots, in *Computing and Graphics in Statistics*, ed. by A. Buja, P.A. Tukey (Springer, New York, NY, USA, 1991), pp. 107–123

108. L. Yang, IEEE Trans. Knowl. Data Eng. **17**(1), 60 (2005). doi:10.1109/TKDE.2005.14

109. W. Peng, M. Ward, E. Rundensteiner, in *IEEE Symposium on Information Visualization. INFOVIS 2004* (2004), pp. 89–96. doi:10.1109/INFVIS.2004.15

110. M. Novotny, H. Hauser, IEEE Trans. Vis. Comput. Graph. **12**(5), 893 (2006). doi:10.1109/TVCG.2006.170

111. A. Pilhofer, A. Gribov, A. Unwin, IEEE Trans. Vis. Comput. Graph. **18**(12), 2506 (2012). doi:10.1109/TVCG.2012.207

112. L. Wilkinson, *SYSTAT for DOS: Advanced Applications* (SYSTAT Inc., Evanston, IL, 1994)

113. D. Borland, R. Taylor, IEEE Comput. Graph. Appl. **27**(2), 14 (2007). doi:10.1109/MCG. 2007.323435

114. R. Rao, S.K. Card, in *Proceedings of the SIGCHI Conference on Human Factors in Computing Systems: Celebrating Interdependence, CHI'94* (ACM, New York, NY, USA, 1994), pp. 318–322. doi:10.1145/191666.191776, http://doi.acm.org/10.1145/191666.191776

115. B. Shneiderman, ACM Trans. Graph. **11**(1), 92 (1992). doi:10.1145/102377.115768, http://doi.acm.org/10.1145/102377.115768

116. B. Shneiderman, M. Wattenberg, in *Proceedings of the IEEE Symposium on Information Visualization 2001 (INFOVIS'01), INFOVIS'01* (IEEE Computer Society, Washington, DC, USA, 2001), pp. 73. http://dl.acm.org/citation.cfm?id=580582.857710

117. B. Shneiderman, C. Plaisant, Treemaps for space-constrained visualization of hierarchies. University of Maryland (2009). http://www.cs.umd.edu/hcil/treemap-history/index.shtml. http://www.cs.umd.edu/hcil/treemap-history/index.shtml, Accessed July 2013

118. A. Cox, All of inflation's little parts. The New York Times (2008). http://www.nytimes. com/interactive/2008/05/03/business/20080403_SPENDING_GRAPHIC.html, http://www. nytimes.com/interactive/2008/05/03/business/20080403_SPENDING_GRAPHIC.html, Accessed July 2013

119. K. Nazemi, M. Breyer, C. Hornung, in *Universal Access in Human-Computer Interaction. Applications and Services*. Lecture Notes in Computer Science, vol. 5616, ed. by C. Stephanidis (Springer, Berlin, 2009), pp. 83–91. doi:10.1007/978-3-642-02713-0_9, http://dx.doi.org/10.1007/978-3-642-02713-0_9

120. K. Nazemi, C. Stab, M. Breyer, D. Burkhardt, T. May, T. von Landesberger, THESEUS CTC-WP5-Innovative Benutzerschnittstellen und Visualisierungen: Alleinstellungsmerkmale der CTC-WP5 Technologien. (THESEUS Core Technology Cluster for Innovative User Interfaces and Visualizations: Unique Features of the CTC-WP5 Technologies). THESEUS Programme Deliverable of the Fraunhofer Institute for Computer Graphics Research (IGD). Nazemi K. (ed.) (2010)

121. M. Rios-Berrios, P. Sharma, T.Y. Lee, R. Schwartz, B. Shneiderman, Gov. Inf. Q. **29**(2), 212 (2012). doi:10.1016/j.giq.2011.07.004, http://www.sciencedirect.com/science/article/pii/S0740624X11001055

122. J.M. Brunetti, in *Proceedings of the 3rd International Conference on Web Intelligence, Mining and Semantics, WIMS'13* (ACM, New York, NY, USA, 2013), pp. 37:1–37:8. doi:10.1145/2479787.2479824, http://doi.acm.org/10.1145/2479787.2479824

123. C. Collins, S. Carpendale, G. Penn, in *Proceedings of the 11th Eurographics/IEEE—VGTC Conference on Visualization, EuroVis'09* (Eurographics Association, Aire-la-Ville, Switzerland, 2009), pp. 1039–1046. doi:10.1111/j.1467-8659.2009.01439.x, http://dx.doi.org/10.1111/j.1467-8659.2009.01439.x

124. C. Stab, M. Breyer, D. Burkhardt, K. Nazemi, J. Kohlhammer, in *Proceedings of SIGRAD 2012. SIGRAD, Linköping Electronic Conference Proceedings; 81* (Linköping University Electronic Press, Linköping, 2012), pp. 83–86. http://bibcd.igd.fraunhofer.de/bibcd/INI_Science/papers/2012/12p121.pdf

125. H. Shin, G. Park, J. Han, in *Proceedings of the 13th Eurographics/IEEE—VGTC Conference on Visualization, EuroVis'11* (Eurographics Association, Aire-la-Ville, Switzerland, 2011), pp. 1131–1140. doi:10.1111/j.1467-8659.2011.01962.x, http://dx.doi.org/10.1111/j.1467-8659.2011.01962.x

126. M. John, C. Tominski, H. Schumann, Visual and analytical extensions for the table lens. in *Proceedings of SPIE*, vol. 6809 (2008), pp. 680907–680912. doi:10.1117/12.766440, http://dx.doi.org/10.1117/12.766440

127. G. Albuquerque, M. Eisemann, D. Lehmann, H. Theisel, M. Magnor, in *2010 IEEE Symposium on Visual Analytics Science and Technology (VAST)* (2010), pp. 19–26. doi:10.1109/VAST.2010.5652433

128. R. Diestel, *Graph Theory* (Springer, New York, 2010). http://diestel-graph-theory.com/basic.html

129. T. von Landesberger, Visual analytics of large weighted directed graphs and two-dimensional time-dependent data. Ph.D. thesis, Darmstadt, TU, Dissertation (2010), 265p

130. K. Nazemi, M. Breyer, A. Kuijper, in *Human Centered Design*. Lecture Notes in Computer Science, vol. 6776, ed. by M. Kurosu (Springer, Berlin, 2011), pp. 576–585. doi:10.1007/978-3-642-21753-1_64, http://dx.doi.org/10.1007/978-3-642-21753-1_64

131. M. Breyer, K. Nazemi (supervisor), D.W. Fellner (supervisor), Benutzerzentrierte Visualisierung von Informationsempfehlungen basierend auf Recommender Systemen. Masterthesis at the Technische Universität Darmstat. Supervised by K. Nazemi (2010), 171 p

132. J. Chen, A. MacEachren, D. Peuquet, IEEE Trans. Vis. Comput. Graph. **15**(6), 889 (2009). doi:10.1109/TVCG.2009.130

133. M. D'Ambros, M. Lanza, in *Proceedings of the 10th European Conference on Software Maintenance and Reengineering, 2006. CSMR 2006* (2006), pp. 10 pp.-238. doi:10.1109/CSMR.2006.51

134. D. Holten, J.J. van Wijk, in *Proceedings of the 10th Joint Eurographics/IEEE—VGTC Conference on Visualization, EuroVis'08* (Eurographics Association, Aire-la-Ville, Switzerland, 2008), pp. 759–766. doi:10.1111/j.1467-8659.2008.01205.x, http://dx.doi.org/10.1111/j.1467-8659.2008.01205.x

135. K. Dinkla, M.A. Westenberg, H.M. Timmerman, S.A.F.T. van Hijum, J.J. van Wijk, in *Proceedings of the 13th Eurographics/IEEE—VGTC Conference on Visualization, EuroVis'11* (Eurographics Association, Aire-la-Ville, Switzerland, 2011), pp. 1141–1150. doi:10.1111/j.1467-8659.2011.01963.x, http://dx.doi.org/10.1111/j.1467-8659.2011.01963.x

136. T. von Landesberger, A. Kuijper, T. Schreck, J. Kohlhammer, J. van Wijk, J.D. Fekete, D.W. Fellner, Comput. Graph. Forum **30**(6), 1719 (2011). doi:10.1111/j.1467-8659.2011.01898.x, http://bibcd.igd.fraunhofer.de/bibcd/INI_Science/papers/2011/11p118.pdf

137. S.G. Kobourov, Spring Embedders and Force Directed Graph Drawing Algorithms (2012). arXiv:abs/1201.3011

138. W.T. Tufte, Proc. London Math. Soc. **3**, 743–768 (1963)

139. L. Shi, N. Cao, S. Liu, W. Qian, L. Tan, G. Wang, J. Sun, C.Y. Lin, in *Visualization Symposium, PacificVis'09. IEEE Pacific* (2009), pp. 41–48. doi:10.1109/PACIFICVIS.2009.4906836

140. B. Gretarsson, J. O'Donovan, S. Bostandjiev, C. Hall, T. Höllererk, in *Proceedings of the 12th Eurographics/IEEE—VGTC Conference on Visualization, EuroVis'10* (Eurographics Association, Aire-la-Ville, Switzerland, 2010), pp. 833–842. doi:10.1111/j.1467-8659.2009.01679.x, http://dx.doi.org/10.1111/j.1467-8659.2009.01679.x

141. T. Crnovrsanin, I. Liao, Y. Wuy, K.L. Ma, in *Proceedings of the 13th Eurographics/IEEE—VGTC Conference on Visualization, EuroVis'11* (Eurographics Association, Aire-la-Ville, Switzerland, 2011), pp. 1081–1090. doi:10.1111/j.1467-8659.2011.01957.x, http://dx.doi.org/10.1111/j.1467-8659.2011.01957.x

142. M. Zinsmaier, U. Brandes, O. Deussen, H. Strobelt, IEEE Trans. Vis. Comput. Graph. **18**(12), 2486 (2012). doi:10.1109/TVCG.2012.238

143. A. Telea, O. Ersoy, in *Eurographics/IEEE-VGTC Symposium on Visualization* (Eurographics Association and IEEE, 2010)

144. G.D. Battista, P. Eades, R. Tamassia, I.G. Tollis, *Graph Drawing: Algorithms for the Visualization of Graphs* (Upper Saddle River, NJ: Prentice Hall, 1998)

145. M. Graham, J. Kennedy, Inf. Vis. **9**(4), 235 (2010). doi:10.1057/ivs.2009.29, http://dx.doi.org/10.1057/ivs.2009.29

146. S. Jürgensmann, H.J. Schulz, in *Poster at the IEEE InfoVis 2010* (2010)

147. D.A. Keim, SIGMOD Rec. **25**, 35 (1996). http://0-doi.acm.org.millennium.lib.cyut.edu.tw/10.1145/245882.245896

148. D. Keim, IEEE Trans. Vis. Comput. Graph. **6**(1), 59 (2000). doi:10.1109/2945.841121

149. M. Wijffelaars, R. Vliegen, J.J. van Wijk, E.J. van der Linden, in *Proceedings of the 10th Joint Eurographics/IEEE—VGTC Conference on Visualization, EuroVis'08* (Eurographics Association, Aire-la-Ville, Switzerland, 2008), pp. 743–750. doi:10.1111/j.1467-8659.2008.01203.x, http://dx.doi.org/10.1111/j.1467-8659.2008.01203.x

150. D. Oelke, H. Janetzko, S. Simon, K. Neuhaus, D.A. Keim, in *Proceedings of the 13th Eurographics/IEEE—VGTC Conference on Visualization, EuroVis'11* (Eurographics Association, Aire-la-Ville, Switzerland, 2011), pp. 871–880. doi:10.1111/j.1467-8659.2011.01936.x, http://dx.doi.org/10.1111/j.1467-8659.2011.01936.x

151. M. Karnaugh, Trans. Am. Inst. Electr. Eng. **9**, 593 (1953)

152. T. May, Modelle und Methoden für die Kopplung automatischer und visuell-interaktiver Verfahren für die Datenanalyse. Models and Methods for Coupling automatic and visual-interactive Approaches for Data Analysis. Ph.D. thesis, Technische Universität Darmstadt (2012)

153. K. Stein, R. Wegener, C. Schlieder, in Advances in *2010 International Conference on Social Networks Analysis and Mining (ASONAM)* (2010), pp. 233–240. doi:10.1109/ASONAM.2010.18

154. G. Andrienko, N. Andrienko, S. Bremm, T. Schreck, T. von Landesberger, P. Bak, D. Keim, in *Proceedings of the 12th Eurographics/IEEE—VGTC Conference on Visualization, EuroVis'10* (Eurographics Association, Aire-la-Ville, Switzerland, 2010), pp. 913–922. doi:10.1111/j.1467-8659.2009.01664.x, http://dx.doi.org/10.1111/j.1467-8659.2009.01664.x

155. D.A. Keim, J. Schneidewind, M. Sips, in *Proceedings of the 1st First Visual Information Expert Conference on Pixelization Paradigm, VIEW'06* (Springer-Verlag, Berlin, 2007), pp. 12–24. http://dl.acm.org/citation.cfm?id=1759363.1759367

Chapter 3
Semantics Visualization

Semantic technologies provide new ways for accessing data and acquiring knowledge. The underlying structures allow finding information easier, gathering meaning and associations of data entities and associating data to users' knowledge. Even though the focus of research in this area is more to provide "machine readable" data, human-centered systems benefit from these technologies too. Especially graphical representations of semantically structured data play a key-role in this chapter. We will give a short overview over the idea of Semantic Web and its technologies. Thus semantic technologies are not in focus of this thesis, the review of the research will be performed in a higher level of abstraction. The goal is to give a common understanding of the term semantics as it is used in Semantic Web. Further an overview of the Semantic Web and the related technologies should give an impression of the way how data and meaning as formalized metadata is investigated. In this context, we will introduce some classifications on semantic formalisms to enable a view on the different levels of formalizing knowledge. The most common languages for describing semantics will be introduced based on the classifications. This allows depicting the continuous spectrum of light-weight to formal semantics. Further it will clarify that information visualization and semantic technologies are contradictory approaches for information acquisition. While semantics is more focusing on providing facts to specific and explicit questions or enhance the search experience, information visualization follows more an exploratory approach of information acquisition. For outlining this issue, models of exploratory search will be introduced in context of human interaction with semantics.

The focus of this chapter is a survey of the existing semantics visualizations. First a definition of semantics in context of visualizations will be given. For a comprehensible illustration of the visualization techniques in context of semantics, a classification of the visualization techniques will be provided. This classification will focus more on the exploratory search approaches, models, and steps. Thereafter the existing semantics visualization systems and approaches will be described based on the acquired classification. Figure 3.1 illustrates the structure of this chapter.

© Springer International Publishing Switzerland 2016
K. Nazemi, *Adaptive Semantics Visualization*, Studies in Computational
Intelligence 646, DOI 10.1007/978-3-319-30816-6_3

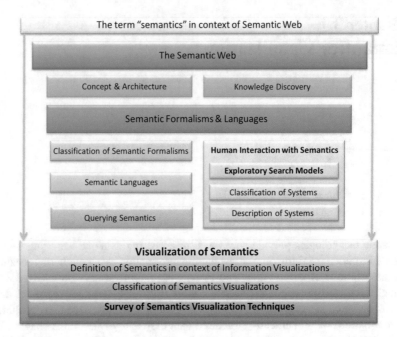

Fig. 3.1 Structure of the chapter semantics visualization

3.1   Terminological Distinction

Semantics is used in various disciplines, e.g. linguistic, logic, and computer science including programming languages and Semantic Web very heterogeneously [1]. It is necessary to have a common understanding of the term semantics in context of this work, in particular in context of semantics visualizations. Therefore this part of the thesis introduces some definitions and investigates different points of view. Here the definition will be given as it will be used for describing the overall idea of Semantic Web and semantic technologies. In the section that describes approaches and techniques for semantics visualization (Sect. 3.5.1), a definition will be introduced in terms and in context of information visualization.

The origin of the term "semantics" lies in semiotics, the science of studying signs. Semiotics is closely related to linguistic and investigates the abstraction, meaning, and rules of languages [1, 2]. Beside the use of the sign studies of semiotics in linguistics, it is a common instrument in logics for describing rules and meanings [1–3]. An early and common definition of semantics in relation to semiotics was proposed by Carnap in a more logical manner:

> If in an investigation explicit reference is made to the speaker, or, to put it in more general terms, to the user of a language, then we assign it to the field of **pragmatics** [...] If we abstract from the user of the language and analyze only the expressions and their designata, we are in the field of **semantics**. And if, finally, we abstract from the designata also and analyze only the relations between the expressions, we are in (logical) **syntax**. The whole science of language, consisting of the three parts mentioned, is called **semiotics**. [2, p. 9]

The definition of Carnap gives not only a linguistic view on semiotics and semantics; it involves the aspect of logic too. Carnap used for his definition of semiotics and the included parts of pragmatics, semantics, and syntax a logical way of description (expression). He outlined in this definition that the context to the user is not given anymore in semantics. Semantics is just the relation between expressions and designata [2]. With other words, semantics is defined by Carnap as the not-user influenced meaning (designatum) or meanings (designata) of expressions. In terms of linguistic these expressions (syntax) can occur as, e.g. words and consists of a logical structure. Hitzler et al. proposed that semantics represents in general the "meaning of words, phrases, and symbols" [1, p. 13]. Further they enhanced that in computer science and in particular in Semantic Web, semantics represents the "meaning" of words and character (strings) with their relation to each other. To provide in this context a more comprehensible understanding of semantics in terms of Semantic Web, they described semantics more precisely. They differentiate in this context between semantics and formal semantics and investigate the term semantics from the viewpoint of mathematical logic, where the focus lies on a correct formal reasoning [1]. The definition is based on the *model-theoretic semantics* [4] and defines semantics as reasoning-relations [1] as follows: a proposition $s \in S$ is a consequence of the sets $S \subseteq S$ (therewith $S \models s$) if each interpretation I, which fulfills each proposition s' from S (therewith $I \models s'$ for all $s' \in S$), is a model of s ($I \models s$ and $\models$ is relational reasoning) [1, p. 92].

Likewise to semiotics, where a sign invokes a concept (identifying an abstract or concrete thing in the world), semantics is used to interpret a data fragment's potential usage [5]. In general, each data fragment has to be interpreted by an appropriate semantics to obtain meaningful results within a consistent system behavior. Representing this data semantics as explicit metadata is the core of research in context of Semantic Web [6]. The research focuses on supporting data re-usability, machine-readability, inference mechanisms and semantic interoperability [7]. The metadata represents the meaning of data [6]. Therewith the term *semantic* in Semantic Web represents a formalized meaning in form of metadata of data and data entities [1, 3, 6, 7].

3.2 The Semantic Web

The World Wide Web (WWW) is a crucial information and communication source of our society. With every day and every hour the Web grows enormously and the access to information is getting more difficult. The growth of data in general and in

Web in particular, is a rising and crucial problem that is investigated in the moment of writing this thesis as the problem of "big data". The Web experiences with each day not only new data, but far more of interest in context of this work are new users of the Web, which make it necessary to research the heterogeneity of the users and provide adaptive systems. It has already changed our way to communicate with each other [8]. We use the Web extensively for searching, exploring, and analyzing information. It is an important resource of today's knowledge acquisition process. Therefore it is more than necessary that Web technologies provide an easy access to the desired information and knowledge. Antoniou and van Harmelen criticized in 2004 that Web technologies with the keyword-based search engines provide a high recall but a low precision, the retrieved results are highly sensitive to vocabulary, and the results are single Web-pages with complementary information that are needed for a search task [8, p. 1].

The idea of Semantic Web was proposed by Berners-Lee [9] as a *Web of Data* to face the above mentioned problems by expressing Web information in a machine readable way [10]. Therefore Semantic Web should be "a Web of data with meaning in sense that a computer program can learn enough about what the data means to process it" [10, p. 5]. Semantic Web enhances the Web with structured and formalized meanings of content (data) [7, 11]. The formalization proposed to make the Web "meaningful" is based on a structured notation of content followed by a structured notation of the underlying structure and provides a rule and meaning inference to make the Web accessible for computer and human [3, 7, 9–11]. To provide such a structure and meaning to the content of Web, a "layer-cake" architecture was proposed [12]. This architecture builds the foundation work on semantic technologies and in particular for the Semantic Web. This construct of Web consists of structured metadata providing well defined meaning described as formal semantics [7]. The following sections will introduce into the general idea of Semantic Web. First the conceptual "layer-cake" will be introduced that give an overall idea of how the enhanced Web works. In terms of getting meaning and structure from the Web-data, an overview of knowledge discovery in context of Semantic Web will be given.

3.2.1   Concept and Architecture of the Semantic Web

The main technique that enables the idea of Semantic Web to become true is XML (Extensible Markup Language [13]), thus most of the Semantic Web technologies were built with the XML family [14]. XML and its derivatives are the foundation of Semantic Web and the technologies and notations that provides those "meaning" that is proposed by the Semantic Web [11, 14]. The starting point of the conceptual structure of the Semantic Web is the Unicode standard [15], an international standard for digitally encoding multilingual character sets. The Unicode standard should enable the development of a multilingual Semantic Web without any language restrictions [14]. Beside the Unicode standard the architecture model illustrates the Uniform

Resource Identifier, which was originally named in as "Universal Resource Identifier" [11] (In this thesis the RFC2396 is applied and uses "Uniform" instead of "Universal"). In combination with XML Namespaces, URIs allow the unification of elements within XML Documents [14]. In Semantic Web URIs (also appearing as URL (Uniform Resource Locator) or URN (Uniform Resource Names)) identify the entities that should be linked to each other and provide with the interlinking structure a meaning [11]. The upper layer of the architecture consists of the already mentioned and introduced XML together with Name-spaces (NS) and xmlschema. The substantial layer of the Semantic Web is formed by XML [14]. This layer builds the main structure that is further formalized in the next steps of the architecture.

The Resource Description Framework (RDF) [16] is the formalization of the meaning of Web data in so called *statements* [6, 11]. The syntax of RDF is commonly appearing in XML, whereas other representations are possible [6]. A statement is an entity-attribute-value triple [6, 27] or as proposed by Berners-Lee a "subject, verb, and object" relation [11]. The entity (or subject) is the resource representing an object of interest and refers to resources on the Web via a URI [6]. The attribute (or verb) in a statement is represented by a property. A property is a special kind of resource and describes relations between resources and can be identified by URIs or located by URLs [6]. The value is either another resource or a literal. Literals appear as atomic values and are reserved identifier in RDF [1, 6]. An RDF statement provides with its triples a "meaning" to the data or entities of interest.

RDF is currently the most common semantic representation on Web. The initiatives of Linked-Data (or Linked-Open-Data) are using commonly the RDF representation and interlink a massive amount of data and information [17–21]. The amount of their representations and the Linked Data cloud is massively growing and provide already today an enhanced and meaningful Web [19].

RDF Schema (RDFS) is in the same layer on the architectural model of Semantic Web and provides already the ability to build simple ontologies [1]. RDFS is an integrated part of the RDF specification and provides appropriate data typing, background information, and schema building for the underlying RDF triples [1, 14, 19]. RDFS defines domain-specific attributes and classes for the resources that are applied in a RDF document [14] and provides universal means of expression for the relations, classes, and the hierarchy [1]. With the ability to build knowledge about the schema, RDFS can be count as an ontology-language, whereas the simple architecture and limitations just allow creating a "lightweight-ontology" [1].

The next layer of the conceptual model is dedicated to build such schema information in form of ontologies. "An *ontology* is an explicit specification of a conceptualization" [22, p. 199]. A conceptualization builds the representation of formal knowledge in form of objects, concepts, and further entities and is thereby an abstracted and simplified view of the world [22]. The term ontology was applied from philosophy and investigates the nature of existence and the type of thing that exist. The most common ontologies appear in Web with the formalization of knowledge in form of term-relations, taxonomies and inference rules for the represented knowledge [3, 11, 14]. At present the most common representation of ontologies in Web is the Web Ontology Language (OWL) [23], which is available in its second version (OWL2) and downward compatible to OWL [1, 14, 23]. While OWL provides many functions to handle and formalize knowledge, inference rules and enable reasoning, XML-schema or the already introduced RDFS can represent the knowledge too in form of ontologies [1, 14]. The difference lies in the formal specification that leads to the differentiation of lightweight and formal ontologies. OWL itself has three degrees of formalization. It may appear as OWL Lite, OWL DL, or OWL Full [1, 23]. The next Sect. 3.3 will investigate in particular the degree of formalization of the ontology and semantic languages and provide an overview of the most common "specifications of conceptualizations".

The logic layer of the architecture is containing logical rules to infer new knowledge from the existing knowledge representations [14]. The proof layer is responsible for the trustworthiness by distinguishing between different Web sources and their trustworthiness [14]. These two layers together with the "digital signature" will lead to a "Web of Trust", where agents act for the user and complete tasks without any human intervention [14]. The aspect of "Web of Trust" is beyond the scope of this thesis, therefore this aspect will not be investigated.

3.2.2 Knowledge Discovery for Semantic Web

Semantic Web premises well-structured and well-defined data in order to provide a meaning to terms that can be processed by machines and used by human. Therefore various methods and techniques are provided that formalizes knowledge and provide such formalized knowledge about a domain. Although, there exist many approaches for formalized documentation and processing of these resources, one main question is how to formalize the given resources that are commonly not well-structured in terms of semantic documents. Thus these documents aim at representing formalized knowledge, it is more than obvious to take the interdisciplinary field of Knowledge Discovery in Databases (KDD) into account. Thus KDD aims at having knowledge as the end product of the data discovery [24, p. 39]. Knowledge Discovery aims at automatically extracting knowledge of large and unstructured databases by using methods and approaches of machine learning and data mining [24, 25]. Grobelnik and Mladenić et al. proposed a methodology for constructing ontologies with KDD methods [25, 26]. Their methodology consists of interrelated phases and is illustrated in Table 3.1.

Commonly the approaches that use KDD for generating semantics are semi-automatic and need the involvement of human in the knowledge and ontology generation process [25]. There are a number of existing approaches and systems [30–33] that enables the construction and generation of ontologies with the involvement of human. These systems make use of different techniques of data mining and

Table 3.1 Interrelated phases of KDD for ontology construction by Grobelnik and Mladenić (adapted from [25, pp. 22])

Phase	Description (and the role of computer vs. human)
Domain understanding	Rely mainly on human to understand the area of investigation. KDD approaches, e.g. information retrieval or focused crawling may assist human in that process
Data understanding	Rely on human to understand the available data and their relation to ontology construction. Information visualization (or Visual analytics) may help to get an overview of the data
Task definition	Rely on human to define tasks to be addressed. The previously steps may help to perform this task
Ontology learning	Semi-automatic process relying on the defined task: use of document collections for extending an existing ontology, use of unsupervised learning for generating ontologies from scratch
Ontology evaluation	Involves human to validate the quality of an ontology. KDD methods can be applied partially [27] or full-automatically [28, 29]
Refinement with human in the loop	Providing the ability for human to refine each phase of the methodology

knowledge discovery, e.g. Topic Models [25, 31]. Further approaches make use of the collective intelligence of users [17–19] or the already exiting structured metadata that contains nearly that information that is necessary for the ontology formalization [21]. The full automatic generation of ontologies without the involvement of human still remains a research topic, whereas the widespread of semantic structures (e.g. Linked-Data or social networks) in Web already give a mass amount of semantic data. With the existence of these data, we are able to say that the Web is already enhanced with semantics and provide meaningful data. The most popular example for the spread and integration of semantics in Web is beside the DBPedia [21] initiative, the integration of semantics in the results of the leading (year 2013) search engine provider. Google's Knowledge Graph provides semantic search results [34], based on the metadata structure of various knowledge platforms. The Knowledge Graph is a product of the collective intelligence that created the various knowledge domains and KDD methods that gather information and organize them to provide a sufficient representation of relationships. Further the Knowledge Graph enables human to validate and correct the presented results. Figure 3.2 illustrates the Google Knowledge Graph with the results for a search for "Edward Tufte". It further illustrates that human are able to claim the correctness of that information: a: the normal view of semantic information with related information such as place and date of birth, written books, pictures, and related people and b: the editable version for validating the information by human.

(a)   **(b)**

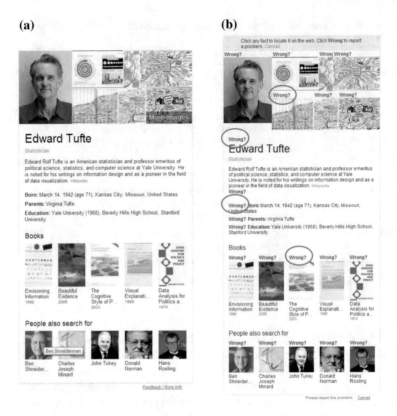

Fig. 3.2 The Google Knowledge Graph illustrating the result for the search query "Edward Tufte" [34]

3.3 Semantic Formalisms and Languages

As the previous sections could already outline, the formal representation of knowledge as ontology or a semantic construct may be modeled in different ways. In general, the formalization of information can be classified into a continuous spectrum from lightweight semantics (informal) to heavyweight-semantics (formal) [3, 14, 35, 36]. The goal of this section is to give an overview of the classifications on the formalization of knowledge as ontologies or semantic constructs for getting an insight of the different viewpoints on this topic. The second part will give a short overview of the most common semantic languages (formalizations). The goal here is to get an insight of the different ways how information can be formalized with semantics.

3.3.1 Classifications of Semantic Formalisms

Obrst introduced a continuous classification of semantics in context of ontologies with the poles of "weak semantics" to "strong semantics" [35]. He characterized the strength of semantics with its expressiveness of meaning from a very simple meaning to an arbitrarily complex meaning expression [35, p. 367]. This classification starts with machine interpretable relational models as schema, which provide taxonomy as sub-classification-of relationships and continues with more and more meaning in form of thesauri in Entity-Relationship (ER) models to conceptual models that define a relationship schema including a "sub-class-of" relationship (RDF/S, XTM, and Extended ER). The higher level includes formalized knowledge in form of *Logical Theory*, which is characterized by transitivity of properties and disjoints "sub-class-of" relationships. Examples for this class may be the Unified Modeling Language (UML), the successor language DAML + OIL, OWL, and Description Logic (DL) [35]. The spectrum further involves Modal Logic and First Order Logic, whereas the formalization does not end at this stage. Future logic constructs may enhance the spectrum to a more formalized knowledge that enables machines to retrieve more logical information about the underlying data. A similar but more focused classification was proposed by Geroimenko in context of Semantic Web visualizations [37]. He classifies XML-Schema as primitive ontologies and the lowest level of formalization of information, whereas the components of Semantic Web, e.g. URI, XML, XML-namespaces are premised for the formalization. His classification continues with the Resource Description Framework (RDF) and its Schema (RDFS). The highest level of his model is built by richer sets of description languages that enable, e.g. Boolean expressions, property restrictions, and axioms. This upper level of semantic includes in the first step the Ontology Inference Layer (OIL) [38] and the DARPA Agent Markup Language (DAML). In the second step the combined version of DAML + OIL [39, 40] provide a more formal representation of knowledge, whereas OWL as the revised version of DAML + OIL that uses the constructs of RDF builds the highest level of his model [37]. Although, Geroimenko does not declare explicitly his model as a categorization of semantic formalisms and language, the categorization is similar to Obrst's model and provides in this context a kind of classification that is useful for understanding the strength of formalisms in particular for XML-based techniques. Figure 3.3 illustrates the models of Obrst and Geroimenko.

Another classification was proposed by Uschold and Gruninger they differentiate between "kinds" of ontologies in terms of their formalization degree and arrange them on a continuum [36]. Their continuum starts with "very lightweight" ontologies that consist of terms with no or little specification of the meaning. With moving along their continuum the amount of specified meaning and the formalization degree increases by reducing the ambiguity [36, p. 60]. Their continuous classification of ontologies ends with formalized logical theories (General Logic). The main difference to the introduced classifications is that the model of Uschold and Gruninger consider Database (DB) Schema as ontologies and illustrate overlapping between ontologies and DB Schema in terms of expressivity [36]. Although, the model of Obrst already con-

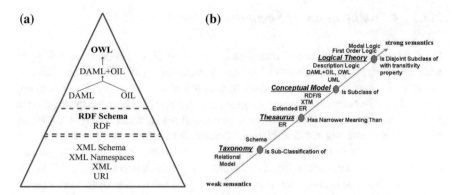

Fig. 3.3 Categorization of semantic formalisms: **a** The genesis OWL model by Geroimenko describing XML-Technologies and their relations (adapted from [37, p. 16] with permission) and **b** the semantic formalism classification by Obrst (from [35, p. 367] with author's permission)

sidered ER models as semantic knowledge representation, the Databases themselves were not mentioned with their expressiveness in terms of objects, properties, aggregations etc. Their model subdivides ontologies into the four classes of *'ordinary' Glossaries, Thesauri, Taxonomies, MetaData, XML Schemas, and Data Models*, and *Formal Ontologies and Inference*.

Guarino et al. applied this model and revised it slightly in order to express the significance of logic in ontologies [5]. Their revised model classifies ontologies based on the continuum of Uschold and Gruninger, but uses for the most formalized category the abstracted term of *Logical Languages* rather than *Formal Languages and Inference*. Their main objective is to classify the logical languages of ontologies into the classes of *higher-order logic, full first-order logic*, or *modal logic* and the stringent subsets of *first order logic*. They argue that the first class with the higher logic is very expressive, but does not allow a complete reasoning. Further they classify the strict subsets of first-order logic into the family of *description logics* (DL), e.g. OWL-DL and *logic programming*, e.g. F-Logic [5]. Figure 3.4 illustrates the described classification by Guarino et al. of the semantics and ontologies.

There exist a variety of further classifications of semantic formalisms and ontologies respectively. This classification is made based on the "subject of the conceptualization" as the introduced examples showed or on the amount and structure of the conceptualization [41, 42]. Further classifications investigate in particular the "heavyweight" ontologies [7, 42]. Heavyweight ontologies or semantic formalisms model the domain in a deeper way and provide more restrictions on domain semantics by adding formal axioms, functions, rules, and procedures in contrast to lightweight formal semantics. There are important relations and implications between the knowledge components (concepts, roles, etc.) used to build the formal semantics, the formal semantics formalisms that represent the components, and the languages that implement the semantic data [7]. Gómez-Pérez et al. propose a differentiation of semantics formalisms that describe the metadata as machine-readable formal semantics (knowl-

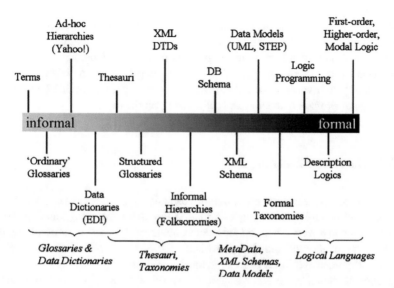

Fig. 3.4 Categorization of semantic formalisms by Guarino et al. (from [5, p. 13] with permission)

edge representation paradigms) into *Semantic Networks, Frame-based Logics*, and *Description logics* [7]. Semantic networks (SN) are one of the oldest knowledge representation formalisms. A SN represents knowledge as a set of nodes connected by labeled links. In such a representation, meaning is implied in the relationship of at least two concepts [43]. A typical and common example for SN is the *Resource Description Framework (RDF)*. Frame-based logic is a knowledge representation, isomorphic to semantic networks. A frame is a named data object that has a set of slots. Each slot is representing a property or attribute of the object. Slots can have one or more values; these values may be pointers to other frames [43]. Description logics (DL) denote a group of logics for knowledge representation that arose from semantic networks. They provide a formal foundation for frame-based systems. Substantially DL constitutes fragments of first-order logic, restricted to a certain complexity class to allow the construction of a high expressive language [1]. The DL model distinguishes between a *terminological box (TBox)* and an *Assertional box (ABox)*. The *TBox* contains intentional knowledge to describe general properties of concepts. The *ABox* contains extensional knowledge, which is specific to the individuals of the discourse domain [7]. The most prominent DL-based languages are OIL [38], DAML + OIL [39], and OWL [3, 42, 44].

The introduced classifications outlined that the spectrum of semantics may start with no or less meaning of termini in an unformalized way [5, 36] to the formalization of knowledge by using higher-order logic. There is a continuous spectrum of formalization, whereas the formalization is the conceptualization of knowledge as proposed by Gruber [22]. In this context various languages for describing ontologies arose that specify a conceptualization in a more or less formal way and provide a meaning to data or information entities. Some of these languages were already mentioned for describing the general idea or the degree of formalization, but there exist many further ways of formal expression to convey information as meaningful knowledge or provide reasoning and inferencing on the data.

3.3.2 Semantic Languages

The introduced classifications of semantic formalisms illustrated an overview of the existing methods for formalizing information to extract "meaning" and enable the idea of Semantic Web. This part introduces some of the most common languages for the formalization of knowledge, whereas the objective of this section is not to give a comprehensive illustration of semantic and ontological formalizations and languages. It should far more be seen as an overview of possible ways to formalize information and to extract meaning from the data. Therefore we will pick up some representatives of the lower level of conceptualization (lightweight) and introduce first the semantic formulated by XML and XML-schema. Currently the most common way of formalizing information is RDF, therefore we will continue with the more formal representation of RDF and RDFS. Thereafter the concept of Topic Maps will be introduced as a technology for building semantic networks above information as a contradictory approach of the resource-centric RDF-technique. To provide an insight into the logical and more formalized languages, we introduce the idea of Description Logic with OWL as its most common representative. This section should not be seen as a handbook for designing formal knowledge representations.

XML and XML-schema
XML [13] is one of the most substantial layers and technologies of Semantic Web [14]. XML was developed as an extensible markup language to provide the ability to annotate information resources or entities with own or predefined "tags" that enable such annotations. The annotations of these documents are commonly called metadata, thus these are information about the underlying data [1]. The data are enclosed with an opening and closing tag, whereas the opening tag may have attributes with values. It is possible to model meaning and relationships in and for data with these rudimentary modeling elements. Let us take the example of a person, who has a working-position in two organization and an academic degree (see Listing 3.1). All the mentioned information about that person can be expressed in sentences and by relationships, as described in Sect. 3.2. In general a XML-document can be seen as a data-model with a tree-structure, thus the elements are nested [1]. With its nested structure a XML-document provides a kind of taxonomy that is similar to a subclass-of relationship. The tags provide a meaning to each data fragment, e.g. the title of the person is "Prof. Dr." and "Fraunhofer IGD" is an organization. Further the relationships of the data fragments are of importance, thus this document allows the extraction of

the information that "*Prof. Dr. Dieter W. Fellner is director of Fraunhofer IGD and professor at the Technische Universität Darmstadt*". In this sentence the assigned relationships are "is director of" and "is professor at".

```
<Person academicDegree="Prof.Dr.">
    <Name>Dieter W. Fellner</Name>
    <Organization position="Director">
        Fraunhofer IGD
    </Organization>
    <Organization position="professor">
        Technische Universitt Darmstadt
    </Organization>
</Person>
```

Listing 3.1 XML example of a "meaningful relationship"

The metadata described here are very similar to semantic relations, but they cannot be processed by machines and computer without further technologies [1, 14]. A machine does not "know" what the strings "Person", "Organization" or "position" mean. Further the context is important, thus the term position, if processed and known by machine, would have various meanings and contexts, e.g. a position in a room or the current position of the moon. Humans are able to derive the information out from metadata as they are really meant based on the context. A first step to face this problem is XML-schema. With XML-schema the meaning of each tag can be annotated to process the information easier. An enhancement to give the machines the ability for describing machine-processable metadata is the Resource Description Language and its schema [3, 14, 16].

RDF and RDFS
The Resource Description Framework (RDF) [16, 45, 46] is currently the most common and important language that builds the foundation of Semantic Web and semantic technologies [1, 3, 14, 17–19]. RDF enables the idea of Semantic Web, as proposed by Berners-Lee that information resources on Web can be expressed by sentences with meanings [11]. RDF can be built with three main concepts: *statements*, *resources*, and *properties* [14, 45], whereas the RDF-statements (rdf:type) are the foundation of RDF and express the meaning and relationships of Web resources in form of "sentences" as *subject* (rdf:subject), *predicate* (rdf:predicate) and *object* (rdf:object) [1, 45]. RDF statements describe directed graphs and thereby a set of nodes that are connected to each other with directed-links [1]. In contrast to XML, where the information is annotated in tree-structures, RDF uses a relational concept and describe far more than just hierarchies. RDF statements allow general descriptions of resource relations in Web [1]. A statement is defined as a RDF-triple (RT), where RT is defined as $RT = (s, p, o)$ and $s \in S$ is a subject referring to a Web recourse, $p \in P$ is a predicate or RDF property defining the relationship and $o \in O$ is a either a resource on Web or a predefined RDF literal [1]. Each element of subject and predicate in a RDF-triple refers to a resource on Web, whereas objects may refer

to resources or to literals. Therefore the concept of URI (and URL) is used to enable a unique identification of resources [1, 45]. Literals are values that may appear as typed or plain [47]. An example for a literal is the numerical value "23". In this case this numerical value can be expressed in different ways, e.g. 023 or 000023, but refer always to the value "23". RDF can be formulated in various ways, e.g. *N3, N-Triples*, or *Turtle* [1], whereas the RDF/XML notation is the most common way of serializing RDF triples [1, 46].

The extraction of semantics as terminological knowledge in RDF is enabled by RDFS [48], the schema language of RDF [1]. RDFS allows creating (lightweight) ontologies by constructing classes, associated properties, and utility properties within a RDF document for a specific knowledge domain [1, 48]. These ontologies are built on the limited vocabulary of RDF and can be treated as RDF documents [1]. The formalization of knowledge with machine-processable schemata is one of the main purposes of RDFS. Therefore a predefined vocabulary is used to assign and annotate information entities as subject or individuals of classes, create class-relations, e.g. as a class hierarchy and annotate properties [1, 48]. RDFS uses therefore a predefined vocabulary commonly starting with the prefix *rdfs:*, e.g. a class or concept in RDFS is annotated as *rdfs:Class*. Further the predefined RDF annotations starting with *rdf:* are used in the schema-level for annotation, e.g. *rdf:Property* or *rdf:Statement*. With the predefined vocabulary a machine-processable ontology can be created that annotated the knowledge within a RDF document. Further information and the specification of RDFS can be found in [48].

Topic Maps
A knowledge-centric approach for generating semantics and ontologies in Web is the *Topic Map* approach [14, 49]. The main concepts of Topic Maps are *topics*, *associations* (or *relations*), and *occurrences* (or *characteristics*) [14, 50]. Almost every knowledge or information entity is represented in Topic Maps as topics. The relationships of topics are modeled through associations. Each topic may have n associations to m occurrences, whereas occurrences refer to resources on Web [14, 49]. Similar to RDF a semantic network can be modeled with the three main components of Topic Maps [49]. The main difference between RDF and Topic Maps is in the knowledge-centric approach of Topic Maps, where each topic represents an abstract knowledge entity and not directly a resource [14]. The resources are linked with the occurrences and model together with the associations a kind of a knowledge map above the Web-resources [14]. Topic Maps are formulated commonly with XML and are specified as XML Topic Maps (XTM) [50].

Description Logic and OWL
RDF and RDFS as today's most common formal representation of semantically annotated data provide already the ability to model information with semantics and provide logical functions for inferencing or reasoning [3]. But RDF has limitations in particular for reasoning and inferencing information and further logical functions [3, 51]. To face these limitations the family of *Description Logic* and more complex information representation structures are provided, which enable more complex logical functions on data [5]. *Description Logic* (DL) is a subset of the first-order logic and the most common logical structure for building formal ontologies in a structured

and formally well-understood way [5, 52]. DL *describes* concepts as unary predicate (atomic concepts) and binary predicate (atomic roles) expressions [52]. The formulation are made by logical constructors, e.g. *negation* ($\neg$) or *existential restriction* ($\exists R.C$) constructors [52]. Let us assume that we want to formalize the concept of "A female professor working at Fraunhofer has three children and all of whom are sons". This concept can be formalized with the concept description as illustrated in the Eq. 3.1.

$$\texttt{Professor} \sqcap \neg\texttt{Male} \sqcap \exists\texttt{working.Fraunhofer} \sqcap (= 3\texttt{hasChild}) \sqcap \forall\texttt{hasChild.Son}$$
(3.1)

The concept example in Eq. 3.1 refers to a simple Boolean construction of the description logic. To model meaningful ontologies DL make use of two further main concepts, the terminological and assertional formalisms [52]. *Terminological axioms* are used for instance to introduce names (abbreviations) or to state constraints. A constraint could be that just human can have human children [52]. The already introduced *TBox* (Sect. 3.3.1) is a set of such terminological axioms [52]. The assertional axiom is used for instance to formalize that individuals belong to a named concept, e.g. the expression `Researcher(Dirk)` formulates that "Dirk" is an individual of the concept "Researcher". The *ABox* is a set of all assertions and the named individuals [52]. Description Logic enables various methods for inferring implicit knowledge from explicitly formalized information. Examples for algorithms that enable such inferencing are the *subsumption* algorithm for inferring the sub- and super-concept of a concept, the *instance* algorithm that determines the individual of concepts, or the *consistency* algorithm that enables the validation of a set of terminological and assertional axioms [52]. The complexity of such algorithms in Description Logic may rise to *undecidable* and unpredictable complexity [52]. The Web Ontology Language (OWL) [23] provides various degrees of logical complexity with different levels of expressiveness to give a choice between logical decidability and expressiveness for the formulation of ontologies and is the most common ontology language for description logic [51].

The Web Ontology Language [23] is built upon RDF and RDFS and has the same syntax [51]. It enhances RDF and RDFS with further logical constructions, e.g. *disjointness of classes, Boolean combination of classes, cardinality restrictions*, and *special characteristics of properties (transitivity, uniqueness, inversion)* [51, pp. 92]. The enhancement of RDF and RDFS with the mentioned logical constructs would lead to a formalized ontological language that is undecidable, unpredictable and uncontrollable [51, 52]. To avoid such an uncontrollable and in particular undecidable language OWL provides three *Species* or sublanguages, namely *OWL-Full, OWL-DL*, and *OWL Lite* [23, 51]. In general the OWL languages are upward compatible as every legal OWL-Lite ontology or conclusion is a legal OWL-DL ontology or conclusion and every legal OWL-DL ontology or conclusion is legal OWL-Full ontology or conclusion [51]. Therewith OWL-Full provide the most complex and expressive language construction of OWL.

OWL-Full enables the construction of ontologies by using all the constructs of the Description Logic, such as *property restrictions, special properties, Boolean combinations*, or *enumerations* in any combinations [51]. OWL-Full is the most expressive language of OWL and the only language that includes and supports the entire constructs of RDFS [1]. It undecidable and only few reasoning and inferencing tools are able to process the complexity of OWL-Full that is not restricted.

To face in particular the limitation of "undecidability" in OWL-Full, OWL-DL exploits the computational tractability and decidability of Description Logic by constraining those aspects that lead to undecidable ontologies [1, 23, 51]. The decidability is important for verifying the ontology for its consistency and deriving information by inferencing in a computable and reasonable manner [51]. Therefore OWL-DL constraints constructs of OWL-Full and Description Logic by: *vocabulary partitioning*, *explicit typing*, *property separation*, *cardinality restrictions on transitive properties*, and *restricted anonymous classes* [51]. In OWL-DL a resource can only be one type of vocabulary (*vocabulary partitioning*), e.g. either class, datatype, or property, further the partitioning of the vocabulary must be explicitly stated (*explicit typing*) [51]. These two restrictions achieve that object and datatype properties are disjoint, this further implies that *inverse*, *functional*, *inverse functional*, and *symmetric* characteristics cannot be specified for properties (*property separation*) further *transitive* properties must not have a cardinality (*cardinality restrictions on transitive properties*) [51]. OWL-DL further constraints the use of anonymous classes. These classes are just allowed in the domain and range of owl:equivalentClass and owl:disjointWith, and in the range of rdfs:subClassOf [51]. With this constraints and restrictions OWL-DL provide a decidable model-theoretical semantics [4] based on Description Logic [52] by using the syntax of RDF/S with a worst-case complexity of *NExpTime* [1, 51]. Antoniou and van Harmelen illustrated in [51] the mapping of the OWL syntax to their Description Logic equivalents.

OWL Lite has further constraints to enable a faster computing of inferences and easier way for describing semantics. It is sub-language of OWL-DL and further restricts OWL-DL. It is for instance not allowed to use the constructors owl:oneOf, owl:disjointWith, owl:complementOf, or owl:hasValue [51]. Further the cardinality statements are limited to 0 and 1 and the owl:equivalent Class statements are not allowed for anonymous classes [51]. OWL Lite as sub-language of OWL-DL is decidable with a maximum complexity of *ExpTime* [1].

The Semantic Web languages were developed to meet the idea of meaningful Web of data [10]. As Berners-Lee explained an idea of Semantic Web, he proposed a Web that is more "intelligent" and provides not only meaning of data but also the integration of agents, which act for human [9]. For such a Web the formal ontologies have to be integrated and spread throughout the Web. And the Web must integrate a Trust layer that was already proposed [12]. In today's Web "semantics" commonly appears as *Linked-Data* [17–19, 53]. Linked-Data appear in various notions with different structures [18], but the most common way of formalizing Linked-Data is the use of RDF and RDFS [17–19, 53]. Therewith RDF and RDFS are today's most common language for the Semantic Web [19].

3.4   Interaction with Semantics

This chapter introduced so far the idea of Semantic Web by explaining its architectural design, the way how data can be annotated semi-automatically with semantics, the different levels of semantic formalisms, and the way how information can be semantically annotated. It is important to not only investigate the formal side of semantics, but how human interacts with semantics and how the interaction with semantic is technically supported. This section gives a short overview of the most common ways how semantics are queried and how human interacts with the responses of the queries. This section does not aim to provide a comprehensive tutorial on semantic data querying or the human interaction with semantics. The goal is to give an insight of what is possible and what is provided.

3.4.1   Querying Semantics

We described in the previous section that the most common formalization of semantics is formulated by RDF and RDFS. The formalization of knowledge has to provide the possibility to query the underlying semantics and get the needed information resources. The most common language for querying semantics from RDF stores and RDF triples is the *SPARQL Protocol and RDF Query language* (SPARQL), a relatively young W3C recommendation [54]. The SPARQL recommendation provides beside the query language, a *SPARQL Service Description*, SPARQL result descriptions in *JSON, CSV and TSV, XML*, and *SPARQL protocols* [54]. This section will just introduce the *SPARQL query language* [55] that is used to query RDF semantics. SPARQL is a language to query (RDF) graphs [1] and uses primitive RDF graphs as query-structure. A SPARQL query commonly consists of a *namespace*, a *query form*, and values for the query form structured as graphs [1, 55]. The SELECT query returns variables and their bindings [55], the syntax SELECT * is used for all variables in a query [55]. Another query form is CONSTRUCT, which returns a single RDF graph by a template, substituting for the variables in the graph template [55]. Such a graph template can contain triples with no variables (ground or explicit triples) [55]. With the ASK query form, the existence of a query pattern solution can be requested, the result is a Boolean expression of the existence [55]. SPARQL provides further the DESCRIBE query form [55]. This form can be used, if the properties (of relevance) in a context are unknown [1]. The result is a single RDF graph containing information about resources [55].

Values for query forms are introduced with WHERE and may contain a simple graph, group graph pattern (with the syntax {}), or complex graph structures [1, 55]. Such complex graph structures can be requested by grouping the queried graph triple with the braces [1]. SPARQL further provides the extended graph declarations of OPTIONAL {*graph*} for requesting an optional graph and UNION {*graph*}, which refers to a logical "or" [1]. The combination of the group pattern ({}), the OPTIONAL pattern, and the UNION pattern may already provide the ability to construct complex query graphs [1]. SPARQL enhances the complex query structure by providing the ability to query for data values, the FILTER keyword that constraints query results, the relational operators (= , < , > , ≤ , ≥ , and ! =) and a set of special operators [1]. The special operator BOUND (A) for instance results with a Boolean true , if the variable 'A' is bounded, the isURI (A) operator indicates if 'A' is an URI [55]. Further special operators are isBLANK (A) (true for an empty node), isLITERAL (A) (true. if 'A' is a literal), STR (A) (maps RDF-literals or URIs to lexical string), LANG (A) (returns a string of the RDF literals), DATATYPE (A) (returns the datatype of 'A'), sameTERM (A, B) (true, if 'A' and 'B' are the same RDF terms), langMATCHES (A, B) (true, if the language tag of 'A' is in the language range of 'B'), and REGEX (A, B (true, if the string 'A' was found in the regular expression 'B') [1, 55]. SPARQL further supports the use of the Boolean operations of *and*, *or*, and *not* and a set of arithmetic operations that enable a variety of complex query structures for the semantic data [1, 55].

The introduced constructs of SPARQL are just a subset of the entire query possibilities of SPARQL. Various constructs, e.g. modifiers, algebraic computing and modifiers and conjunctive queries were not introduced [1, 56]. SPARQL is an easy to use and intuitive query language in particular for querying RDF graphs, but with its several constructs and functions, SPARQL is a very powerful query language. It can lead to very complex queries that affect the computing time for such queries and leads to very long response times. The performance aspect is important for human-computer interfaces, where the response time and thereby the reaction time may lead to inefficient systems. In particular visualization systems are affected, thus the visual interfaces try to build an interactive picture of the entire or a subset of the semantic structure. For providing efficient and performant visualizations, approaches have to be developed that lead to a nearly "real-time" querying of the complex semantic structure. Therefore we will introduce in our conceptual design (see Sect. 6.1.1.1) an approach for a more efficient querying of semantics in distributed semantic resources.

3.4.2 Human Interaction with Semantics

The introduced methods, languages, approaches, techniques, and technologies for formalizing information, conceptualizing knowledge, and querying the domain knowledge enable theoretically a variety of solving heterogeneous information and knowledge acquisition tasks. The previous sections introduced the model and techniques of Semantic Web by introducing the different levels of formalisms, the formal languages, and the access to the information resources by querying the information space. The meaningful relationships in semantics provide, as introduced, not only accessing and querying explicitly formalized knowledge, it further enables to infer new and implicit knowledge with the underlying logical descriptions and functions. It is likely that the semantic conceptualization of knowledge provide a wide range of search and knowledge acquisition possibilities, in particular for human. The human access to the formalized information plays a key-role, thus the human stands at the end of the chain.

This section introduces various human semantics interfaces. We define as human semantic interface, each human computer (or machine) interface that accesses and queries semantically annotated data. The level of formalism is not of great importance in this context. The focus lies far more on the human informational benefit in interacting with such systems. To enable a better categorization and definition of the informational tasks two search models will be introduced. Search includes complex cognitive activities that are associated with human knowledge and information acquisition, and learning, which is supported by perceived information [57]. The goal is to identify and classify the different existing systems. Based on the models a classification will be performed exemplary for today's human semantics interfaces. The exemplary systems do not claim to give a comprehensive overview of today's human access to semantics. It should far more work out what the role of visualizing semantics is or could be in the process of knowledge acquisition and search. In this context the search and models that describe and define exploratory search are of great interest.

Marchionini introduced a search model based on Bloom's taxonomy [58] with three kinds of search activities, *Lookup*, *Learn*, and *Investigate* [59]. He proposed that the activities are overlapping and searchers may be involved in more than one activity in parallel. *Lookup* is the lowest level of a search activity and thereby the basic step in his search model. The *Lookup* search activity results in discrete and well-structured information. Commonly the results are facts that answering the questions of *who*, *where*, and *when* in contrast to *why* or *how*. The formulation of such queries premises domain knowledge of searchers [57, 59]. Marchionini proposed that a carefully specified query results with a precise set of knowledge items [59]. The next step of his search process model is *Learn*, which is assigned together with *Investigate* as the exploratory activities. This search activity involves multiple iterations of searching and result evaluation to enhance the knowledge about a certain domain or topic. The main aspect here is that the searcher develops new knowledge by e.g. comparing, interpreting, or making qualitative judgments. The returned results that may come from various media require cognitive processing and interpretation. "Learning" is used in a general manner that includes self-directed, directed, or professional learning and aims to achieve beside knowledge acquisition, the comprehension of concepts and skills, interpretation, comparison, aggregation, and user development [57, 59]. The exploratory character of the *Learn* activity leads to reformulate and precise the search queries and develop strategies for searching and enhancing knowledge.

The most complex cognitive activity in the search process is *Investigate* that includes tasks as analysis, synthesis, and evaluation. This activity requires enhanced knowledge about the topic or domain of interest. This search activity includes not only finding and acquiring new knowledge and information. It involves analytical tasks such as discovering gaps in the knowledge domain. Marchionini further proposes that serendipitous browsing is a kind of investigative search too. Serendipitous browsing stimulates analogical thinking [59], where users relate their experiences

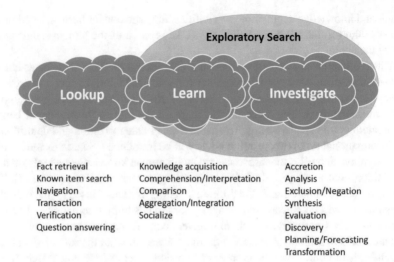

Fact retrieval	Knowledge acquisition	Accretion
Known item search	Comprehension/Interpretation	Analysis
Navigation	Comparison	Exclusion/Negation
Transaction	Aggregation/Integration	Synthesis
Verification	Socialize	Evaluation
Question answering		Discovery
		Planning/Forecasting
		Transformation

Fig. 3.5 Exploratory search model by Marchionini with the three overlapping search activities (adapted from [59, p. 42], with permission)

and internalized knowledge from another knowledge tasks to a related one [57, 59]. In this process of knowledge discovery they construct knowledge by investigating various sources and ideas [60]. Therefore it is important to support this process of search and learning by maximizing the number of possible relevant objects (recall) rather than minimizing the number of possible irrelevant information (precision). Marchionini proposed that the active involvement of users in the search process is important for providing the full range of his search activity model. The matching of queries to the most relevant documents does not solve the entire range of search and information tasks [59]. Figure 3.5 illustrates the exploratory search model of Marchionini with the described three activities.

White and Roth proposed a model of exploratory search behavior by considering the *problem context* and the *search process* [57]. Based on the introduced model of Marchionini [59] they suggest that exploratory search can be defined either by the motivation and intention of search (problem context) or by the process how the search is conducted (search process). The problem context is motivated by an incompleteness or problematic situation of the searcher that leads to the "desire" for information [57]. The search starts with a certain state of user's knowledge with a gap between the known and desired information. The "desire" may be the answer to an unknown question which can be responded by facts, but it may be that the desire of the user is personal development or the wish to learn about a certain topic in general. In exploratory search, the problem context often involves complex situations, is ill-structured, and requires additional information for clarifying the goals and activities. The lack of prior domain knowledge leads to unclear and unsystematic activities through the information step [57]. With the lack in prior domain knowledge the searcher the uncertainty of the user is higher. The problem solving process is more related to discovery and exploratory activities. During the exploratory steps the knowledge if the searcher increases and thereby the ability to verbalize her desire more precisely. The problem solution is constructed from information within knowledge accumulation within the knowledge domain. The searchers' verbalization ability for more precise query statements and the ability to identify relevant information resources increases [57].

The *search process* in the model of White and Roth includes two main steps, *exploratory browsing* and *focused search* [57]. Based on the problem context, which can be abstracted defined as prior domain knowledge, desire, and level of uncertainty, the searcher is involved in one of the steps [57], whereas the steps may overlap during the search process. *Exploratory browsing* is performed by searchers as movement in a connected space to better define their information needs and promote new ideas [57]. Searchers may exhibit exploratory interaction behavior, such as "wayfinding" to traverse the information space without prior knowledge [57]. This kind of searching involves the activities of knowledge acquisition to improve topic or domain knowledge [57]. Browsing is characterized by examining multiple documents and maybe generate hypothesis about a cause of an observed phenomena or the best way to solve an information problem [57]. With the increased topic or domain knowledge the searcher is able to formulate his information need in a more precise way while his uncertainty in the information problem solving process decreases [57]. In the *focused search* step the search query and the search result examination are in close proximity of search results [57]. During focused search the user has a clear sense of the information goal with extant prior knowledge to verbalize his information problem [57]. The proposed model visualizes the exploratory search behavior in temporal manner, where the searcher starts with no or low domain knowledge, high uncertainty, and high frequency of search interactions. During the process of exploratory search his domain or topic knowledge increases, the uncertainty and the frequency of search interactions decrease [57].

The introduced search and search behavior models illustrated the wide range of informational tasks related to search. A precise formulated search query leads to maximize the amount of relevant topics or documents. It requires extant searcher knowledge about the knowledge domain and is commonly answering questions of *what*, *who*, and *when*. The prior domain knowledge or the constraint problem context leads to specify the formulation of such queries. In contrast to a precise or focused search the exploratory search includes more the aspect of discover and situational constructing. The problem context is often complex and can include the personal development or the wish of general knowledge in a topic. The searchers learn during the search interaction, acquire domain knowledge and develop higher-level intellectual capabilities. The search process itself consists of browsing through interlinked resources and knowledge and involves at highest cognitive level aspects of analysis, synthesis, or evaluation.

It is likely that the formal conceptualization of knowledge as semantics and ontology should not only enable the different kind or types of search and knowledge acquisition, it should support both the entire range of informational problem context and the entire range of the search process including the different activities. For evaluating exemplary the human access to semantics, we introduce a three-fold classification based on the previously described models, by differentiating *question-answering systems*, *key-word based non exploratory search systems*, and *exploratory search systems*. *Question-answering systems* translate the human language into a formalized query to extract declarative, fact-knowledge. Such systems are able to answer questions based on human natural language processing. An example for such a question may be "Where was Barack Obama born?". The systems are able to disambiguate the explicitly formalized knowledge and provide a fact as response, in our example just the "place-of-birth" of the entity "Barack Obama" would be presented. The question-answering systems are equivalent to the *Lookup* activity in Marchionini's model and the *focused search* in the search process of White and Roth. Deines and Kreschel proposed a system that is translating the human natural language into SPARQL queries and provides facts as responses of domain ontologies [61]. Their system makes use of text-based and semantic-based similarities for identifying resources in ontologies and generates an adequate SPARQL query that retrieves the right fact with a precision and recall of 66 % [61].

Similar systems that partially further transforms spoken language to textual queries before formalizing them into a query language are proposed in [62–66]. A survey of *question-answering systems* based on semantics can be found in [67].

The class of *key-word based non-exploratory systems* imitates traditional search engines as they are available on Web. It is likely that such systems can be used in the process of search to explore knowledge too, but the goal of these systems is not to actively support exploratory search approaches. These systems provide based on the search query a result set that can be examined by users, who may enhance their knowledge based on the result and refine their queries. An active support, e.g. by providing alternatives, actively minimizing the precision, or enable a "wayfinding" or browsing approach is not integrated. The results are commonly ranked and illustrated as lists. The formalized knowledge base enables the possibility to provide facets for refining or constraining queries, but the use of faceted approaches is commonly assigned to exploratory search systems that go beyond a ranked illustration of the structured knowledge [68, 69]. A prominent example for such a user interface is *Freebase* [70], a community curated semantic database. Freebase provides different APIs [71] that can be used to implement different kinds of search interfaces in particular exploratory interfaces. But the integrated user interface and search limits the use of semantics. The searcher is able to type a query and get a set of ranked results as a list. He then can retrieve the information. Another approach that illustrates in particular the imitation of traditional search to semantic formalized knowledge is proposed by Fazzinga et al. [56]. They reduced the processing of Semantic Web search queries to standard Web search queries by compiling the TBox of offline ontology and generating ABox terms, which are then searched in traditional search engines [56]. Their conjunctive query approach consists of an on offline ontology reasoning step, where semantic annotations are completed by entailed membership axioms, and an online reduction to standard Web search, where the axioms are transformed to key-words [56]. Another example that makes use of the conjunction of standard and semantic information to provide a key-word based search is the introduced (Sect. 3.2.2) Google Knowledge Graph [34]. Prominent people, places, and things are enriched with a "semantic view" that shows relations to other resources on Web (see Fig. 3.2) and provide thereby a kind of browsing through among the relations to enhance the knowledge in particular topic [34]. The example of Google's Knowledge Graph illustrates clearly that a strict bisection between the classified systems is not really possible. Far more the aspect of providing search abilities for the entire range of search activities is of importance. Google Knowledge Graph, for instance does not provide an overview on the entire queried information space as proposed by Shneiderman [72]. An overview to detail approach is consequently not actively supported.

In contrast to the previously introduced classes of search systems and human interactions with semantics, the *exploratory search systems* provide explicitly the ability to browse, compare, investigate, or analyze semantic data and structure. They make use of the "structural information" of semantics and provide in best case both, a bottom-up-approach that starts with a set of semantic entities and can be enhanced to an overview illustration, and a top-down approach that starts with the overview on the topic and provides the possibility to retrieve detailed information. As the model of White and Roth showed, a mix of the approaches would further help to reduce the user's uncertainty [57]. There are a variety of systems that claim to support the exploration of data, for instance by providing faceted search [68, 69], but in many cases the entire range of a bottom-up and top-down views on the information space and the formalized knowledge is not supported. White and Roth propose that information visualization is important for exploratory search, thus different tasks, such as understanding and analyzing, are supported [57]. But they criticize that information visualization does not support the information seeking process. We illustrated in Sect. 2.4 those different approaches of information visualization support solving a variety of tasks, including information seeking in various forms. Further, information visualization provides different approaches to illustrate structural information, overview information and support detailed views on the data entities. Information visualization that includes the process of information seeking and provide, both bottom-up and top-down approaches for semantic search is the best suited tool to support the entire range of information seeking, in particular the exploratory steps.

We will investigate the aspect of information visualization of semantics more detailed in the next section. Therefore we will first introduce a definition of Semantics Visualization to have a common understanding of the term, followed by a classification that will be used to categorize the introduced systems.

3.5   Visualization of Semantics

The conceptualization of knowledge as semantics or ontology provides various enhanced features for retrieving information. This information can be explicitly modeled in a semantic knowledge base or implicitly inferred by logical functions. Semantics as formalization of information has experienced wide dissemination in Web, research, and industry. In particular the RDF-based Linked-Data formalizations have experienced a wide acceptability and dissemination in Web. The human access to semantics is commonly performed by various kinds of search, whereas answering simple questions about facts in form of "who", "what", "where" and "when" seems to be the focus of the semantic search approaches. We outlined that specifying such questions needs prior domain knowledge or a constrained problem context. The user needs the ability to verbalize such specific questions, which can be constructed by browsing and exploring for acquiring domain knowledge. We could further outline that key-word based approaches often do not provide an exploration of the information space, whereas the exploration is of great importance for acquiring knowledge and performing complex cognitive tasks. These tasks can be supported by visualization systems that graphically present the structure of data and the entities (resources) with their detailed information. The visualization systems should provide further a bottom-up and top-down approach for supporting the entire search process. With these functions the search process would profit from the underlying semantics and semantics would profit from the exploratory approaches of information visualization and semantics visualization respectively.

This section introduces the state of art and technology in semantics visualizations. To obtain a clear picture of existing systems and approaches, we will first define the term 'semantics visualization'. Thereafter we will introduce a classification for providing a comprehensible picture of the existing systems. The classification will be used to introduce the existing approaches and systems for visualizing semantics.

3.5.1   Definition of Semantics in Context of Information Visualization

Semantics provide formalized information of a certain domain knowledge. We have introduced a variety of different levels of formalisms that builds a spectrum of "simple meanings" of terms to a decidable and predictable subset of first-order logic as formalizations by description logic. Semantics provide in different system, applications and knowledge domains various degrees of formalization. It depends commonly to the informational task to be solved, what semantics is in each case. In this thesis the aspect of human interaction with semantics plays an important role. We outlined that commonly semantic is used for human-computer interaction to solve information seeking tasks. Information seeking or search may have various steps or involve different activities as we described with the exploratory search models. The formalized characteristic of semantics provides high precision for question-answering systems within a knowledge domain. But the verbalization of such specific questions premises prior domain knowledge that is not actively supported by such systems. The construction of knowledge can be actively supported by exploration and discovery of the information space [57–60]. Information visualization is predestinated and suitable for exploration tasks, whereas the aspect of information seeking is not supported actively [57]. Even in context of semantics, visualization techniques commonly aim at visualizing the formalized structure of the conceptualized knowledge domain and refer more to ontologies. In this context commonly the term *ontology visualization* is used [73]. *Ontology Visualizations* aims at visualizing the semantic relationships between concepts or instances within formal domains of knowledge. Visualizations were designed to illustrate the formalized structure of ontologies and provide primarily a view for validating and overviewing a formal modeled domain. Some of these technologies provided further functionalities for editing or annotation. User-centered approaches for solving information seeking tasks by exploring the information space, retrieving overview and detailed views, and enabling the "investigation" of the domain knowledge as proposed by Bloom [58] or Marchionini [59] were commonly not the focus of research and development.

The term *Ontology Visualization* constraints interactive visualizations to graphical representations of formal descriptions and definitions of ontologies. With the upcoming light-weight semantics, in particular in form of Linked-Data or social networks, the visualization of these knowledge concepts would not fulfill the requirements to belong to *Ontology Visualizations*. Semantics in a lower formalization represents meaning of terms, resources, or entities. Thus information visualization is defined by Card et al. [74] as "The use of computer-supported, interactive, visual representations of **abstract** data to amplify cognition" [74, p. 6] it is likely to adapt in particular the "amplifying cognition" more precisely in terms of semantics. This would lead to investigate the entire search process of Marchionini [59], which includes *Lookup*, *Learn*, and *Investigate* with the various activities that amplify cognition. But the question of what semantics is in context of information visualization still remains. We could illustrate that a well-structured semantic annotation can be performed by a directed graph, as a meaningful relationship between two resources, terms, or knowledge entities. These relationships are commonly designed as triples, e.g. as a *subject*, *predicate*, and *object* triple in RDF. But if we take a look at the *reference model of information visualization* (Fig. 2.2) [74], the first transformation step is the *data transformation* to *data tables* with a set of relations defined by a set of tuples (see Sect. 2.1) [74]. And if we assume that a table can be defined as triple of *row*, *value*, and *column*, where row and column are named, human can retrieve meaning from this structured table. Thus in information visualization human and his informational perception plays an important role (see Sect. 2.2) we can define *Semantics* as a meaningful interrelation of at least two information or data entities, to provide in best case a disambiguated meaning of interlinked data. The data entities or knowledge representations may have different natures, e.g. topic, concept, resource, or just values. The important aspect is that human can retrieve in best case a disambiguated meaning from the underlying relationships.

With this definition of semantics, *ontology visualizations* are a subset of semantics visualization, which includes the visualization of any meaningful data relation.
Semantics visualization supports actively the information seeking process with the
underlying steps of *Lookup, Learning*, and *Investigate* to amplify cognition in the
search process. The main goal is to provide user-centered interactive graphical representations for solving visual tasks with semantics and supports in best case the entire
search process. Therewith semantics visualization bridges the three dimensional gap
between users, tasks and data. Based on the criteria introduced above, we define
semantics visualizations as:

> Semantics visualizations are computer-aided interactive visualizations for effective explo-
> ratory search, knowledge domain understanding, and decision making based on semantics.

- *Whereas semantics is defined as data with meaningful relations of at least two
  information or data entities, to provide in best case a disambiguated meaning,*
- *and exploratory search is defined by* Bloom [58] *and* Marchionini [57, 59] *and
  includes the activities of Lookup, Learn, and Investigate with the various sub-
  activities, e.g. analyze, synthesize, and compare.*

With this definition semantics visualization supports actively the search process
as proposed by Marchionini [59], and White and Roth. The visualization of structural
information of the underlying knowledge domain, e.g. the hierarchy, visual patterns,
or relationships amongst entities supports the search process and amplifies cognition
to learn, investigate, and decide. Therefore semantics visualizations should support
both a top-down approach, as proposed by Shneiderman in his *Visual Information
Seeking Mantra* [72] and a bottom-up approach: from detailed-view to the abstract
semantic relations of a knowledge domain. For supporting the visual tasks it is
important to provide visual information on entities, the hierarchical structure, and
the arbitrary relations of the data.

3.5.2 Classification of Semantics Visualizations

There exist a variety of classifications for information visualization based on different criteria. We have introduced in the previous chapter various classifications
based visual interactions, visual tasks, data, and visualization techniques. On the
semantics level, there exist various classifications too. In this context we introduced
classifications based semantics formalisms, search process, and human access to
visualizations. Although, the introduced classifications already cover a huge range
of the aspects that are supported by semantics visualizations, an explicit classification of semantics visualization can scarcely be found. Katifori et al. introduced a
survey and overview of ontology visualizations [73, 75]. In this context they categorized the existing systems based on the characteristics of ontologies [73]. They
characterized an ontology (O) based on the definition of Amann and Fundulaki
[76] as $O = (C, S, isa, I)$, where $C = \{c_1, c_2, \ldots c_m\}$ is a set of classes,

where c_i represents a real world object. $S = \{s_1, s_2, \ldots, s_n\}$ is a set of slots, where s_i represent either the property of class or a binary relationship of two classes. $isa = \{isa_1, isa_2, \ldots, isa_p\}$ is a set of inheritance relationships, where isa_i represents the inheritance of classes. $I = \{i_1, i_2, \ldots, i_q\}$ is a set of instance, where i_w is an instance of the class $c_x \in C$ [73, pp. 10:2–10:3].

Based on this definition, they identified relevant elements of ontologies that should be visualized by ontology visualization. They identified and differentiated the following elements: *Classes, Instances, Taxonomies (isa relations), Multiple Inheritance, Role Relations*, and *Properties* [73]. The identification of these elements is the foundation of a two-stepped classification. They differentiate in the first classification step between *Indented list, Node-link and tree, Zoomable, Space-filling, Focus + context or distortion*, and *3D Information landscapes* [73]. The second step differentiates the visualization according their number of space dimensions into 2D and 3D methods, whereas they apply the term $2\frac{1}{2}$D to 2D visualizations with a perspective view [73]. In context of this work their second classification step is not of interest, thus the space dimension is a projection of the semantics to the display. We will investigate the first step of their classification in order to get an overview of the aspects that are relevant for classifying ontology visualizations. *Indented list* visualizes the hierarchy of an ontology commonly on the class or schema level. Classes are presented as indented, expandable lists in a tree-structure, where a subclass is placed under the parent-class [73]. Multiple-inheritance is illustrated is multiple occurrences of the same class under different parent classes. Further information, e.g. role relations or properties can be visualized in separated windows, but are not supported by indented lists [73]. *Node-link and tree* visualizes the hierarchy of ontologies by graph-based techniques, either with a top-down or a left-right metaphor. Users are enabled to interact with the nodes and get more information about the selected nodes, e.g. sub-nodes or properties [73]. *Zoomable* ontology visualizations include all visualization techniques that present the child-nodes of hierarchical class structure as nested nodes within the parent classes [73]. Users are able to *zoom-in* into the child nodes and enlarge them in order to get more information, whereas the child-nodes may be subclasses in the nested visual representation or instances [73]. *Space-filling* ontology visualizations use the entire space of the screen and subdivide the available space amongst its children [73]. Commonly classes with more sub-classes or instances use more space of the screen to indicate the higher amount of included entities [73]. In many systems the space-filling approach uses the nested model too, as proposed by Shneiderman [77]. *Focus + context or distortion* visualizations focus on the visual presentation of objects (e.g. classes or instances) and their neighborhood relations [73]. The selected object is commonly placed on the center of the visual representation, while the other objects are placed around it with a reduced size [73]. The distance to the centered object is the metric for the size, until the objects with higher distance are not visualized anymore [73]. *Information landscapes* make use of size- and color-coded objects and place them in three-dimensional "landscape metaphor" [73].

Katifori et al. investigate in their classification and in their survey of existing system in particular the formal aspects of ontology. Their foundation of classification is the formalization of knowledge in its highest degree of formalism. Their focus of investigation is e.g. the ability of visualization for presenting multiple-inheritance or role-relations. It is a fact that these aspects are of great interest for a formal view on ontologies. But the way how users are supported in their search process is not in focus. Further, their classification seems to be mixture of classifications of interaction-techniques (e.g. Zoomable or Focus + context) and visualization techniques. It is possible to abstract visualizations that targets visualizing conceptual hierarchies, as *hierarchical visualizations*, instead of defining three classes of visualizations that fulfill the same task. In the definition of semantics visualization we described that ontology visualizations are a subset of semantics visualization. The formal view on ontologies is not of great interest in context of this work, thus we focus more on adaptation issues. Our foundation is the exploratory search model of Marchionini [59] with the three elementary activities of *Lookup*, *Learn*, and *Investigate*. Further we map the *Visual Information Seeking Mantra* of Shneiderman [72] with the steps *overview*, *zoom*, and *details-on-demand* to the search model. Thus the information seeking process can be started by users with prior domain knowledge or a constraint problem context, the order of the seeking mantra is neglected. We assume that semantics visualization should support in best case both, a top-down approach as proposed by Shneiderman [72] and a bottom-up approach as implemented for instance by van Ham and Perer [78]. For this case the structure of the modeled knowledge plays an important role, as Katifori and colleagues already worked out. The third aspect that we consider in our classification is the semantic data structure. As we already worked out in our definition of semantics and semantics visualization, the structure can be described as a directed graph and this can describe any arbitrary relation or a hierarchical relation. According to the introduced criteria, we categorize the existing approaches for semantics visualizations into: (1) *hierarchical semantics visualization*, (2) *relational semantics visualizations*, and (3) *Entity-based semantics visualization*. (1) Hierarchical semantics visualizations are focused on visualizing hierarchical aspects of semantic information, e.g. the concept taxonomies or inheritance structures. They provide an exploratory overview in terms of taxonomies. Users are able to view the entire hierarchical structure or the hierarchical structure of focused entities and retrieve information about the hierarchical domain. (2) Relational semantics visualizations use the meaningful relations of semantics to represent correlations between semantic entities. They provide in particular a neighborhood relationship that can be used to browse a knowledge domain. Users are able to retrieve contextual information about a knowledge domain with the relational visualizations. (3) Entity-based semantic visualizations support the bottom-up approach in search. The starting point is always a performed search with a set of result entities. Users are able to choose one or more entities and navigate through the relations or hierarchies. These categories are not mutually exclusive. There are also approaches that can be assigned to multiple categories (e.g. visualizations that are presenting hierarchical and relational aspects). Therefore we use the dominant characteristic of visualizations for its classification.

3.5.3   Survey of Semantics Visualization Techniques

This section will illustrate the state of art and technology for semantics visualization. For a clearer picture of the topic we will introduce the existing systems and approaches based on the above defined classification. The focus of investigation will be the last decade, whereas some early fundamental examples will be mentioned to provide a clearer picture of the development history. In our investigation, we found that the term "semantic visualization" was already used in 1998 by Chase and colleagues [79].

3.5.3.1   Hierarchical Semantics Visualizations

Hierarchical structures form the foundation of each knowledge domain. Already light-weight semantics may contain hierarchical structures and provide a comprehensible view on the knowledge. Hierarchies provide the opportunity to categorize domain-specific resources in inherited concepts and allow a topic-related access to the modeled domain knowledge. Users are able to locate a search topic in the hierarchical structure and get thereby a starting point to abstract his query by querying the parent node or precise his search by choosing a sub-node. Vice versa the overview on hierarchical structures provides the ability to find the starting point and locate or verbalize the knowledge-resource of interest. In contrast to arbitrary knowledge structures, hierarchies can be visualized in various ways, thus we perceive nested [80], treemap [81, 82], indented [83], or top-down and left-right structures intuitively as hierarchies. Common approaches for visualizing hierarchical structures that are also used in the *Protégé Class Browser* [83] and *OntoEdit* [84] are tree-based visualizations. Indented lists are often used for navigating file systems. Because of their familiarity, indented lists are easy to use, allow high performance in semantics exploring [73] and provide a clear view of entity labels and the concept hierarchy. Further approaches may include special graph-visualizations, such as hyperbolic-trees [85] to visualize hierarchical structures.

An example for a graph-based hierarchical semantics visualization is *OntoTrack* [86] that focuses on visualization and editing of formal ontologies formalized as OWL [23]. *OntoTrack* visualizes either classes or properties as directed acyclic graphs [86] based on *SpaceTree* [87]. The hierarchies can be visualized in a top-down, left-right, bottom-up, or right-left graph [86]. The visualization is designed to show only direct subsumption relationships and hide all redundant relationships [86]. The formal character of *OntoTrack* with OWL-editing possibility, a direct interfacing to a reasoner, and the validation ability is an appropriate visual tool for ontology development and editing [86], the search process is not in focus and not supported adequately.

An example for a nested visual metaphor is the *CropCircles* ontology visualization [88, 89], which represents the class hierarchy tree as a set of concentric nested circles. *CropCircles* aims to gives users intuitions on the complexity of a given class hierarchy at a glance. Nodes are given the appropriate space in order to guarantee enclosure of all the sub-trees. If there is only one child, it is placed as a concentric circle to its parent. Otherwise the child-circles are placed inside the parent node from the largest to the smallest. In order to navigate the ontology structure, the user may click on a circle to highlight it and see a list of its immediate children on a selection pane [89]. The selection pane let users drill down the class hierarchy level-by-level and also supports users' browsing history. The user may also select which top level nodes to show in the visualization [88, 89]. The user interface consists of two areas: the main view is the visualization of hierarchies as nested circles and a smaller panel on the left, which serves for navigation and detailed views [89]. *CropCircles* does not provide a search functionality, is more designed for an overview of a formal ontology.

A space-filling approach for visualizing light-weight ontological hierarchies was brought by *Kriglstein* and colleagues with the *Knoocks* (Knowledge blocks) visualization [90, 91]. In their first version [90] the main view of their application that visualizes classes as rectangles was introduced. They visualized classes next to each other (from left to right) with their sub-class-of relationship. The instances were nested with text within each class [90]. The visual metaphor had the advantage that the rectangles did not overlap and the users were not overcharged. But the space of the screen was not used in an efficient way. Further the visualization did not provide any search or exploration approaches. It was dedicated to visualize ontological hierarchies as side-by-side rectangles instead nested rectangles as proposed by Shneiderman [77]. Their visualization was limited to the hierarchical relations of classes and was even to able to visualize between class relations. In their revised version they enhanced their visualization with a multi-view approach [91] consisting of an overview of class relations, the already introduced hierarchical space-filling visualization, and window for toolbox for searching, filtering, and history illustration [91]. In particular the overview visualization, which enabled to see the relations of all classes, was a missing enhancement that enriched their visualization. It used a similar rectangular metaphor for the classes with interlinked with each other [91], whereas instance relations or resource relations were not considered. The search was limited to the visualized ontology. In any case the entire ontology was visualized, whereas the user could bookmark instances of relevance [91]. Beside the limited search capabilities, the visualization was not able to illustrate the resource contents. The entire search, navigation, and exploration process was performed on the abstract level of an OWL-Lite ontology.

In context of collaboration Allemang et al. introduced the *Cove* tool [92] that is assigned as visualization but provides far more a collaborative approach for engineering ontologies. The main goal is to provide a platform for collaboration and evolution of ontologies, whereas the evolution is illustrated in hierarchical graph visualization. Therefore they use the sub-class-of relations of ontologies to visualize a graph that presents the highest hierarchical level on the left with linked children nodes on right. The approach does not provide any search or exploration capabilities in terms of search. It provides comprehensible view on the abstract level of the ontology hierarchy on class-level. An enhanced approach targeting the collaboration aspect with more than one view is proposed by FU and colleagues in their *BioMixer* [93]. The goal of *BioMixer* is to provide a collaborative approach for engineering ontologies too, but they provide a set of different visualization layouts with minor text-information to support this process [93]. Both introduced examples are designed for collaboration tasks among generating or editing ontologies, whereas *BioMixer* allows the visualization of two or more ontologies at the same screen by using an ontology-mapping approach [93]. The techniques do not focus on search or learning tasks and therefore a search is not possible.

de Souza et al. introduced the visualization of ontology hierarchies as *hypertrees* [94]. They proposed that the use of hypertrees for ontologies provide a site map and a navigation tool. In addition the hypertrees would allow a better contextualization of information. Their approach was designed for ontologies, with references to the entity-resources, whereas the underlying domain ontology was formalized as Dublin Core XML [94]. A double-click on an entity in the visualization opened the linked resource in a separated window and the user was able to investigate that resource [94]. The main advantage of their system was the comprehensible view on the contextual, commonly hierarchical information with a direct link to the resources. Their approach used just one type of visualization with one level of detail. An overview to detail or vice versa approach was not possible. Further the system opened the resources in separated windows, which led to lose the context of the information entity that was investigated.

Buntain introduced an approach for three-dimensional visualization for ontologies with multiple views on the domain ontology [95]. The main and dominant aspect of his approach was using three-dimensional rendering techniques for visualizing the topology of a given ontology in a tree-like fashion. He proposed three different views on the ontology, *Structural View*, *Document Map View*, and a *Result View*. The *Structural View* allowed users to interact in a three-dimensional fashion with the hierarchical structure of an ontology. Users were able to rotate the tree structure and had a larger space to click for navigating through the hierarchical structure. The *Document Map View* enabled users to see how a set of document maps into a selected ontology. The used metaphor was assigned as *information molecules*, whereas the work on integration was not completed [95]. The *Result View* visualized the result of a user query on a concept level in the *information molecules* metaphor. The *information molecule* metaphor organized concept and their relationships in transparent circles to intend their relationships. The main aspect of Buntain's visualization approach was the topological view on the ontological structure. It allowed, for a limited amount of concepts, a comprehensible view on the structure.

Samper et al. introduced *OntoService*, an ontology visualization tool for creating Semantic Web service profiles [96]. The visualization of ontologies is based on two main visual components that are both aiming at visualizing the ontology hierarchy. The user interface consists of an indented list visualization with a folder metaphor for classes and a hierarchical node-link visualization that visualizes the concept hierarchy as rectangular nodes with directed links. The main contribution of their approach is the capability of the visualization to cache the latest user query for increasing the speed [96]. Their system is targeted for service profile creation and does not provide an exploratory search approach. Although, the caching of user queries is interesting, thus the reaction time of visualizations decreases, the search and visualization capabilities are limited to ontological hierarchies.

Motta et al. introduced with *KC-Viz* an ontology visualization and navigation tool, claiming to support the exploration process [97, 98]. *KC-Viz* is an integrated visualization tool for the ontology engineering environment *NeOn Toolkit* [99]. The integrated visualization aims at providing a visual environment for hierarchical semantic structures by integrating a *key-concept approach* [97, 98]. *KC-Viz* uses the algorithm for *key concept extraction* (KC), [98] to compute an importance score of each class in a particular ontology. The importance factor of a concept is computed based mainly on two main criteria, the notions *density* refers to concepts that are richly characterized with properties and taxonomies (sub-class-of relations) and the notion *coverage* refers to key-concepts related in a taxonomic manner (again hierarchical) to a concept. *KC-Viz* visualizes the hierarchy of the computed key concepts as a node-link graph. Each node represents a key-concept tagged with the name of the class, followed by the amount of direct subclasses followed further by the amount of indirect subclasses. For instance the class *Region* with 16 direct subclasses and 386 indirect subclasses is labeled in the visualization as Region [16, 386] [98]. The classes are interlinked with directed graphs using an arrow metaphor, whereby the link is not labeled due the limitation to a *sub-class-of* or *is-a* relation and the arrow indicates the course of the super- or parent-class. The amount of the visualized nodes are chosen by the user, whereas a high-level (commonly *thing*) class is added to provide a closed graph-structure. The claimed exploration process is supported by a "flexible set of options" [98]. The user is able to get more information about a certain class. *KC-Viz* opens after a right-click a context-menu with several information about the class, e.g. type, URI, amount of direct super- and sub-classes, amount of indirect super- and sub-classes, and the amount of individuals (instances) of the class. Further the user is enabled to expand a class and get the direct subclasses of the concept. *KC-Viz* provides in its current implementation the visualization of concepts in a hierarchical manner by using an importance factor that is computed by the amount of sub-classes and properties. The visualization provides an interesting top-down approach, thus the key-concepts are not explicitly annotated in an ontology. It is similar to page-ranking and relevance measure approaches that compute the relevance of an entity based on the number of interlinked objects and references. The top-down approach enables a view on the taxonomy with relevant classes, whereas the relevancy is canonical and not based on any preferences of the user. A search or a bottom-up approach or functionality is not provided.

Hierarchical structures are fundamental components of semantic data and formalisms. These structures provide the ability to visualize semantics on an abstract level in an adequate and meaningful way to navigate through the classes or concepts of interest and retrieve information. Further the overview on the entire retrieved data plays an important role. The hierarchical view on data with different perspectives enables to gain an insight on an entire knowledge domain or a relevant sub-part. With the semantically labeled clusters of instances or information resources, the concept hierarchy is an essential part of visualizing formalized data. Although, many systems and approaches provide sufficient visualization techniques for hierarchical structures, the main task of semantics, namely providing an efficient human information acquisition by exploratory search and learning is not actively supported by the introduced systems. The semantics visualization approaches are focusing far more on the overview aspect, rather than on navigation through the hierarchical concepts for gathering the information entities (instances). Most of the systems are designed for ontology engineering and do not support search. They remain on the abstract level of ontological concepts and do not provide a real top-down or bottom-search paradigm. And even if search is supported, the only search paradigm is a top-down search that enables viewing classes and subclasses and navigating through the hierarchical structure. A bottom-approach is missing in the hierarchical visualization of semantics at all.

3.5.3.2 Relational Semantics Visualizations

Relational visualizations aim to visualize the semantic context of information and provide navigation and browsing abilities within an information space. Common approaches for visualizing semantic relationships are usually based on graph-based visualization techniques and provide navigation through the nodes and semantic neighborhoods. Alani proposed with *TGVizTab* [100] such a graph-based relational visualization of formal ontologies based on the *TouchGraph* [101] visualization as a tab plug-in for Protégé [83]. The user interface of *TGVizTab* consists of an ontology schema that illustrates class taxonomies as indented list provided by the *Protégé* class browser, which is the initial point of interaction. After selecting a class in the indented list of the class panel, the main visual interface provides a relational visualization as arbitrary node-links in the main panel [100]. The main graph-visualization provides a gradually visualization and navigation through the relational structure of the formal ontology. Users are able to choose focal points, zoom geometrically into the information space and change the number of shown relations. The arrangement of the visualization in the user interface cannot be changed. The main panel is a relational view that is designed for ontology experts.

Godehardt and Bhatti introduced with the *SAP-TM-Viewer* [102] a relational visualization for knowledge-based Topic Maps [49]. The *SAP-TM-Viewer* (Topic-Map Viewer) offers fields or sectors, which can be extracted from the underlying topic maps. The concepts in each field are represented as areas with different levels of semantic relationships, separated visually by lines in so called *sectors*. The knowledge entities are visualized as arbitrary graphs, with iconic linked nodes. The entities of each concept have an own iconic representation, to provide an easier differentiation between the knowledge entities [102]. The concept level can be enlarged manually to provide a kind of zooming by enlarging the surrounding space. The *SAP-TM-Viewer* further provides the clustering of objects to hide uninteresting related objects. A search query not provided in the visualization, thus the system is an embedded visualization of a Web-based system. Far more of interest is the loading-on-demand support [102]. Users' interactions result in loading more and more knowledge enti-

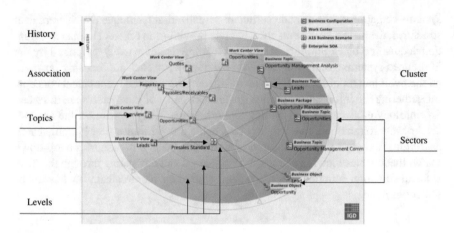

Fig. 3.6 Relational semantics visualizations: the *SAP-TM-Viewer* with the concepts sectors and associations between the iconic entities (from [102, p. 5])

ties from a predefined knowledge base. Figure 3.6 illustrates the user interfaces of the introduced system *SAP-TM-Viewer* with their different areas and functionalities.

Fluit et al. introduced the *Cluster Map* visualization approach for visualizing instances of a set of selected classes based on their hierarchy and organized by their classifications [103, 104]. Therefore they cluster instances from the same class into a grouped cluster in a balloon-shape. Each class of a *Cluster Map* may have a color, which differentiates the class from the others. The overlapping of classes is illustrated as color-overlapping. The layout algorithm used for *Cluster Map* is a spring-embedder and visualizes the semantic closeness of the classes and the included instances, where semantically related classes are placed next to each other. The *Cluster Map* was included in different user interfaces for different tasks [103, 104]. The main aspect of this visualization is the comprehensible overview of the classes and the included instances, whereas the sub-class-of relations are not visualized as nested clusters. *Cluster Map* supports the Top-down search approach with the overview of relations, whereas the initial design was dedicated for hierarchical views. The idea of *Cluster Map* for relational visualization of semantic structures was applied in various further applications [103–105]. An enhanced example is *VISCover* [106] commonly in combination with other visualization tools for supporting the discovery process. A further similar approach was proposed by Bhatti, who used the force-directed [107] algorithm for visualizing knowledge spaces [105].

OntoViz is a plug-in visualization for Protégé ([83] in [108]), based on the AT&T visualization technology *GraphViz* [109]. *OntoViz* visualizes the structure of a formal ontology as an arbitrary graph, where each node represents either a class or an instance and nodes represent the semantic relationship [108]. It provides the ability to visualize just parts of an ontology by picking a set of classes. The user is able to set colors for nodes and edges. With a limited set of classes, *OntoViz* supports some closure operations, e.g. sub-, or super-class relations [108]. *OntoViz* is another visualization example for formal ontologies: it does not support the search process and is designed as Protégé plug-in for ontology experts.

Lanzenberger and Sampson introduced the multiple view ontology visualization *AlViz* for aligning ontologies visually [110]. The user interface consists of four windows, two for each ontology. The main window illustrates in a node-link diagram as a clustered graph the entities of an ontology (either classes or instances) connected by selectable mutual properties, e.g. *isA*, *isPart*, or *isDistinct*, whereas only one property type can be visualized at same time. A smaller window illustrates the hierarchy of the ontology as indented list. The main focus of the visualization tool is to locate and align ontology elements. Therefore the two ontologies are visualized on the screen. The use of color indicates similarities or equivalences of entities [110]. Thus the task to be solved with *AlViz* is limited to visual ontology alignment, a search or exploration approach is not provided. Furthermore, the visualization is limited to one type of relation in a formal ontology.

A further plug-in visualization for Protégé was introduced by Bosca et al. [111]. They visualized in their *OntoSphere* tool, the structure of formal ontologies in 3-dimensional hyper-space, where information is presented on a 3D view-port enriched by several visual cues (as the color or the size of visualized entities). Although, taxonomies are visualized in *OntoSphere* the dominant aspect of visualization is the relational view based on the ontological structure. Therefore *OntoSphere* displays three different views on the ontology, the *RootFocus Scene*, the *TreeFocus Scene*, and the *ConceptFocus Scene*. The *RootFocus Scene* presents the semantic relations of classes on the surface of "earth-like" sphere [111, p. 7]. This view targets on presenting ontological primitives, e.g. the root classes and provide an overview. The *TreeFocus Scene* illustrates the hierarchical structure of a chosen class and the relations that do not constitute a sub-class-of or instance-of relationship [111]. The *ConceptFocus Scene* is a kind of detail-on-demand view. It visualizes all available information about a single class, as a node-link visualization. Users are able to manipulate the three-dimensional view on the ontology by rotating, panning and zooming. It is strongly bound to the "one hand" interaction paradigm, allowing to browse the ontology as well as to update it, or to add new concepts and relations. Ontology elements are represented with different shapes, concepts are visualized as spheres, instances are depicted as cubes, literals are rendered as cylinders, and the relationships between entities are symbolized by arrowed links, where the arrow itself is constituted by a cone [111]. Bosca et al. introduced a revised and enhanced version of their *Ontosphere* renamed in *Ontosphere3D* [112] with the capabilities to visualize the OWL logic constraints, e.g. restrictions, cardinalities, or disjointness. *OntoSphere* provides with the different views and levels of detail an enhanced approach for visualizing and editing formal ontologies. The three-dimensional view may have certain surpluses, but is confusing in search and learning tasks [74]. Further *OntoSphere* does not provide any search capabilities and is limited to the strong formalisms of formal ontologies.

Kerrigan introduced the *WSMOViz* ontology editing and visualization tool as an integrated tool of WSMO (*Web Service Modeling Ontology*) toolkit based on

JPowerGraph [113, 114]. The visualization of ontologies is performed with a spring-embedder algorithm that visualizes concepts, relations and instances as an arbitrary node-link graph [113]. The visualization is enhanced with a dynamic legend that updates the list of the node-types of the visualized graph. For reducing the ontological complexity the semantic level is reduced by removing certain class or ontology information, e.g. class properties. *WSMOViz* is designed to support users through the ontology engineering process. A search or learning capability is not implemented and was not the focus of the system. Another example for a relational visualization with embedding editing functionality is the *IsaViz* visual authoring tool in its currently third version [115]. It provides a visual environment for browsing and authoring RDF ontologies represented as directed graphs [73, 115]. The graph is presented in a $2\frac{1}{2}$ dimensional visualization with zooming and navigation functionalities in RDF graphs [115]. It further allows the creation and editing of ontologies in RDF CML. Searching, exploration and learning are not supported and not indented by *IsaViz*. It is designed to create or edit ontologies.

Deligiannidis et al. introduced with their semantic analytics visualization (*SAV*) a three-dimensional graph-layout visual analytics tool for semantic enriched data [116]. Their visualization approach is based on the *GraphViz* library [109] that makes use of the three-dimensional graph-visualization. It is coupled to data analytical techniques and designed for experts to interactively investigate complex relationships between various information and sources. For visualizing the complex structures of the relationships in huge semantic data, they partitioned the visualization space into two volumes, a foreground and a background [116]. The foreground visualizes the entities and the relations amongst the entities, whereas the background refers to the documents. With other words this visualization is one of those that are able to visualize both, the entity, or semantic relationships and the related sources and documents. The *query* in *SAV* can be performed by selecting concepts or entities form the visualized ontology. The relations to the selected objects are visualized and highlighted. The approach of this visualization has the main surplus that the real sources and documents can be visualized and retrieved too. It provides a top-down approach, with the foreground-view, the users are able to view the information space of the semantic structure and the background illustrated the visual representations on the document or resource level. *SAV* is designed for experts, who are able to perceive and interact with masses of visualized information. Although, the system makes use of two view-level, the complexity of the information and their relationships is still high. An abstracted view on the different levels is not provided. The search process by verbalizing queries and retrieve a more focused part of the ontology is not provided too. The querying with the system is performed by navigating through a complex information structure and selecting nodes of interest, which may be entities or resources depending on the view-level [116].

Another approach for exploring RDF graphs with incremental loading of RDF graphs proposed by Deligiannidis et al. too [117]. They introduced with the *Paged Graph Visualization* (*PGV*), a semantic visualization tool that provided an incremental view on the RDF sub-graphs based on a first user query. *PGV* consists of two main components, the *PGV explorer* and the *PGV pager*. The *PGV explorer* is the visualization component of their system that uses a concentric radial algorithm for visualizing sub-graphs of RDF semantics. It consists of a *starter subsystem* and a *visualizer subsystem* [117]. The *starter subsystem* is the initial point of a user query. Therefore the user selects a resources from the RDF schema representation and is able to place a query. The user gets a pre-formulated SPARQL-query and is enabled to enhance it with a keyword or even reformulated it with his knowledge about the query language. The *visualizer subsystem* visualizes based on the initial query the direct neighborhood of the RDF graph instead of visualizing the entire graph. As common in concentric radial algorithms, the queried resource is placed in the center and the direct neighborhood is presented with directed and labeled links. By selecting a neighborhood, the direct neighbors of the selected resources are visualized, while the original graph remains. The color of the chosen nodes changes during the interaction to illustrate that the node was explored. If a chosen resource has too many neighbors Deligiannidis et al. call them "hot spot". They propose the use of *Ferris Wheel* metaphor that is just visualizing a part of the neighbors at once. If the user rotates the concentric radial visualization, other neighbors are revealing, while the already visualized ones disappear. This technique is the paging approach of *PGV* [117]. Responsible for this task is the *PGV pager* that enables a loading on demand techniques [117]. *PGV* proposes a bottom-up approach for exploring the semantic relationships of a RDF graph. It makes use of an interesting loading and demand technique, but the paging of the visualization may lead to lose the context, the many resources are hidden. Further the approach does not provide any top.-down search and exploration approaches, e.g. by visualizing in a separated window the classes of RDFS. The search itself seems to be a not intuitive way as the SPARQL query notion is not common for every user. Further the amount of the visualized graphs is still a remaining question: what if the user clicks on the third or fourth neighbor?

da Silva et al. introduced *OntoViewer*, an ontology visualization tool considering the degree of interest [118, 119]. *OntoViewer* aims at reducing the complexity of relational visualizations by calculating the degree-of-interest (DOI) and providing a multiple view on the underlying ontology. The degree-of-interest is automatically calculated based on a main concept that is chosen by a user. Based on this main concept, vertices (representing concepts) and edges (representing relations) of interest are computed that correspond to the *semantic distance* to the main concept. Therefore the percentage of the amount of the involved instances id calculated to infer the semantic relation of the concepts. Based on the semantic distance, the distance between the main concept and the most similar and related concepts are calculated to reduce the amount of visualized entities. The calculated classes of interest are visualized in a user interface with three different visualizations. The main visual element visualizes the arbitrary relations between classes in a $2\frac{1}{2}$D radial tree visualization [118]. The main concept is placed in the center, whereas the computed semantic distance is visualized by the distance to the center of the visualization [118, 119]. The classes are connected with links and illustrate the type of relationship by using different colors, whereas the hierarchical relations are visualized as plain 2D links. A smaller visualization in the *OntoViewer* user interface illustrates only the hierarchical structure as two-dimensional hyperbolic tree. To reduce the complexity in particular for the hierarchical relations, a treeview visualization is used to illustrate the class hierarchy [118]. Further the user interface provides two control panels for the visualization and the degree of-interest adjustments, where the linkage between visualization can be chosen or the degree of relations can be assigned with sliders [118]. A further revised version provides a view on instances by a two-dimensional icicle tree and a pixel oriented visualization [119].

The aspect of the degree-of-interest and the multiple views are interesting and useful approaches for exploring the information space. The *OntoViewer* follows the *visual information seeking mantra* of Shneiderman [72] and provides a top-down visual metaphor. A real search is not provided by *OntoViewer* and a bottom-up approach is missing too. Further the visualizations are fixed; the user is not able to interact with the hierarchy in a focused manner. Investigative tasks, like comparing parts of the information space by visualizing two or more parts of the ontology are not supported. *OntoViewer* is designed to view and interact with one information space at the same time. An investigation of different ontologies is only serially possible.

Bach et al. introduced with *OntoTrix* a relational visualization approach dedicated for visualizing the instances of an underlying ontology [120]. *OntoTrix* is designed to enable the visualization of large amount of instances by a hybrid approach that makes use of node-links and adjacency matrix representations. Their approach is inspired by *NodeTrix*, a hybrid visualization approach for social networks proposed by Henry and *Fekete* [121]. The overall structure of the ontology is visualized as an undirected graph of matrices. The grouping of instances in the matrices is performed in three ways, by *Density*, *Global class membership*, *Local class membership*, or *Property type*. The grouping by *Density* clusters the instance by taking into account all object properties, which are further differentiated by using various adjustable colors. The *Global class membership* uses the structure of the class relation in the ontology in terms of *instance-of* [120], while the *Local class membership* represents a trade-off between the first introduces grouping methods. This grouping method applies first the *Density* clustering of the instances and in a second step the *Global class membership*, which may results in matrices that correspond to the same class. The *Property type* grouping clusters the instance according to their object properties, whereas irrelevant properties can be deselected [120]. The *OntoTrix* is further enhanced by a Birdseye-view that enables the view on the entire visualized structure [120]. Further a separated node-link view visualizes the class hierarchy and a further separated window illustrates the property hierarchy as radial tree.

OntoTrix provide an interesting approach for visualizing semantics, thus not only the explicitly modeled structure of the ontology is visualized. It provides other clustering methods using the semantics for enabling a different view on the data. An underlying search paradigm cannot be identified in *OntoTrix*. The visualization targets at detecting certain patterns on instance level with a high degree of abstraction. The hybrid view allows viewing the correlations of instances with different grouping algorithms, but does not provide a detailed view on the instance or a resource of interest rather than the content itself.

Voigt et al. proposed a semantics visualization system that enhances the existing approaches by the use of multiple visualization view on different levels and perspectives [122]. They considered in their approach in particular the "lay-user" to provide a visualization system that enables the task solving process. Their visualization approach makes use of a context-aware information visualization workflow for semantic data on Web that depicts the human and system interaction as a foundation of visualizing semantics [122, p. 8]. The human side consists of the steps *data upload*, *data pre-selection*, *data and vis selection*, *visualization configuration*, and *perception and internalization* [122]. These human interactions are directly related to system interaction without an iterative character. The system reacts or should react on the human interactions with *data augmentation, data reduction, visualization recommendation, visualization integration*, and *knowledge externalization*. The entire process of visualizing semantics starts with *data upload* supports not a bottom-up exploratory search approach. Voigt et al. focus far more on a top-down approach that is recognizable by the stapes of their workflow, thus after *data upload*, the user preselects data related with the system by *data reduction* [122]. The most interesting aspect of their visualization approach is the visualization recommendation as a response of the human *data and vis selection*, whereas the visualization configuration does not consider the visual appearance, but far more the visualization integration, which is similar to the visualization selection. The recommendation process is based on rankings of the semantic faceted weights. The user interface of their system provides various visualizations and techniques to interactively present the semantics, whereas the resources and thereby the real content is not investigated. Thus the main visualization targets at a relational view on data, we categorize their system as relational visualization, although some aspects of hierarchies and facets are presented too.

Nunes et al. proposed with *Cite4Me* a relational visualization of the *LAK* digital library content [123]. Their goal was to interlink the digital library content with sources of the Linked-Open-Data Cloud and provide an exploratory search and visualization approach for the semantically enriched data. In particular the support of students in the learning process was targeted by the visual approach. *Cite4Me* uses standard vector space models (*tf-idf*) for indexing and retrieving documents to provide *cosine similarity measure* and provide similar content instead of exact term matching [123]. Beside the *free text search*, an *exploratory search* and a *semantic search* is supported by *Cite4Me*. The *exploratory search* is implemented by enriching the digital library documents with the semantics of Linked-Data. Therefore the DBPedia API [21] is used to extract entities, entity types, and entity categories. The *semantic search* is implemented by computing a *tf-idf* score for the entities of their interlinked semantics instead of a calculated score for the terms. Another feature of *Cite4Me* is the recommendation of papers, where the number of paths and the distance between given entities are computed to recommend related documents [123]. The search and recommendation functionalities provide interlinked data that are visualized in *Cite4Me* with a relational graph view for exploring the entire data set of the library. Beside the relational view on the entire data set, the search functionalities enable a bottom up search, where semantically related documents can be recommended. The main advantage of *Cite4Me* is the enabling of both, the bottom up search by free-text, semantics, and recommendations and the top-down approach by providing a visual overview of the entire graph. The exploratory search paradigm is not interactive, so the user is not able to interact through the node-link overview and get direct details on demand. She has to inform herself about the domain and use the bottom-up approach of searching to get more detailed information. *Cite4Me* further does not support investigative and comparative search capabilities due to its one visualization per user interface paradigm.

Relational visualizations play an essential role in visualizing semantics. The natural structure of semantics is predestinated for visualizing semantic relationships and thereby a structural view on the domain of given knowledge. There are many ways to visualize these semantic structures; we have introduced the most common visualization methods with respect to their search and interaction ability. Further systems that were introduced for visualizing these relations [124–128] have applied the relational visualizations in different domains or for various sources, e.g. text corpus or scientific relations. A relational view on semantics is essential, due to the ability to get an overview of the entire knowledge space or see the relations of resources to interact and browse for a knowledge path. However, the requirements of exploratory search cannot be fulfilled with relational visualizations only. It is far more necessary to provide multiple visualization views on the same or on different information spaces, provide details about the entities, and support the knowledge investigation by viewing the content itself. Relational visualizations are one essential view on semantics, but not the only one.

3.5.3.3 Entity-Based Semantics Visualizations

In contrast to relational or hierarchical semantics visualizations that utilize the underlying structure of the semantics information, entity-based semantics visualizations focus on the queried set of entities and provide either an exploratory approach by navigating through a visual structure or retrieving information by a specifically defined query. This class of semantics visualization is designed for search tasks. The result visualization as a set of entities is the main focus. Further approaches for graphically representing the relationships or hierarchies are secondary and are commonly used to support the search process.

A visualization approach that is not based on formal semantics but investigates semantics as defined in this thesis is *PaperVis*, introduced by Chou and Yang [129]. *PaperVis* allows search and exploration within a bibliographic system, whereas the search function is more a filtering of the entity-based visualization depending on the chosen mode. The semantic in *PaperVis* refers in their approach by citations of a paper and a kind of categorization based on the resulted keywords of a searched set of papers [129]. *PaperVis* provides three modes for searching and exploring scientific papers of a bibliographic system, the *Citation-Reference Mode*, *Keyword Mode*, and *Mixed Mode*. In the *Citation Reference Mode* the users specify a level scope of the citation network as starting point for exploring papers. *PaperVis* loads the bibliographic entries based on the selected levels and scopes and visualizes them in a concentric radial visualization using a modified radial space filling and a Bullseye view technique [129]. The nodes represent the result entities and links illustrates the citations relationship. The users are able to type a query that leads to a filtering of the results. In this separated window a certain paper can be selected to get the citation links to other papers. The detailed information about the paper is visualized in another separated window, containing the title, source and year of publication, and authors. The user interface enables the choice of the mentioned three modes. In the *Keyword Mode* the initial starting point of the search is a user's query by a keyword. Based on this keyword all the keywords that are assigned and annotated as keywords are searched and co-occurrence keywords are assigned as categories as far as they have a certain amount of occurrences [129]. The results are then visualized in a radial space-filling visualization enabling users to refine the query by choosing one subcategory. A double-click in a sub-category places the key-word in the center of the visualization and the relevant papers around it. The relevance is defined by the citation, which builds the main semantic relationship in *PaperVis*. The *Mixed Mode* premises at least one keyword [129]. If a user selects a paper in this mode, additional papers are loaded based on their citations but visualized as in the *Keyword Mode*. *PaperVis* allows searching and exploring entities that have a weak semantics as a XML-Schema in bibliographic entries, which is the main advantage of the system. The *Citation-Reference Mode* provides a top-down search capability, whereas the *Keyword Mode* targets a bottom-up search. The semantic visualization of *PaperVis* is limited to the constrains of relevance and importance of citations and keywords. A real visual interactive behavior is not provided by the system due to the different search modes that have to be explicitly chosen. The user may lose the context of his search during the changes of the modes. Further the *PaperVis* tool is limited to one visualization type with three layouts. If a user is not satisfied with the circular visualization or is overcharged, the system does not provide any alternatives. The focus point of the *PaperVis* system is always one main visualization without the ability to change it in size or reconfigure and personalize the visualization. But the main shortcoming is the gap in supporting investigative search tasks. Users are not able to place two visualizations juxtaposed to compare any issue.

Schenk et al. introduced *SemaPlorer*, an entity-based search and visualization tool for heterogeneous distributed semantic data on Web [130]. *SemaPlorer* integrates and leverages various semantic databases with Linked-Data, e.g. *DBPedia* [21], *GeoNames*, *WordNet*, and *FOAF*. The Linked-Data repositories and search engines are used to retrieve in particular semantic data about *Locations, Time, Persons*, and *Tags*, whereas the *Tags* refer to the non-semantic formalized database *Flickr* [130]. *SemaPlorer* aims at conducting complex data exploration tasks. As an example of *complex data exploration task* they introduce the example of searching street art in a city. They propose that searching these kind of information is "almost impossible" with current search engines [130, p. 2]. For solving these *exploratory* tasks *SemaPlorer* uses data federated from various data bases using faceted, blended browsing, querying those data bases simultaneously. The defined facets are the semantic data, retrieved from the databases. The user has the choice to search for a query and refine his search by choosing of one of the tabs *Location, Time, Person*, or *Tags*. *SemaPlorer* uses the semantic data for an entity-based illustration, consisting of three visualization panels. A left vertical panel illustrates the search results for the queried term in the as listed terms categorized by the facets. The main panel in the center of *SemaPlorer* visualizes initially the entities of the search on a geographic map, whereas no semantic relations are illustrated. This main panel can further visualize the tagged media, with one entity at the center and a horizontal list at the bottom [130]. The panel on the right illustrates "contextual" information about the chosen entity in terms of more detailed information about a certain location. Further it provides to choose one of the four assigned facets with tabs [130].

In contrast to many other semantic visualization techniques, *SemaPlorer* does not provide any relational or hierarchical view on the semantics. The search paradigm is a bottom-up search, where the user starts with a query and may find some unknown related entities. Although an exploratory approach is proposed, the exploration itself is not supported by the system. The user may learn some aspects about the queried term but there is no significant difference between *SemaPlorer* and a search engine that does not rank the results. A top-down or comparative view is not supported at all. Further the visual panels are fixed; the user is not able to change the size of a visualization panel to retrieve more information at a glance. *SemaPlorer*'s most interesting approach is the integration of various data-bases simultaneously and the aspect of bottom-up entity search that is scarcely supported by other visualization techniques.

Petrelli et al. proposed that the visual support of exploratory tasks cannot be performed by a single visualization type [131]. It is necessary to provide different perspectives on the semantic to support exploration and knowledge discovery in particular by space and time contextualized semantics visualization [131]. For supporting the exploration process they introduced a set of visualization and interaction techniques that claimed to be complementary and provide different perspectives on the semantics. Thus their main approach targets at visualizing entities of semantics in spatial and temporal way, we classify their visualizations as entity-based semantics visualizations. Although, they follow the *Visual Seeking Mantra* of Shneiderman [72] and propose a visualization type for the hierarchical structure, the main contribution is an entity-based visualization. Their *GeoPlot* visualization depicts semantic resource entities based on RDF triple on a geographic world-map and provides an entity-based view similar to *SemaPlorer*. The entities are place on the map with icons and the number of instances correlated to a spatial area. A further entity-based visualization uses the temporal annotations of RDF-triples to visualize them in a temporal manner with their *TimeLine*. The *TimeLine* illustrates the temporal spread of the RDF-resources and entities using a two-dimensional *Scatterplot*. This view further provides filters for *dynamic querying* in terms of filtering the visualized set of entities and a tree-based hierarchical visualization with a folder-metaphor. A further hierarchical view, the *TopologicalPlot* uses an *engine-metaphor* to provide overview-information about the ontology that conceptualizes domain knowledge about engines.

Petrelli et al. proposed in their work that the aspect of knowledge discovery and exploration is of great interest in semantics visualizations. For this purpose they introduced the different perspectives on the semantics enhanced by filters. Their approach follows a top-down search paradigm, whereas the users always interact on the top-level. Details about the visualized entities cannot be requested on demand.

The most interesting aspect is that the visualization approaches are focusing primarily at the entities of RDF triples. Furthermore, the relations and an adequate overview of the relations is missing for the exploratory approach as they claim. The visualization perspectives cannot be used at the same time and are not interlinked. Comparative or investigative tasks are consequently not supported. The idea of different perspectives on the same data is interesting, however a real perspective view on the semantic information is not provided.

With the upcoming importance of search in information visualization, entity-based visualizations gain more and more importance. Commonly the goal of these visualizations is not to provide a visual pattern and overview on the data. Entity-based visualizations aim at giving a kind of interactive "picture" on resources, documents, text [132], or other kinds of entities that builds the content of the underlying visualized data. In particular in semantics visualizations, where a visualized entity refers to a resource on Web or to another data-base, the investigation of the resource itself could enhance the search and exploration process. The described entity-based visualizations aimed at providing a kind of investigation of the resources themselves [129, 130]. They mapped the entity to certain contextual information, e.g. time, location, or semantic similarity. The shortcoming of these systems are commonly that they either provide an abstract view on the schema of the semantics or a schema-less view on the entities. The exploratory search process as described by Marchionini [59] or White and Roth [57] is not actively supported. The idea of providing contextual information for certain semantic entities and support the entire search process is promising, but not yet solved.

3.6   Summary and Findings

While information visualization aims at giving insights of data and amplify the human cognition to retrieve visual information patterns, semantic technologies in context of human computer interaction aim primarily at providing answers to questions, which can be specifically verbalized by human. This chapter introduced semantics visualization as an approach for bridging the gap between the contradictory concepts of semantic technologies and information visualization. We first introduced the term "semantics" as it is used in context of Semantic Web and formalization of knowledge. Thereafter the general idea of Semantic Web was introduced. Thus the Web provides a crucial information resource, Berners-Lee proposed the idea of a Web of data that enables the access to these resources with sense of "meanings" as *Semantic Web* [9, 10]. We could emphasize that the main idea of Semantic Web was to formalize the data and information in a way that is machine-readable [10]. The formalization aims at making the Web "meaningful" based on formalized notation of content followed by a formalization of the underlying structure and provides a rule and meaning inferencing to make the Web accessible for computer and human [3, 7, 9–11]. In this context we introduced the "layer-cake" architecture of Semantic Web [12] to give an idea how this Web of "meaning" was proposed to be realized. The architectural model builds the foundation for semantic technologies and in particular for the Semantic Web. This construct of Web consists of structured metadata providing well-defined meanings described as formal semantics [7]. The formalization in this context may consist of "light" description in form of markups and tags, or investigate formal logic. In any case the information on Web has to be formalized to enable the idea of semantics. The way from data to formalized information or knowledge may be performed by application and approaches of Knowledge Discovery in Databases (KDD). We introduced in this context the idea of KDD for semantic generation. We outlined that commonly the applications that use KDD for generating semantics are semi-automatic and need the involvement of human in the knowledge and ontology generation process [25]. A more promising and disseminated way of knowledge formalization is the use of *collective intelligence* and the already existing metadata to interlink data entities with each other in the so called Linked-Data cloud [17–19]. With the existence of these data, we are able to say that the Web is already enhanced with semantics and provide meaningful data with various degrees of formalization [21, 34]. We further described that the formalization of information can be classified into a continuous spectrum of lightweight semantics (informal) to heavyweight-semantics (formal) [3, 14, 35, 36]. In this context we introduced various models for classifying the formalization degrees and the related description languages. A dedicated section introduced the most common formalization languages. Therefore we picked up some representatives of the lower level of conceptualization (lightweight) to more formal languages as *Description Logic* or the Web Ontology Language (OWL). We further worked out that most prevalent language is the *Resource Description Framework* (RDF) that is commonly used for the Linked-Data approach.

After introducing the general idea of Semantic Web and semantic with its formalizations, degree of formalizations, and the architecture, we investigated the access to semantic data. Therefore we first introduced the most common query language for RDF to give an idea how this information can be accessed. The next important question was for what are this information accessed by human. We identified that the most common task solved with semantics is the process of *search* with all the related tasks, e.g. learning or exploring. To provide a comprehensible view on *search* as a human information retrieval task, we introduced two exploratory search models. These models built the foundation to review existing exemplary systems of *human-semantics-interaction*. To provide a comprehensible picture of the existing systems, we categorized the systems based on the introduced search models into *question-answering systems*; *key-word based none exploratory search systems*, and *exploratory search systems*. By reviewing the existing systems, we outlined that commonly systems are designed to answer specific questions that require prior-domain knowledge. The exploration process is more activated by information visualization approaches of semantics and semantic data.

The main goal of this chapter was to give a comprehensive and comprehensible state of art and technology for semantics visualizations. For obtaining a clear picture of existing systems and approaches, we first defined the term semantics visualization as *computer-aided interactive visualizations for effective exploratory search, knowledge domain understanding, and decision making based on semantics*, where semantics is defined as data with meaningful relations of at least two information or data entities. With this definition of semantics *ontology visualizations* are a subset of semantics visualization and were thereby part of our review. Thereafter we introduced a classification for providing a comprehensible picture of the existing systems, thus no sufficient classification for semantics visualization in terms of information search was found. Our classification categorizes semantics visualization in the three categories of *hierarchical semantics visualizations*, *relational semantics visualizations*, and *entity-based semantics visualizations*. These categories are not mutually exclusive. There are also approaches that can be assigned to multiple categories (e.g. visualizations that are presenting hierarchical and relational aspects), therefore we used the dominant visualization of a system for its classification.

Our state of the art review investigated systems and approaches of the last decade, whereas some fundamental previously introduced systems were mentioned too. We introduced the systems based on our classification with respect to their search and exploration activation as proposed in the introduced search models. In this context the formal view on ontologies was not of great interest. The foundation of our review was performed based on the exploratory search model of Marchionini [59] with the three elementary activities of *Lookup*, *Learn*, and *Investigate*. Further we mapped the *Visual Information Seeking Mantra* of Shneiderman [72] with the steps *overview*, *zoom*, and *details-on-demand* to the search model. Thus the information seeking process can be started by users with prior domain knowledge or a constraint problem context, the order of the seeking mantra was neglected. We premised that semantics visualization should support in best case both, a top-down approach as proposed by Shneiderman and a bottom-up approach as implemented for instance by van Ham and Perer [78]. The mapping of the two paradigms to the semantics with enriched labeled hierarchies, relations and meanings builds a triangle of the way how we used to review the visualizations.

Based on the categorization and the defined criteria the review of the existing systems was performed. Our survey investigated the entire range of semantics visualizations, including ontology visualization and those visualizations that make use of any kind of semantic relationship. The focus was reviewing the systems based on their active support for exploratory search. There exist an enormous number of visualization techniques, approaches, and methods. Many of them are using more than one visualization technique to provide a visual interactive picture on the semantics. As already assumed in the classification, the categorization was not always exclusive, there are approaches that make use of both relational and hierarchical view on semantics. A huge number of the found systems were designed for ontology engineering. Contrary to our assumption the aspect of exploration and search did not play any role for these systems. In contrast to these systems, we found approaches that claimed to support the exploration process. Some interesting approaches were found that makes use of a top-down approach on a semantic level, but non of them fulfilled the criteria of exploratory search. They either did not support the search process at all, or even if they provided search functionality, this was commonly limited and did not provide a sufficient search process. In case of hierarchical visualization, we found interesting approaches that provide an overview on the data. The hierarchical view on data with different perspectives enables to get an insight of the entire knowledge domain or relevant sub-parts. The semantics visualization approaches are focusing far more on the overview aspect, rather than on navigation through the hierarchical concepts to get the information entities (instances). They remain on the abstract level of ontological concepts and do not provide a real top-down or bottom-search paradigm. And even if search is supported, the only search paradigm is a top-down search that enables viewing classes and subclasses and navigating through the hierarchical structure. A bottom-approach is missing in the hierarchical visualization of semantics at all. The category of relational visualizations plays an essential role in the visualization of semantics, thus the structure of semantics is predestinated for visualizing semantic relationships and a structural view on the domain knowledge. There are many ways to visualize these semantic structures. We introduced the most common visualization methods with respect to their search and interaction ability. A relational view on semantics is essential, due to the ability to get an overview of the entire knowledge space or see the relations of resources and interact and browse a knowledge path. However, the requirements of exploratory search cannot be fulfilled with relational visualizations only. It is far more necessary to provide multiple visualization views on the same or on different information spaces, provide details about the entities, and support the knowledge investigation by viewing the content itself. Relational visualizations are one essential view on semantics, but not the only one. We were not able to find a system that fulfills the requirements of the entire search process, even in those systems, which visualizes the data with multiple visualization methods. The category of relational semantics visualizations provides the most interesting approaches, but a sufficient view on the semantics with different and separated perspectives was not found. In the category of entity-based visualization, the number of existing systems is limited. These approaches claim to support the search process and provide contextual knowledge on entity-level. It was very interesting to find visualization systems that focus primarily on the search based on semantic entities. In particular in case of semantics visualization, where a visualized entity refers to a resource on Web or to another data-base, the investigation of the resource itself could enhance the search and exploration process. The described entity-based visualizations aimed at providing a kind of investigation of the resource itself. They mapped the entity to certain contextual information, e.g. time, location, or semantic similarity. The shortcoming of these systems are commonly that they either provide an abstract view on the schema of the semantics or a schema-less view on the entities. The exploratory search process as described by Marchionini [59] or White and Roth [57] is not actively supported. The idea of providing contextual information for certain semantic entities and support the entire search process is promising, but not yet resolved.

Our review on the existing systems could clearly outline that none of the existing systems support the level of *Investigate*, the highest level of cognitive search tasks, as proposed by Marchionini [59]. Therewith tasks like comparison of data-sets or data-parts, or analysis and synthesis are not supported by today's semantics visualization systems. Further the view on different perspectives is not provided in a sufficient way. Although, systems exist that provide different views [122, 131], a separation of the complex structures on relational, hierarchical, and entity-based level is not yet given. Further the aspect of personalization is not sufficiently investigated. The existing visualizations are commonly static with one main panel on one property of semantics.

References

1. P. Hitzler, M. Krötzsch, S. Rudolph, Y. Sure, *Semantic Web: Grundlagen (Semantic Web: Foundations)* (Springer, Berlin, 2008)
2. R. Carnap, *Introduction to Semantics*, Studies in Semantics, vol. 1 (Harvard University Press, Cambridge, Massachusetts, 1948). http://de.scribd.com/doc/146390494/Carnap-Introduction-to-Semantics
3. S. Staab, R. Studer (eds.), *Handbook of Ontologies* (Springer, Heidelberg, 2009)
4. A. Tarski, Philos. Phenomenol. Res. **4**, 341 (1944)
5. N. Guarino, D. Oberle, S. Staab, in *Handbook of Ontologies*, ed. by S. Staab, R. Studer (Springer, Heidelberg, 2009), pp. 1–17
6. G. Antoniou, P. Groth, F. van Harmelen, R. Hoekstra, *A Semantic Web Primer*, 3rd edn., Cooperative Information Systems (The MIT Press, Cambridge, Massachusetts, 2012)
7. A. Gómez-Perez, M. Fernandez-Lopez, O. Corcho, *Ontological Engineering: With Examples from the Areas of Knowledge Management, e-Commerce and the Semantic Web* (Springer, 2007)
8. G. Antoniou, F. van Harmelen, *A Semantic Web Primer*, Cooperative Information Systems (The MIT Press, Cambridge, Massachusetts, 2004)
9. T. Berners-Lee, Semantic web roadmap, (W3C Online Publishing, 1998). http://www.w3.org/DesignIssues/Semantic.html
10. T. Berners-Lee, *Weaving the Web: The Original Design and Ultimate Destiny of the World Wide Web* (HarperBusiness, 2000)
11. T. Berners-Lee, J. Hendler, O. Lassila, Sci. Am. (2001). http://www.scientificamerican.com/1999/0599issue/0599bosak.html
12. E. Miller, in *5th European Conference on Research and Advanced Technology for Digital Libraries* (2001)
13. L. Quin, K. Ercim, Extensible markup language (xml) (2013). http://www.w3.org/XML/
14. V. Geroimenko, C. Chen (eds.), *Visualizing the Semantic Web—XML-Based Internet and Information Visualization* (Springer, New York, 2006)
15. T.U. consortium. The unicode standard, version 6.20 (2013), http://www.unicode.org/. Accessed Aug 2013
16. RDF Working Group: Resource description framework (rdf) (2013), http://www.w3.org/RDF/
17. C. Bizer, T. Heath, K. Idehen, T. Berners-Lee, in *Proceedings of the 17th International Conference on World Wide Web, WWW 2008*, Beijing, China, 21–25 Apr 2008, pp. 1265–1266
18. C. Bizer, T. Heath, T. Berners-Lee, Int. J. Semantic Web Inf. Syst. **5**(3), 1 (2009)
19. T. Heath, C. Bizer, *Linked Data—Evolving the Web into a Global Data Space*, Synthesis Lectures on the Semantic Web: Theory and Technology (Morgan & Claypool Publishers, 2011)
20. C. Bizer, P. Boncz, M.L. Brodie, O. Erling, SIGMOD Rec. **40**(4), 56 (2012). doi:10.1145/2094114.2094129. http://doi.acm.org/10.1145/2094114.2094129
21. P.N. Mendes, M. Jakob, C. Bizer, in *Proceedings of the Eight International Conference on Language Resources and Evaluation (LREC'12)* (Istanbul, Turkey, 2012)
22. T.R. Gruber, Knowledge Acquis. **5**(2), 199 (1993). http://dx.doi.org/10.1006/knac.1993.1008. http://www.sciencedirect.com/science/article/pii/S1042814383710083

23. M. Smith, I. Horrocks, M. Krötzsch, B. Glimm, *Owl 2 Web Ontology Language: Conformance*, 2nd edn. (2012) http://www.w3.org/TR/2012/REC-owl2-conformance-20121211/. Accessed Aug 2013
24. U. Fayyad, G. Piatetsky-Shapiro, P. Smyth, in *Advances in Knowledge Discovery and Data Mining*, ed. by U.M. Fayyad, G. Piatetsky-Shapiro, P. Smyth, R. Uthurusamy (MIT Press, Cambridge, Massachusettes, 1996)
25. D. Mladenic, M. Grobelink, B. Fortuna, M. Grcar, in *Semantic Knowledge Management— Integrating Ontology Management, Knowledge Discovery, and Human Language Technologies*, ed. by J. Davis, M. Grobelink, D. Mladenic (Springer, 2009), pp. 21–36
26. M. Grobelnik, D. Mladenć, *Knowledge Discovery for Ontology Construction* (Wiley, 2006), pp. 9–27. doi:10.1002/047003033X.ch2. http://dx.doi.org/10.1002/047003033X.ch2
27. D. Mladenic, M. Grobelnik, in *29th International Conference on Information Technology Interfaces, ITI 2007* (2007), pp. 547–551. doi:10.1109/ITI.2007.4283830
28. J. Brank, M. Grobelnik, D. Mladenić, in *Natural Language Processing and Text Mining*, ed. by A. Kao, S. Poteet (Springer, London, 2007), pp. 193–219. doi:10.1007/978-1-84628-754-1_11. http://dx.doi.org/10.1007/978-1-84628-754-1_11
29. J. Völker, D. Vrandečić, Y. Sure, A. Hotho, Appl. Ontol. **3**(1–2), 41 (2008). http://dl.acm.org/citation.cfm?id=1412417.1412422
30. M. Grobelnik, D. Mladenic, in *27th International Conference on Information Technology Interfaces* (2005), pp. 188–193. doi:10.1109/ITI.2005.1491120
31. B. Fortuna, M. Grobelnik, D. Mladenic, in *Proceedings of the 9th international multi-conference information society IS-2006* (Ljubljana, Slovenia, 2006). http://kt.ijs.si/blazf/papers/OntoGen2_SIKDD2006.pdf
32. N. Bhatti, Visual semantic analysis to support semi automatic modeling of service descriptions. Ph.D. thesis, Diss., Darmstadt, Technische Universität, 2011, 205p. http://bibcd.igd.fraunhofer.de/bibcd/INI_Science/dissertations/diss_Bhatti.PDF
33. T. Kamps, Conweaver—Make your data work, http://www.conweaver.de/en/product-explanation-what-is-conweaver (2013). Accessed Aug 2013
34. Google Press Center: The Knowledge Graph (2013), http://www.google.com/intl/en/insidesearch/features/search/knowledge.html. Accessed Aug 2013
35. L. Obrst, in *Proceedings of the twelfth international conference on Information and knowledge management CIKM '03*, (ACM, New York, 2003), pp. 366–369. doi:10.1145/956863.956932. http://doi.acm.org/10.1145/956863.956932
36. M. Uschold, M. Gruninger, SIGMOD Rec. **33**(4), 58 (2004). doi:10.1145/1041410.1041420. http://doi.acm.org/10.1145/1041410.1041420
37. V. Geroimenko, in *Visualizing the Semantic Web—XML-Based Internet andInformation Visualization* ed. by V. Geroimenko, C. Chen (Springer, 2006)
38. D. Fensel, I. Horrocks, F. Harmelen, S. Decker, M. Erdmann, M. Klein, in *Knowledge Engineering and Knowledge Management Methods, Models, and Tools*, vol. 1937, Lecture Notes in Computer Science, ed. by R. Dieng, O. Corby, (Springer, Berlin, 2000), pp. 1–16. doi:10.1007/3-540-39967-4_1. http://dx.doi.org/10.1007/3-540-39967-4_1
39. van Harmelen, F., Horrocks I. (eds.), Reference description of the daml+oil ontology markup language. Technical report, DAML Program (2001), http://www.daml.org/2000/12/reference.html. Accessed Aug 2013
40. P. Patel-Schneider, I. Horrocks, F. van Harmelen, in *Proceedings of AAAI'02* (2002)
41. N. Guarino, Int. J. Hum.-Comput. Stud. **46**(2–3), 293 (1997). doi:10.1006/ijhc.1996.0091. http://dx.doi.org/10.1006/ijhc.1996.0091
42. A. Gómez-Perez, M. Fernandez-Lopez, O. Corcho, *Ontological Engineering: with examples from the areas of Knowledge Management, e-Commerce and the Semantic Web*, 3rd edn. (Springer, 2010)
43. D. Fensel, J. Hendler, H. Lieberman, W. Wahlster (eds.), *Spinning the Semantic Web: Bringing the World Wide Web to Its Full Potential* (MIT Press, Cambridge, Massachusettes, 2003)
44. J. Davis, M. Grobelink, D. Mladenic (eds.), *Semantic Knowledge Management—Integrating Ontology Management, Knowledge Discovery, and Human Language Technologies* (Springer, 2009)

45. F. Manola, E. Miller, RDF primer, W3C Recommendation, 10 Feb 2004, http://www.w3.org/TR/2004/REC-rdf-primer-20040210/. Accessed Aug 2013
46. D. Beckett, RDF/xml syntax specification (revised), W3C Recommendation, 10 Feb 2004, http://www.w3.org/TR/REC-rdf-syntax/. Accessed Aug 2013
47. G. Klyne, J.J. Carrol, Resource description framework (RDF): concepts and abstract syntax, W3C Recommendation, 10 Feb 2004, http://www.w3.org/TR/rdf-concepts/dfn-URI-reference. Accessed Aug 2013
48. D. Brickley, R. Guha. RDF vocabulary description language 1.0: RDF schema, W3C Recommendation, 10 Feb 2004, http://www.w3.org/TR/rdf-schema/. Accessed Aug 2013
49. M.S. Lacher, S. Decker, Markup Lang. **3**(3), 313 (2001). doi:10.1162/109966201753750333. http://dx.doi.org/10.1162/109966201753750333
50. S. Pepper, G. Moore. Xml topic maps (xtm) 1.0. topicmaps.org specification (2001), http://www.topicmaps.org/xtm/. Accessed Aug 2013
51. G. Antoniou, F. van Harmelen, in *Handbook of Ontologies* ed. by S. Staab, R. Studer (Springer, Heidelberg, 2009)
52. F. Baader, I. Horrocks, U. Sattler, in n *Handbook of Ontologies* ed. by S. Staab, R. Studer (Springer, 2009)
53. H. Stuckenschmidt, J. Data Semant. **1**(1), 1 (2012). doi:10.1007/s13740-012-0003-z. http://dx.doi.org/10.1007/s13740-012-0003-z
54. C.B. Aranda, O. Corby, S. Das, L. Feigenbaum, P. Gearon, B. Glimm, S. Harris, S. Hawke, I. Herman, N. Humfrey, N. Michaelis, C. Ogbuji, M. Perry, A. Passant, A. Polleres, E. Prud'hommeaux, A. Seaborne, G.T. Williams, SPARQL 1.1 overview, W3C Recommendation, 21 Mar 2013, http://www.w3.org/TR/sparql11-overview/Acknowledgements (2013)
55. S. Harris, A. Seaborne, SPARQL 1.1 query language, W3C Recommendation, 21 Mar 2013, http://www.w3.org/TR/2013/REC-sparql11-query-20130321/
56. B. Fazzinga, G. Gianforme, G. Gottlob, T. Lukasiewicz, in *Foundations of Information and Knowledge Systems*, vol. 5956, Lecture Notes in Computer Science, ed. by S. Link, H. Prade (Springer, Berlin Heidelberg, 2010), pp. 153–172. doi:10.1007/978-3-642-11829-6_12. http://dx.doi.org/10.1007/978-3-642-11829-6_12
57. R.W. White, R.A. Roth, *Exploratory Search: Beyond the Query-Response Paradigm*, vol. 1, Synthesis Lectures on Information Concepts, Retrieval, and Services, ed. by G. Marchionini (Morgan & Claypool Publishers, 2009). doi:10.2200/s00174ed1v01y200901icr003. http://dx.doi.org/10.2200/s00174ed1v01y200901icr003
58. B.S. Bloom, *Taxonomy of Educational Objectives* (David McKay Co. Inc, New York, 1956)
59. G. Marchionini, Commun. ACM **49**(4), 41 (2006). doi:10.1145/1121949.1121979. http://doi.acm.org/10.1145/1121949.1121979
60. J.S. Bruner, Harv. Educ. Rev. **31**, 21 (1961)
61. I. Deines, D. Krechel, in *Semantic Technology*, vol. 7774, Lecture Notes in Computer Science, ed. by H. Takeda, Y. Qu, R. Mizoguchi, Y. Kitamura (Springer, Berlin Heidelberg, 2013), pp. 278–289. doi:10.1007/978-3-642-37996-3_19. http://dx.doi.org/10.1007/978-3-642-37996-3_19
62. N. Reithinger, G. Herzog, A. Blocher, KI - Künstliche Intelligenz 2/07, 30 (2007)
63. D. Sonntag, in *Human Interface and the Management of Information. Interacting in Information Environments*, vol. 4558, Lecture Notes in Computer Science, ed. by M. Smith, G. Salvendy (Springer, Berlin Heidelberg, 2007), pp. 645–654. doi:10.1007/978-3-540-73354-6_71. http://dx.doi.org/10.1007/978-3-540-73354-6_71
64. W. Wahlster, in *Informatikforschung in Deutschland* (Springer, Berlin, 2008), pp. 300–311. doi:10.1007/978-3-540-76550-98
65. V. Lopez, A. Nikolov, M. Sabou, V. Uren, E. Motta, M. d'Aquin, in *Knowledge Engineering and Management by the Masses*, vol. 6317, Lecture Notes in Computer Science, ed. by P. Cimiano, H. Pinto (Springer, Berlin, 2010), pp. 193–210. doi:10.1007/978-3-642-16438-514. http://dx.doi.org/10.1007/978-3-642-16438-5_14
66. D. Damljanovic, M. Agatonovic, H. Cunningham, in *The Semantic Web: ESWC 2011 Workshops*, vol. 7117, Lecture Notes in Computer Science, ed. by R. Garc-Castro, D. Fensel, G.

Antoniou (Springer, Berlin, 2012), pp. 125–138. doi:10.1007/978-3-642-25953-111. http://dx.doi.org/10.1007/978-3-642-25953-1_11

67. V. Lopez, V. Uren, M. Sabou, E. Motta, Semant. Web **2**(2), 125 (2011). doi:10.3233/SW-2011-0041. http://dx.doi.org/10.3233/SW-2011-0041

68. A. Wagner, G. Ladwig, T. Tran, in *Database and Expert Systems Applications*, vol. 6860, Lecture Notes in Computer Science, ed. by A. Hameurlain, S. Liddle, K.D. Schewe, X. Zhou (Springer, Berlin, 2011), pp. 303–319. doi:10.1007/978-3-642-23088-222. http://dx.doi.org/10.1007/978-3-642-23088-2_22

69. S. Ferré, A. Hermann, in *The Semantic Web at ISWC 2011*, vol. 7031, Lecture Notes in Computer Science, ed. by L. Aroyo, C. Welty, H. Alani, J. Taylor, A. Bernstein, L. Kagal, N. Noy, E. Blomqvist (Springer, Berlin, 2011), pp. 177–192. doi:10.1007/978-3-642-25073-612. http://dx.doi.org/10.1007/978-3-642-25073-6_12

70. Freebase Consortium, Freebase—a community-curated database of well-known people, places, and things (2013), http://www.freebase.com/. Accessed Aug 2013

71. Freebase Consortium, Freebase api. build intelligent apps with freebase data, (2013), https://developers.google.com/freebase/. Accessed Aug 2013

72. B. Shneiderman, in *VL* (1996), pp. 336–343

73. A. Katifori, C. Halatsis, G. Lepouras, C. Vassilakis, E. Giannopoulou, ACM Comput. Surv. **39** (2007)

74. S.K. Card, J.D. Mackinlay, B. Shneiderman, *Readings in Information Visualization: Using Vision to Think*, 1st edn. (Morgan Kaufmann, 1999). http://www.amazon.com/exec/obidos/redirect?tag=citeulike07-20&path=ASIN/1558605339

75. A. Katifori, E. Torou, C. Vassilakis, G. Lepouras, C. Halatsis, in *Second International Conference on Research Challenges in Information Science, RCIS 2008* (2008), pp. 133–140. doi:10.1109/RCIS.2008.4632101

76. B. Amann, I. Fundulaki, in *ECDL-99: Research and Advanced Technologies for Digital Libraries*, Lecture Notes in Computer Science (Springer, Berlin, 1999), pp. 234–253

77. B. Shneiderman, ACM Tran. Graph. **11**(1), 92 (1992). doi:10.1145/102377.115768. http://doi.acm.org/10.1145/102377.115768

78. F. van Ham, A. Perer, I.E.E.E. Trans. Vis. Comput. Graph. **15**, 953 (2009)

79. P. Chase, R. D'Amore, N. Gershon, R. Holland, R. Hyland, I. Mani, M. Maybury, A. Merlinoa, J. Rayson, in *Coling-ACL Workshop on Content Visualization and Intermedia Representation (CVIR'98)*, ed. by J. Pustejovsky, M.T. Maybury (1998), pp. 52–62

80. M. Storey, M. Musen, J. Silva, C. Best, N. Ernst, R. Fergerson, N. Noy, in *Workshop on Interactive Tools for Knowledge Capture* (2001)

81. E. Baehrecke, N. Dang, K. Babaria, B. Shneiderman, BMC Bioinformatics **5**(1), 1 (2004). doi:10.1186/1471-2105-5-84. http://dx.doi.org/10.1186/1471-2105-5-84

82. S. Zillner, T. Hauer, D. Rogulin, A. Tsymbal, M. Huber, T. Solomonides, in *21st IEEE International Symposium on Computer-Based Medical Systems, CBMS '08* (2008), pp. 296–301. doi:10.1109/CBMS.2008.11

83. N. Noy, R. Fergerson, M. Musen, in *Knowledge Engineering and Knowledge Management Methods, Models, and Tools*, vol. 1937, Lecture Notes in Computer Science, ed. by R. Dieng, O. Corby (Springer, Berlin Heidelberg, 2000), pp. 17–32. doi:10.1007/3-540-39967-42. http://dx.doi.org/10.1007/3-540-39967-4_2

84. Y. Sure, J. Angele, S. Staab, in *On the Move to Meaningful Internet Systems 2002: CoopIS, DOA, and ODBASE*, vol. 2519, Lecture Notes in Computer Science, ed. by R. Meersman, Z. Tari (Springer, Berlin, 2002), pp. 1205–1222. doi:10.1007/3-540-36124-376. http://dx.doi.org/10.1007/3-540-36124-3_76

85. P. Eklund, N. Roberts, S. Green, in *Proceedings of the First International Symposium on Cyber Worlds (CW'02)*, (IEEE Computer Society, Washington, DC, USA, 2002), pp. 0405. http://dl.acm.org/citation.cfm?id=794192.794831

86. T. Liebig, O. Noppens, in *The Semantic Web at ISWC 2004*, vol. 3298, Lecture Notes in Computer Science, ed. by S. McIlraith, D. Plexousakis, F. Harmelen (Springer, Berlin, 2004), pp. 244–258. doi:10.1007/978-3-540-30475-318. http://dx.doi.org/10.1007/978-3-540-30475-3_18

87. C. Plaisant, J. Grosjean, B. Bederson, in *IEEE Symposium on Information Visualization, INFOVIS 2002*, pp. 57–64. doi:10.1109/INFVIS.2002.1173148
88. B. Parsia, T. Wang, J. Goldbeck, in *Proceedings of the 4th International Semantic Web Conference* (2005)
89. T.D. Wang, B. Parsia, in *Proceedings of the 5th international conference on The Semantic Web ISWC'06* (Springer, Berlin, 2006), pp. 695–708. doi:10.1007/1192607850. http://dx.doi.org/10.1007/11926078_50
90. S. Kriglstein, R. Motschnig-Pitrik, in *12th International Conference on Information Visualisation IV '08* (2008), pp. 163–168. doi:10.1109/IV.2008.16
91. S. Kriglstein, G. Wallner, in *2010 International Conference on Complex, Intelligent and Software Intensive Systems (CISIS)*, pp. 950–955. doi:10.1109/CISIS.2010.55
92. D. Allemang, I. Polikoff, R. Hodgson, P. Keller, J. Duley, P. Chang, in *2005 IEEE Aerospace Conference*, pp. 1–10. doi:10.1109/AERO.2005.1559633
93. B. Fu, L. Grammel, M.A.D. Storey, in *ICBO'12* (2012), pp. -1-1
94. K.X.S. de Souza, A.D. dos Santos, S.R.M. Evangelista, in *Proceedings of the Latin American conference on Human-computer interaction CLIHC '03* (ACM, New York, USA, 2003), pp. 251–255. http://dl.acm.org/citation.cfm?id=944519.944551
95. C. Buntain, in *Proceedings of the 46th Annual Southeast Regional Conference on XX ACM-SE 46* (ACM, New York, USA, 2008), pp. 204–208. doi:10.1145/1593105.1593158. http://doi.acm.org/10.1145/1593105.1593158
96. J. Samper, V. Tomas, E. Carrillo, R.P. do Nascimento, IEEE Trans. Knowl. Data Eng. **20**(1), 130 (2008). doi:10.1109/TKDE.2007.190698
97. E. Motta, S. Peroni, N. Li, M. D'Aquin, in *Proceedings of the EKAW2010 Poster and Demo Track* (2010)
98. E. Motta, P. Mulholland, S. Peroni, M. d' Aquin, J.M. Gomez-Perez, V. Mendez, F. Zablith, in *The Semantic Web at ISWC 2011*, vol. 7031, Lecture Notes in Computer Science, ed. by L. Aroyo, C. Welty, H. Alani, J. Taylor, A. Bernstein, L. Kagal, N. Noy, E. Blomqvist, (Springer, Berlin Heidelberg, 2011), pp. 470–486. doi:10.1007/978-3-642-25073-630. http://dx.doi.org/10.1007/978-3-642-25073-6_30
99. NEON Consortium, NEO Toolkit (2012), http://neon-toolkit.org/wiki/Main_Page. Accessed Aug 2013
100. H. Alani, in *Workshop on Visualization Information in Knowledge Engineering Knowledge Capture (K-Cap'03)*, 26 Oct 2003. http://eprints.soton.ac.uk/258326/
101. TouchGraph, TouchGraph Navigator (2013), http://www.touchgraph.com/navigator. Accessed Aug 2013
102. E. Godehardt, N. Bhatti, in *Scaling Topic Maps*, Lecture Notes in Computer Science (LNCS) (Springer, Berlin, 2008), p. 6. http://bibcd.igd.fraunhofer.de/bibcd/INI_Science/papers/2008/08p111.pdf
103. C. Fluit, M. Sabou, F. van Harmelen, in *Handbook on Ontologies*, ed. by S. Staab, R. Studer International Handbooks on Information Systems (Springer, 2004), pp. 415–434
104. C. Fluit, M. Sabou, F. van Harmelen, in *isualizing the Semantic Web. XML-Based Internet and Information Visualization*, 2nd edn., ed. by V. Geroimenko, C. Chen (Springer, London, 2006)
105. N. Bhatti, in *Proceedings of ED-Media 2008* (Association for the Advancement of Computing in Education (AACE), 2008), pp. 312–317. http://bibcd.igd.fraunhofer.de/bibcd/INI_Science/papers/2008/08p129.pdf
106. T. Liebig, O. Noppens, F. von Henke, in *IEEE Symposium on Visual Analytics Science and Technology, VAST 2009*. doi:10.1109/VAST.2009.5333946
107. S.G. Kobourov, arXiv:1201.3011
108. M. Sintek, Ontoviz tab: Visualizing protégé ontologies (2007). http://protegewiki.stanford.edu/index.php/OntoViz. Accessed Aug 2013
109. J. Ellson, E.G.Y. Hu, A. Bilgin, D. Perry, Graphviz—Graph Visualization Software, Drawing graphs since 1988 (2013), http://www.graphviz.org/About.php. Accessed Aug 2013

110. M. Lanzenberger, J. Sampson, in *Tenth International Conference on Information Visualiza-tion, IV 2006*, pp. 430–440. doi:10.1109/IV.2006.18
111. A. Bosca, D. Bonino, P. Pellegrino, in *Proceedings of the 2nd Italian Semantic Web Workshop SWAP 2005*, Trento, Italy (2005)
112. A. Bosca, D. Bonino, M. Comerio, S. Grega, F. Corno, in *Proceedings of the twelfth inter-national conference on 3D web technology, Web3D '07* (ACM, New York, USA, 2007), pp. 89–96. doi:10.1145/1229390.1229405. http://doi.acm.org/10.1145/1229390.1229405
113. M. Kerrigan, in *Tenth International Conference on Information Visualization, IV 2006*, pp. 411–416. doi:10.1109/IV.2006.135
114. M. Kerrigan, Jpowergraph (2013), http://sourceforge.net/projects/jpowergraph/
115. E. Pietriga, Isaviz: a visual authoring tool for rdf (2007), http://www.w3.org/2001/11/IsaViz. Accessed Aug 2013
116. L. Deligiannidis, A. Sheth, B. Aleman-Meza, in *Intelligence and Security Informatics*, vol. 3975, Lecture Notes in Computer Science, ed. by S. Mehrotra, D. Zeng, H. Chen, B. Thurais-ingham, F.Y. Wang (Springer, Berlin Heidelberg, 2006), pp. 48–59. doi:10.1007/11760146. http://dx.doi.org/10.1007/11760146_5
117. L. Deligiannidis, K.J. Kochut, A.P. Sheth, in *Proceedings of the ACM first workshop on CyberInfrastructure: Information Management in eScience CIMS '07* (ACM, New York, USA, 2007), pp. 39–46. doi:10.1145/1317353.1317362. http://doi.acm.org/10.1145/1317353.1317362
118. I. da Silva, G. Santucci, C. del Sasso Freitas, in *EuroVA 2012: International Workshop on Visual Analytics* (Eurographics Association, 2012), pp. 91–95. doi:10.2312/PE/EuroVAST/EuroVA12/091-095. http://diglib.eg.org/EG/DL/PE/EuroVAST/EuroVA12/091-095.pdf
119. I.C.S. da Silva, C.M.D.S. Freitas, G. Santucci, in *Proceedings of the 2012 BELIV Workshop: Beyond Time and Errors—Novel Evaluation Methods for Visualization, BELIV '12* (ACM, New York, USA, 2012), pp. 2:1–2:7. doi:10.1145/2442576.2442578. http://doi.acm.org/10.1145/2442576.2442578
120. B. Bach, E. Pietriga, I. Liccardi, G. Legostaev, in *Proceedings of the 20th international conference companion on World Wide Web WWW '11* (ACM, New York, USA, 2011), pp. 177–180. doi:10.1145/1963192.1963283. http://doi.acm.org/10.1145/1963192.1963283
121. N. Henry, J. Fekete, M. McGuffin, IEEE Trans. Vis. Comput. Graph. **13**(6), 1302 (2007). doi:10.1109/TVCG.2007.70582
122. M. Voigt, S. Pietschmann, K. Meissner, in *Semantic Models for Adaptive Interactive Systems*, ed. by T. Hussein, H. Paulheim, S. Lukosch, J. Ziegler, G. Calvary, Human (at) Computer Interaction Series (Springer, London, 2013), pp. 83–107. doi:10.1007/978-1-4471-5301-65. http://dx.doi.org/10.1007/978-1-4471-5301-6_5
123. B.P. Nunes, B. Fetahu, M.A. Casanova, in *Proceedings of LAK Data Challenge, Third Con-ference on Learning Analytics and Knowledge (LAK 2013)*
124. H. Lin, J.A. Rushing, T. Berendes, C. Stein, S.J. Graves, in *ACM Southeast Regional Confer-ence*, ed. by H.C. Cunningham, P. Ruth, N.A. Kraft (ACM, 2010), p. 38. http://dblp.uni-trier.de/db/conf/ACMse/ACMse2010.html#LinRBSG10
125. S. Sen, H. Charlton, R. Kerwin, J. Lim, B. Maus, N. Miller, M.R. Naminski, A. Schneeman, A. Tran, E. Nunes, E.I. Sparling, in *Proceedings of the 16th International Conference on Intelligent User Interfaces IUI '11* (ACM, New York, USA, 2011), pp. 457–458. doi:10.1145/1943403.1943497. http://doi.acm.org/10.1145/1943403.1943497
126. F. Kboubi, A.H. Chaibi, M.B. Ahmed, Semantic visualization and navigation in textual cor-pus, Int. J. Inf. Sci. Tech. (2012). arXiv:1202.1841. http://dblp.uni-trier.de/db/journals/corr/corr1202.html#abs-1202-1841
127. F. Bertault, W. Feng, A. Krastins, L. Yi, A. Verza, in *Proceedings of the 2011 Joint Inter-national Conference on the Semantic Web JIST'11* (Springer, Berlin, 2012), pp. 411–416. doi:10.1007/978-3-642-29923-031. http://dx.doi.org/10.1007/978-3-642-29923-0_31
128. A. Panchenko, P. Romanov, O. Morozova, H. Naets, A. Philippovich, A. Romanov, C. Fairon, in *Advances in Information Retrieval*, vol. 7814, Lecture Notes in Computer Science, ed. by P. Serdyukov, P. Braslavski, S. Kuznetsov, J. Kamps, S. Rger, E. Agichtein, I. Segalovich,

E. Yilmaz (Springer, Berlin Heidelberg, 2013), pp. 837–840. doi:10.1007/978-3-642-36973-597. http://dx.doi.org/10.1007/978-3-642-36973-5_97

129. J.K. Chou, C.K. Yang, in *Proceedings of the 13th Eurographics/IEEE—VGTC conference on Visualization EuroVis'11* (Eurographics Association, Aire-la-Ville, Switzerland, Switzerland, 2011), pp. 721–730. doi:10.1111/j.1467-8659.2011.01921.x. http://dx.doi.org/10.1111/j.1467-8659.2011.01921.x

130. S. Schenk, C. Saathoff, S. Staab, A. Scherp, Web Semant. **7**(4), 298 (2009). doi:10.1016/j.websem.2009.09.006. http://dx.doi.org/10.1016/j.websem.2009.09.006

131. D. Petrelli, S. Mazumdar, A.S. Dadzie, F. Ciravegna, in *Proceedings of the 8th International Semantic Web Conference ISWC '09* (Springer, Berlin, 2009), pp. 505–520. doi:10.1007/978-3-642-04930-932. http://dx.doi.org/10.1007/978-3-642-04930-9_32

132. Y. Wu, T. Provan, F. Wei, S. Liu, K.L. Ma, in *Proceedings of the 13th Eurographics/IEEE—VGTC conference on Visualization EuroVis'11* (Eurographics Association, Aire-la-Ville, Switzerland, Switzerland, 2011), pp. 741–750. doi:10.1111/j.1467-8659.2011.01923.x. http://dx.doi.org/10.1111/j.1467-8659.2011.01923.x

Chapter 4
Adaptive Visualization

Human-centered adaptation has been subject of research and development for more than two decades. In particular adaptive learning systems built the foundation of this research area. Content-based adaptation, navigation support, and help-systems are just example for the broad range of adaptive hypermedia and learning systems. The developed approaches commonly focus on users' knowledge in a particular domain of interest, e.g. teaching programming languages. In contrast to those systems, adaptive visualizations are a relatively young area of research. Although the main goal, namely to support users in information and knowledge acquisition processes is quite similar, a linear way of learning is not focused in adaptive visualizations. Visual environments contain various variables that can be adapted and thereby support or confuse users. This chapter investigates the state-of-the art in adaptive visualizations. To have a common understanding of the term "adaptive", we will introduce definitions for adaptation in a general manner and derive from these a definition for adaptation for this work. Then, influencing works in context of adaptation will be introduced to provide a comprehensible view on the general process of adaptation. Based on this general process a number of statistical methods will be introduced that deal in particular with uncertainty in the adaptation process. Thereafter, we will investigate in particular the adaptation process in information visualization. It is important to understand the differences for adapting visualization systems. To provide a comprehensible way for conveying these differences, we introduce three main aspects: *influencing factors* by means of "to what can visualizations be adapted", *knowledge modeling* that refers to the way how the influencing factors can be formalized (represented) and which factors may play a role for the adaptation process, and *human interface adaptation* that refers to visualization and their capabilities for adaptation.

The main focus of this chapter is a survey of existing adaptive visualizations. To have a comprehensible view on the introduced systems, we first introduce a definition for adaptive visualizations. Thereafter different classifications for adaptive visualizations and adaptive systems will be investigated to derive a classification for adaptive visualizations. The introduced classification will focus more on the human interface and visualization adaptation capabilities. The existing adaptive

© Springer International Publishing Switzerland 2016
K. Nazemi, *Adaptive Semantics Visualization*, Studies in Computational
Intelligence 646, DOI 10.1007/978-3-319-30816-6_4

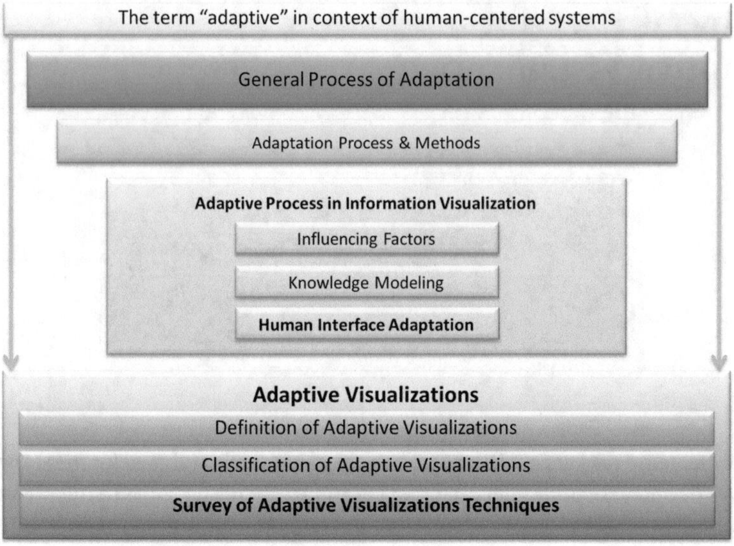

Fig. 4.1 Structure of the chapter adaptive visualizations

visualization systems will be introduced based on the derived classification. This provides a more comprehensible view on the systems, their strengths, and the identification of gaps. Figure 4.1 illustrates the described structure of this chapter.

4.1 Terminological Distinction

The term adaptation in computer systems is used in various contexts ambiguously. Even if the application in such a system provides a human interface that is adaptive, the term shows some diversity that has to be clarified at least in context of this thesis. The review on literature illustrates various terms that are used for similar adaptive functionalities. In this context terms like *user-adaptive* [1], *adaptive* [2–6], *intelligent* [7–10], *personalized* [11, 12], *adaptable* [1, 13], *personalization* [14], and *customization* and *customizable* [15] are used for the different systems. Commonly the terms are chosen based on application domain, e.g. in context of learning systems (adaptive tutoring, adaptive hypermedia [16, 17] the term *adaptive* is dominantly used, while in the domain of e-commerce the term *personalization* is used dominantly [18]. *Stephanidis* et al. defined adaptation as a process of tailoring the user interface based on user abilities, skills, requirements, and preferences during the application use [2]. These tailoring and changes may be performed by users or the system itself. The main aspect is that *adaptive* is denoted to changes during the system use. In contrast to that, *adaptable* is the process of change and tailoring the system before starting the usage. It is more the initial adaptation that is performed before using a system and does not change (and is not changeable) during system use [2]. Kobsa et al. proposes a different definition of adaptation. Although, they use in their definition the term *personalized*, their definition denotes to the adaptation process, thus the definition of Stephanidis is criticized as a "different sense" understanding adaptive and adaptable system behavior, where the adaptable characteristic refers more to configurable system [11, p. 3]. Kobsa et al. defined *personalized hypermedia applications* as hypermedia systems that adapt content, structure, and presentation to the individual user characteristics, usage behavior, and usage environment [11, p. 3]. Further they identified basic types of adaptation, depending on the users' control on the adaptation process. They differentiated between adaptation *initiator*, adaptation *proposer*, adaptation *selector*, and adaptation *producer*. The main difference between adaptable and adaptive systems is that in adaptable system the users have the control on the identified basic types, whereas the system may support the user in choosing the adequate basic type [11, 13]. With this definition adaptable and adaptive systems co-exist in applications [11]. Further the user control on the adaptive system behavior can be on different levels: on a general level by enabling and disabling the adaptation at all, on a type-level by approving or disapproving certain types of adaptations, or on the case-by-case level [11]. Another definition of adaptive systems, in particular of *user-adaptive* systems was proposed by Jameson [1], that a user-adaptive system can be defined as:

An interactive system that adapts its behavior to individual users on the basis of process of user model acquisition and application that involve some form of learning, inference, or decision making [1, p. 2].

Further he distinguishes adaptable systems as those on which the individual user can explicitly express and adjust their preferences. Adaptable systems provide therewith users with the opportunity to specify and tailor desired properties on the user interface [1]. Adaptable systems do not adjust any parameter automatically. Fink and Kobsa used the term *personalization* in the domain of e-commerce, because of the predominance of the term [18]. They proposed that personalization is a generic term that refers to user-adaptive systems and user modeling. This phenomenon is often given in literature, where user-adaptive systems that adapt certain information or user interface parameters to the individual user's requirements, skills, or any other criteria are denoted as *personalized system* or *personalization* [11, 12, 14, 18–21]. Summarized personalization or personalized systems can be defined as user-adaptive systems that are based on users' individual requirements, preferences, skills etc. In contrast to that, the term customization (or customizable, configurable) can be denoted to systems that enables users to take control of changes in functionalities and presentation aspects of a systems [15]. Therewith and according to Kobsa [11] customization and configurable systems can be summarized to adaptable systems, thus the user has the control on the adaptable functionalities and presentation parameters. Another often used term in this context is *intelligent* [7]. It is commonly used synonymously to *adaptive* in particular in context of intelligent user interfaces (IUI) [7, 22]. Álvarez-Cortéz and Zárate-Silva proposed that intelligent or adaptive user interfaces are characterized as personalized systems that support users in solving tasks, reduce the information overflow and provide help on using new and complex applications [7]. The main difference between intelligent and adaptive may be the use of approaches from artificial intelligence in intelligent user interface that are not necessarily part of adaptive systems [7, 9]. With the introduced definitions the terms personalization (or personalized systems), intelligent (in particular intelligent user interfaces), user-adaptation (user-adaptive systems), customization (or customized systems), and adaptable systems should be clear. It relies on the automatic and systems-driven tailoring based on the requirements of individual users (user-adaptive). Adaptable systems (including customizable and configurable) provide a more user-driven control on the parameter to be adapted. The system is commonly not actively adapting any parameter to users. It is the users' actions that lead to changes in system behavior. In this context we have constrained the definition to user-adaptive and not to a general adaptation of a system. Thus commonly user-adaptive or personalized systems incorporate the users' individual preferences, knowledge, skills, etc. for the adaptation process. As Jameson proposed, the general schema of a user-adaptive system starts with the information about the user, which is further used for the user model acquisition [1]. Based on the user model the system performs adaptations to the individual user.

The general schema of Jameson illustrates very clearly that the main factor for influencing the adaptation process in user-adaptive system is the (individual) user. The model incorporates a user model that acquires information about a certain user. Based on the user model an application predicts or decides about certain adaptive feature for the user. As we already introduced with the definition of Kobsa, who proposed to acquire more than just knowledge about a user, e.g. the usage environment [11], an adaptive system can be seen as a generalization of a user-adaptive system. In that sense the influencing factors depends on the application, tasks, and users of an adaptive system and is not limited to the user. Kobsa proposed in his early works that worthwhile adaptation is application specific and depends on the requirements on an application [23]. Further influencing factors beside the usage environment [11] may be the environmental context [9], tasks [3, 7], data [24], and any other influencing factor that supports users in system use. The main focus of adaptation still remains the user support, even if contextual or environmental information are used. An adaptive system is therewith always user-centered, even if it is not user-adaptive. To define adaptation in context of this work, we take the introduced definitions of Kobsa and James and generalize them to:

> Adaptation in human-computer interfaces is the automatic and system-driven changes on content, structure, and presentation of system-behavior that involve some form of learning, inference, or decision making based on one or many influencing factors to support users.

Further we enhance the definition that influencing factors are commonly modeled in knowledge repositories, e.g. user models or context models. The aspects of learning, inferencing, and decision making are not combined in one adaptive system. If the learning component is given, the term intelligent can be used too. The main aspect in this definition is that the adaptation to users' needs is not anymore coupled directly to the existent of user models, whereas we believe that commonly user model are essential parts of adaptive systems and user interfaces respectively.

4.2   Adaptation in Computational Systems

One main goal of adaptive systems is to help users in achieving their intended tasks faster, easier, or with better results [25]. In this context a number of classifications, methods and applications arose that should give in this section a short overview on some main topics. The goal of this section is not to give a survey on adaptive systems. It is more intended to introduce the idea of adaptive systems for classifying and reviewing adaptive visualizations.

An early and influencing work on adaptive systems was brought by Brusilovsky with the dimensions of adaptive systems [17]. Although, the pioneering work was mainly proposed to provide an insight on Adaptive Hypermedia Systems (AH), the dimensions and depended questions can be applied to the general idea of adaptation and was applied in various further works [26, 27]. Brusilovsky introduced with his dimensions four elementary questions, which we generalize to adaptive systems: *Where* can adaptivity be applied, *To What* can be adapted (influencing factors), *What* can be adapted, and *Why* are certain methods or techniques used [17]. He listed six areas, where adaptation helps users to solve tasks in a more efficient way. We pick-up one area that is closely related to visualization of abstract information, namely "online information systems" to refer on the surpluses of adaptation in information systems. He proposed that the main goal of such systems is to provide a reference access to information, rather than an educational process. The subjects working with these systems differ in their knowledge-level. The knowledge-level of users is criterion enough to provide adaptation to their level knowledge and support the "free" learning process. But there are more user features, the adaptation can be influenced by. Brusilovsky lists four elementary dimensions of users that influence the adaptation process (To What), users' knowledge, goals, background and experience, and preferences [17]. The next dimension is what can be adapted to the influencing factors (in this case users) Brusilovsky identifies two general levels of adaptation: (1) adaptive presentation and (2) adaptive navigation. In this classification, (1) the presentation refers to the content that can be adapted based on the acquired user model. Presentation regulates the amount and complexity of the content to be presented to the user. Further he proposed the adaptation of (2) navigation support. The main idea of the navigation support is provide a path-finding by guiding, sorting, hiding, annotating and providing (global and local) maps to the user. The dimensions of Brusilovsky are for the work in adaptive hypermedia systems. But they can be applied to provide the general idea and limitations of adaptive systems in general.

Kobsa et al. proposed a general model of adaptive system addressing three main *tasks* that are usually performed to provide adaptation: (1) *acquisition*, (2) *representation*, and (3) *production* [11]. *Acquisition* (1)) is the task that identifies available information that can or should be used for the adaptation process. It may include data about users, usage behavior or usage environment. The process of acquisition can be performed implicitly by observing users, analyzing interaction etc. or explicitly by asking for users demographic data or any other information that are useful for the adaptation process, whereas the process of implicit data acquisition is less disturbing and annoying [11]. This task is further responsible for the initial model of the acquired information. Kobsa and colleagues list *user data*, *usage data*, and *environmental data* with various facets as potential input data that can be used to acquire *User Models* (UM), *Usage Models*, and environmental models [11]. The task (2) *representation* or *secondary inferencing* is responsible for expressing the acquired information in a formal and machine-processable way [11]. Further it incorporates *secondary* assumptions about the acquired information, e.g. by integrating information from various sources. Kobsa et al. list *deductive reasoning, inductive reasoning (learning), analogical reasoning*, and *hybrid approaches* for formalizing and inferencing information about the input data [11]. Deductive reasoning contain logic-based representations and inferencing that formalizes the information in a logical manner (e.g. as ontology) and provide logical inferencing on the modeled data. It further may make use of probabilistic methods on features (e.g. Bayesian Networks, where the inferencing is the conditional probabilities associated with the network links [11]. Inductive reasoning involves observing and monitoring users' interactions with the adaptive system to enable "learning about users" [11, p. 25]. Their introduced examples made use of neural networks, Bayesian networks, nearest neighbor algorithms, and probabilistic models, whereas the results of assumption are very similar. Analogical reasoning make use of the "similar behavior" [11, p. 28] of certain users by using of *Clique-based filtering* (collaborative filtering) methods in order to find similar neighbors, select a comparison group of neighbors, or compute prediction based on weighted representations of selected neighbor [11, pp. 28–29]. Further user clustering approaches are proposed to identify similar groups of users. Hybrid approaches combine the formal representation of, e.g. as ontology with prediction models. The last step of the adaptation process is (3) production, which translates the acquired models for an adaptive output. Kobsa et al. differentiate in this context between content adaptation, presentation and modality adaptation, and structure adaptation. Content adaptation refers to presentation adaptation of Brusilovsky and adapts the content in an appropriate way to the user. The presentation and modality adaptation changes the format and layout of certain hypermedia objects. In particular the change of modalities plays an important role, e.g. from text to audio or from video to single images [11, p. 37]. This type of adaptation is often combined with content adaptation. The adaptation of structure refers to the way how "the link structure of hypermedia documents or its presentation to users is changed" [11, p. 39]. In this context aspects link sorting, link annotation, hiding and unhiding, link disabling and enabling, link removal and addition, link recommendation, and the navigation support through links (as proposed by [17] too) are listed. Although, the adaptation process in the introduced model of Kobsa et al. is more similar to traditional recommendation systems, it provides a good foundation to understand and comprehend the process of adaptation. In particular the *production* and therewith the adaptation *output* goes one step beyond the proposed model of Brusilovsky [17] and is still the foundation of many works of adapting user interfaces (e.g. Bunt07, ahn10, AhnDiss10, feigh12). Figure 4.2 illustrates the high-level steps of the introduced adaptation process.

Kobsa et al. and Brusilovsky shaped the understanding and definition of adaptive systems in context of information-use still today. Although, the steps in Kobsa

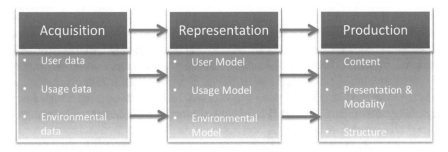

Fig. 4.2 General process of adaptation (based on the proposed model of [11])

and colleagues' processes has changed in the way what acquisition includes and what the part of modeling or representation is, the process itself did not change [1, 6, 28]. One further elementary classification and model was proposed by Jameson [1]. He distinguished adaptive systems in general for supporting two main processes: (1) *system use* and (2) *information acquisition*. (1) System-use refers to the adaptive support to operate a system successfully and reduce user's efforts. This may be solved by overtaking routine tasks by system instead by users, or adapting the user interface in a way that fits better to the user's way of working. Further adaptation for system use may be giving advices about how to use a certain system, e.g. as a mentor or supporting tutor [29] or supporting users by natural language dialog [1].

The second main area of adaptive systems according to Jameson is (2) information acquisition, thus the amount of information on Web is growing with each day and the access to these data is getting more difficult, Jameson proposed that one way to support the information acquisition process is helping users in finding information [1]. He introduced in this context an example that provides three main kind of helping users: (i) *support for browsing*, (ii) *support for query-based search or filtering*, and (iii) *spontaneous provision of information* [1]. *Support for browsing* (i) refers to the adaptive functionality of "recommending or selecting promising items or directions of search" [1, p. 6], *support for query-based search or filtering* (ii) refers to the use of user models from other knowledge domains to enrich the search and filtering, and *spontaneous provision of information* (iii) refers to system-driven information suggestion during a task, e.g. providing relevant links while the user is typing in a word processor. Another way of supporting users by adaptive information acquisition according to Jameson is *tailoring information*. Based on user's degree of interest, user's knowledge, user's preferences and needs, and the capabilities of the computing device the degree of information can be tailored to the individual user. Further areas, where the information acquisition process is supported by adaptive systems are *product recommendations*, *collaboration*, and *learning* [1].

Adaptive systems can be found for both areas: information acquisition and system-use. Commonly the adaptive systems process contextual, user, and further relevant data to improve the human computer interaction [30]. Thereby the reduction of information, personalized views on information, supporting users in their tasks, and providing help is commonly performed by systems to improve the human computer interaction [7]. Most of the adaptive systems are part of research in learning and knowledge acquisition [3, 4, 11, 16, 17, 23, 31] in so called adaptive hypermedia, adaptive hypertext, intelligent or adaptive tutoring system etc. Further the area of recommendation systems profit in particular from the information acquisition in adaptive systems [32, 33]. Adaptive systems investigate more and more further areas, e.g. the environmental context [11, 34, 35], sensors, mental and cognitive states of users' [28], or mobile devices [9, 36].

4.3   Adaptation Process and Methods

The adaptation process aims at providing user interfaces and systems that support users in their task, make the work with a system more attractive and support the information acquisition process. As we already outlined the main process starts with gathering information about users, usage, environment etc., acquire some kind of machine-readable formal representation of the data [11]. These formal representations are then used to create adaptive interface, information reduction, filtering and recommendation, or help and tutoring systems [16]. This section will introduce the general process of adaptation by giving a view on how features and influencing factors can be modeled to formal representations that further leads to adaptive system behavior. Our goal in this section is to give a short overview of possible methods that make use of these techniques.

4.3.1   The Adaptation Process

Frias-Martinez et al. introduced a survey of data mining approaches for user modeling in particular for adaptive hypermedia systems [37]. Complementary to the already introduced model of Kobsa et al. [11], their survey introduces steps for generating automatic user models in adaptive systems. Frias-Martinez et al. proposed that the automatic generation of user models, the inferencing from the modeled, and the validation start with *Data Collection*. This step incorporates the collection of needed data for the adaptation process and may include data from users' interaction with the system, data about the environment, and explicit feedback and data from users [37]. The second step in their process is *Preprocessing and Extraction*. This step "cleans" the data from noises, in particular the identification of users and session reconstruction belong to this step. The third step, *Pattern Discovery*, applies machine learning and data mining techniques in order to model the user's behavior. This steps leads to learning about user's behavior and interest and provides knowledge determining adaptations. In the last step of *Validation and Interpretation* the obtained structures are analyzed and interpreted to validate them [37].

Frias-Martinez et al. introduced machine learning methods for gathering and inter-preting user and usage information. The machine-learning methods can generally be classified in two main categories: *supervised* and *unsupervised* methods [37]. Supervised learning requires a preclassification of the training data in order to label the training items for signifying the class to which they belong [37]. In case of user interaction analysis, the process of learning have to be observed to define for each interaction an analogue output. Thus, user interactions occur during the entire process of interacting with the system to learn and refine the model of the user, these methods are in our opinion not appropriate for the implicit user interaction analy-sis. Examples for supervised methods are decision trees [37, 38], neural networks [37, 39], or SVMs [37, 40]. Unsupervised learning does not require a feedback, validation, or preclassification of the trained data [37] and are thereby more appro-priate for implicit interaction analysis. One main class of unsupervised learning is clustering algorithms [37]. Clustering algorithms are able to find, group and cluster training in concepts data based on similarities between the unclassified instances. The most common techniques for clustering are *k-means clustering* and *self-organizing maps* (SOM) [37, 41]. The clustering is commonly based on distance and similar-ity measurements (e.g. cosine algorithm or pearson similarity) [37]. Based on the *distance* training data can be compared and classified. The classification of user interaction data with clustering algorithms is commonly constrained, thus the user interactions are occurring as series of interactions and the subset that should be com-pared is relatively fuzzy to identify [42, 43]. Another group of unsupervised learning are *probabilistic learning* (or predictive statistical) methods [41], which are used to describe predictions between various variables and states [44]. One main advantage is that these methods are able to handle uncertainty appearing in user models and the wealth of data that are produced as the result of information gathering [41]. Thus users' interactions with systems produces both, uncertainty and wealth amount of data as consequent of the users' interaction [25, 41, 45] the next section will intro-duce common predictive statistical models. The main goal is to provide an idea of how the commonly well-known unsupervised machine-learning methods may help to model knowledge, in particular about users as user models.

4.3.2   Predictive Statistical Methods

Modeling a formal and machine-readable representation of knowledge, which enables reasoning of further information, is one of the main steps in developing adaptive systems [11, 37, 41]. Beside logical representations and inferencing, the involvement of predictive statistical methods gained more and more importance in modeling such knowledge [46]. This section provides an overview of some unsu-pervised predictive statistical methods. The goal is to provide a short insight into the modeling and usage of the techniques. This section was partially published in [42, 43, 47] and relies on these works.

Bayesian Networks

Bayesian Networks are directed acyclic graphs, where the nodes represent real-world objects and events, which are usually are described as variables [48]. Directed links (arcs) represent the causal relationships between nodes in a Graph Model (GM). A graph model is a triple of $GM = (K, E, P)$, where K is a finite set of nodes, E is a finite set of links, and P is a set of conditional probability distribution [49]. In Bayesian Networks the directed links may refer to causal relation between the nodes (commonly as parent-child relation) and each node is associated with conditional probability distribution [41]. Therewith Bayesian Networks combine graph- and predictive theories and are also called *belief networks* [50, 51] or *causal networks* [41, 48]. Inferencing in Bayesian Networks is the computing of probabilities for each value of a variable, if the values of other variables are known [49]. Mihajlovic et al. proposed that inferencing can be performed as an "exact probability propagation in a singly connected network" or as an approximation of the inferencing algorithms [48, p. 6]. If the structure or variable values in Bayesian Networks are not known, learning algorithms such as *sample statistics, EM or gradient ascent, search through model space*, or *structural EM* help to fill the unknown information [52]. On the other hand, if the structure and all variables are known, the conditional probability distribution can be determined, e.g. by using the *Likelihood-method* [49]. Dynamic Bayesian Networks are enhancements of BNs and provide the ability to model temporal changes of variables. Detailed information for further readings about Bayesian Networks can be found in [48, 53].

Probabilistic Relational Models

Bayesian Networks do not provide the possibility to represent rules or dependencies between similar objects [54]. Although, they are still used in various applications [46] for different purposes, they have limitations in particular for representing large and complex domains [55]. Bayesian Networks of given domains involve a prespecified set of random variables, where the relationships are fixed [55]. Therewith Bayesian Networks cannot be used in domains with varying number of entities and a variety of configuration possibilities [55]. Thus visual interfaces commonly have these varieties, knowledge modeling and inferencing via Bayesian Networks are therewith not appropriate in visualization applications. In contrast to that, *Probabilistic Relational Models* (PRM) combine frame-based logical representations with probabilistic semantics based GMs in particular Bayesian Networks [55–58]. PRMs are enhancements of Bayesian Networks and are commonly used to determine probability distribution over databases [55, 58, 59]. One main advantage is that the relational structure of data is consistent in contrast to Bayesian Networks. Objects of a PRM are divided in set of concepts [58]. Each concept or class contains a set of attributes with relations and dependencies to other attributes of the same and different classes [58]. PRMs premise a relational schema, a set of classes $X = X_1, \ldots, X_n$, where each class contains a set of *descriptive attributes* $A(X_i)$ and a set of *reference slots* $R(X_i)$, which associates the objects [55]. A class can be seen as equivalent to a table in a relational database, with descriptive attribute representing the columns and reference slots representing secondary keys [55, 58]. PRMs are able to model uncertainties

of missing values in structures [56] and enhances Bayesian Networks with probability distribution to fill-out the structure [55]. Inferencing in PRMs are equivalent
to those in Bayesian Networks [54], but the often existing huge amount of classes
in PRM does not allow exact inferencing [60]. Therefore approximated inferencing
algorithms are commonly used [55, 58]. Learning in PRMs can be applied to support
attribute uncertainty (e.g. parameter estimation, Likelihood, Bayesian parameter estimation), structural uncertainty (e.g. reference uncertainty, Existence uncertainty), or
learning PRMs with class hierarchies [55]. Similar to Dynamic Bayesian Networks,
Dynamic Probabilistic Relational Models (DPRM) enhance PRMs with the ability
to model temporal changes of variables. Detailed information for further readings
about PRMs can be found in [55, 58].

Markov Models
Markov Models (MM) are used to describe sequences of events and predict possible future events [41]. Their relatively simple structure is due to the assumption that
occurrences of the next events depends on a fixed number of previous events (*Markov
Assumption*) [41]. Thereby the next event is predicted from the probability distribution of the events which have to be followed after the fixed number of observed
previous events [41]. In case of only one single previous variable (state) a *first-order
Markov* specifies the probability of each state and the transition of one state to another
[59]. There exist further variants of Markov Models, e.g. Hidden Markov Model that
contain two (one hidden and one observable) or more variables [59]. A *Markov Chain*
for instance is defined by a set of states $S = \{s_1, s_2, \ldots, s_n\}$, a transition probability
matrix P and an initial probability distribution π of the states in S [61]. A transition
from state s_i in s_j occurs with the probability p_{ij} from P [61]. The probability of the
next state s_j is therewith depending on the given state s_i (Markov Assumption) [61].
The initial probability distribution π over the states S, is defined by the start state
and therewith with state of highest probability distribution [61]. Markov Chain can
be expressed by the triple $MK = (S, P, \pi)$ [61]. The states of Markov Chains are
all observable [59, 61, 62]. Some scenarios require the investigation of unobservable states. *Hidden Markov Models* (HMM) are enhancements of Markov Chains,
where unobservable states occur [63]. A Hidden Markov Model can be expressed
as quintuple $HMM = (S, M, P, B, \pi)$, where S is a set of unobservable states
($S = \{s_1, s_2, \ldots, s_n\}$), M is a set of observable states ($M = \{o_1, o_2, \ldots, o_m\}$), P is
the probability transition matrix ($a_{ij} = P(q_{t+1} = s_j | q_t = s_i)$, $1 \leq i, j \leq n$), B
is a set of probability distributions, and π is the initial probability distribution [62].

The introduced Markov Models premise that the states are trained independently and therewith the probability of a state can be determined with occurrence
of the state itself in the data [59]. For computing the probability huge amount of
data are necessary to get high precision [59]. Anderson et al. proposed that *Relational Markov Models* (RMM) enable statements about states that not occur in the
training data [59]. A Relational Markov Model can be expressed as a quintuple
$RMM = (D, R, Q, P, \pi)$, where D is a set of domains representing an abstraction hierarchy of values, R is a set of relations. Each argument x of a relation
$r \in R$ is an element of the domain d. Q is a set of states, P is the probability

transition matrix, and π is the initial probability distribution [59, p. 3]. The relations and domains (abstraction hierarchies) are used to define sets of states as abstractions over Q [59]. Elements of these sets relations and values often occur in the same sub-trees, while states with strong differing values occur together in general abstractions [59]. Learning in Markov Models is performed by computing the transition probability and the initial probability distribution [59]. Commonly high amount of training data are needed to enable a precise prediction. In many cases the states are not or rarely defined in the training data. Relational Markov Models provide with their generalization the ability to infer states, even if they are not defined in the data [59]. Detailed information of learning in Relational Markov Models can be found in [59].

LEV- und KO-Algorithm

Künzer et al. proposed with their *LEV* and *KO* algorithms two predictive models that enable the probability prediction of interaction events [64]. They premised that an alphabet A with all possible interaction events and a sequence $S = s_1, \ldots, s_n$ as training data with $s_{1,\ldots,n} \in A$ are given [64]. Based on the sequence S the probability distribution in a is computed to predict the next event. The *LEV* algorithm computes the probability distribution of all interaction events by searching similar sequences in the training data [64]. For measuring the similarity, the *Levenshtein-Distance* [65] is used [64]. For each interaction event $a \in A$ the sequence $S_a = s_{n-markovOrder+1}, \ldots, s_n, a$, which is equivalent to the last elements of the length $MarkovOrder$ concatenated with the interaction event a, is searched in the sequence $S^* = s_1, \ldots, s_{n-markovOrder}$ and the occurrence is counted [64]. If for all sequences S_a in S^* an exact matching is not found, the Levenshtein-Distance is used to search for similar sequences [64]. Once the given distance is found the probability distribution is computed based on the quantity of the interaction events a [64]. In contrast to the LEV-algorithm, the KO-algorithm does not use similarity measurements [64]. The KO-algorithm searches for identical sequences with different length and weights their occurrence based on their length [64]. Thereby the length of a sequence is limited by the parameter $SearchDepth$ [64]. For each interaction event $a \in A$ the occurrence of the sequences $S_a = s_{n-i+1}, \ldots, s_n, a$ for all i between 1 and $SearchDepth$ in $S^* = s_1, \ldots, s_{n-i}$ counted and based on i weighted [64]. Künzer et al. proposed for computing the weight w, $w(i) = i^{19}$ [64]. With the weighted occurrences for each interaction event $a \in A$ the probability distribution of the next event is computed [64]. Thereby they propose a maximum length of 3 for the sequences $SearchDepth$ [64]. Their proposed initiation of the KO algorithm with the weight $w(i) = i^{19}$ and the maximum length of 3 is therefore commonly called *KO3/19-algorithm* [64]. Their proposed algorithm was evaluated and showed better results in the mean prediction probability than other predictive statistical models. The detailed algorithm and pseudo-code can be found in [64].

In general the adaptation process is related to the generation of knowledge about certain influencing factors, in particular the user. The knowledge generation process can be performed in various ways. One way still remains the use of predictive statistical models. The variety of these models has advantages and disadvantages. For each case and each scenario the knowledge generation process has to be considered and in best case evaluated. We have introduced in this section various predictive statistical models and illustrated for example that Bayesian Networks model dependencies and variables with directed acyclic graphs. They enable expressions about variables, which are not observable or not contained in the training data. The main challenge here is the modeling of the Bayesian structure that contains dependencies of the variables. This can be done with accurate knowledge about the domain and is domain-dependent. Probabilistic Relational Models are enhancements of Bayesian Networks and extend BNs with relational structures. In contrast to BNs Probabilistic Relational models can be used in varying domains with different configurations. The effort of inferencing in PRMs can rise enormously with huge amount of object. Approximated inferencing reduces the effort in PRMs. Markov Models are used to describe sequences of events and predict events. For reliable predictions in Markov Models huge amount of data are necessary. The amount can be reduced by using relational structures, such as Relational Markov Models. Hidden Markov Models provide further the expression about not observable events. The LEV und KO algorithms enable the prediction of interaction events. In contrast to other algorithms the KO-algorithm provide a better mean prediction probability. All these algorithms were used in adaptive systems with different scopes.

4.4 Adaptive Process in Information Visualization

We introduced so far the general idea and process of adaptation and included some examples of how implicit information can be gathered. The general process of adaptation in computational systems involves three main steps, data acquisition that influences the adaptation process, representation of the data in a formal and machine-readable way, and the production of the adaptation output [11]. One main goal of this work is to apply these ideas and processes for visualizing information in particular semantic information. Therefore this section will give a more focused overview on the adaptation process in information visualization. Based on the general model proposed by Kobsa [11], we will investigate three aspects to describe the process. First, those influencing factors will be introduced that have the most impact on visualization systems. In this context we focus on visualizations used in standard situations by common users and computer systems. The investigation of further influencing factors, like environmental factors would go beyond the focus of this work. Thereafter a short description of how the gathered data can be modeled will be given. Thus this work focuses on visualization adaptation, we will investigate the human interface adaptation in particular for adapting visual parameters.

4.4.1 Influencing Factors

Influencing factors refers to the question *to what* a visualization system should or could be adapted to. In case of information visualization, we have already outlined the major aspects that could be used for influencing the characteristics of visualizations (see Chap. 2). The process of information visualization aims at amplifying human cognition by using interactive visual representation [66]. Based on this assumption, we define three main influencing factors that are best suited to serve as influencing factors: *human* (or users), *data*, and *tasks*.

Human as Influencing Factor
The use of interactive information visualization systems is commonly coupled with users' interactions, e.g. by selecting, zooming, or navigating. Rich proposed in her early work on user modeling that human's information and information about human can be gathered in two ways, by implicitly gathering users' interaction with the system and by explicitly asking users for the required information [67]. In the explicit user data gathering, the users actively provide information about themselves [7]. Commonly demographic data , e.g. age, sex, or profession are captured in this way, whereas asking users in a computer software that should support users could be conceived as intrusive and is in any case time consuming [25]. Rich claimed further that the information provided by users is not per se valid [25]. Human are not always reliable information sources [67] and the self-description may depend on social or personal context [68]. Consequently the main way to get information about users, their behavior and further needed aspects, are the users interactions with a computational systems [11, 16, 25, 41, 69]. This information is the result of the natural interaction with a system to achieve a goal or solve a task. Commonly the users' interactions with the system are observable, at least by the system and provide the possibility to infer unobservable information from the data [41]. The inferencing could lead in particular to information that can be predicted based on users' past behavior [25]. The implicit gathering of users' information has the advantages that users are disturbed in their working process and the information about the users are reliable, thus these are the way they work [30]. Further these data may contain information that users are not able to provide by self-observation [30]. The implicit data gathering of users is the consequence of users' interactions with the system. Based on these data a system could be enabled to model users' behavior, interests and further aspects that may be relevant for visualizations.

Data as Influencing Factor
Adaptive systems commonly use the knowledge about users to filter, rank or adapt data (content) [6, 11, 32], in information visualization data itself can be considered as an influencing factor for the adaptation process. The transformation model of Card et al. [66] illustrates clearly that data with its properties, values, and dimensions plays an essential role for visualizing information. We outlined in Sect. 2.5 Data Foundations that data can be characterized in context of information visualization in various ways.

We identified the data dimension as one main classification factor [70] that influences the way of visualization. Further the *data value* plays an essential role for the choice of various visual variables and visualization types to provide a meaningful way of graphical representations. Although, all visualization systems commonly investigate the data as an essential part of the visual design, *data as influencing factor* for the adaptation process was investigated rarely in the design and creation process. The early and pioneering work of Mackinlay has applied the idea of investigating data characteristics as influencing factors for the design of static two-dimensional presentation of information [24]. With the limitation to two-dimensional data and no-interactivity, his proposed system *A Presentation Tool* (APT) focused on data values and their ability to be ordered [24]. He proposed to transform these values to sentences of graphical languages with a precise syntax and semantic and defined that visual design is codified by *expressiveness* and *effectiveness* [24]. Thereby *expressiveness* identifies the graphical languages with a formal syntax and semantic that express the intended information and *effectiveness* identifies the most effective language to exploit the capabilities of the output and human visual system [24]. According to Mackinlay a set of facts is *expressible* if the sentence encodes all the facts and nothing else as the given facts [24, p. 119].

Task as Influencing Factor
The term task in information visualization is defined very heterogeneously as we worked out in Sect. 2.4. In that sense a task can be the basic selection of a visual item representing a data entity [71] or the entire exploration process that includes a number of users' interactions [72–74]. In context of adaptive visualizations, we define a visual task as users' intention to achieve a certain goal. This goal may be known by the user such as finding certain patterns or information or to learn about a certain topic, refine the knowledge and get a better understanding. Both, a goal-direct or exploratory intention of users may influence the visualization of information. While exploratory approaches can be designed by visual system through an open-information paradigm with no or rare constraints on content and structure, the goal-directed tasks may lead to a kind of navigation-support through changes in visual appearances. However, tasks are just like information about human and information about data, just data that have to be gathered though an adaptive system. Thus, this is a special kind of user information; the acquisition of information about the tasks can be performed similar to the users' information. One way is to explicitly ask users for their intended task [29, 75]. Another way is to analyze the users' interaction and find certain patterns or repeated interaction events [76]. A third way is the combination of implicit and explicit information gathering [77, 78]. In the mixed case the probability distribution determined a task with an uncertainty and the user is explicitly asked, if she is doing this task. The most popular example in that case is the *Microsoft Office* assistant of the *Lumiére Project* [75, 77, 78]. The user is for example explicitly asked if he writes on a letter and want to have letter template, if the system recognizes a pattern that is similar to a letter-writing pattern [77].

4.4.2   Knowledge Modeling

The influencing factors serve as input-data for adaptive systems to behave differently. The acquisition of these data are essential for the adaptation process, but the
way how they are modeled as knowledge makes them useful for a system to infer
information that leads to the adaptation process. Different types of information have
different effects on the visual system and provide together a knowledge-repository
for the systems. We mean with the term knowledge the formalization of the input-
data in any way to infer the adaptation effect. This section gives an overview of the
representation, meaning, and the modeled issues of input-data that effect the adaptation. Thus commonly the user and her abilities are in focus of investigation [16, 79,
80], the focus lie on user, whereas the modeling of tasks is commonly part of user
modeling.

User Modeling
In context of adaptive systems, the user was and is in focus of the adaptation process
[16, 23, 31, 67, 69, 79–82]. Rich proposed that user models can be classified through
three main dimensions [67]. The first dimension is the canonical user versus individual user, a second dimension is a way how information is gathered, explicitly or
implicitly, and the third dimension dresses the temporal aspect of user models, is
there very specific short-term information modeled or more general long-term [67].
Canonical user models contain information about groups of users [83]. They describe
in contrast to individual user models the canonical behavior of all or groups of users
and can be investigated as the common way how users (or user groups) interact
in general with a system. Individual user models investigate model the behavior of
one individual user and represent the behavioral pattern of the observed user [81].
The differentiation between short-term and long-term information about user may
be investigated to model different aspects of users. Short-term information reflects a
situation and more task-oriented representation of the user. This kind of user models
are commonly used to identify the current tasks, goals, and needs, while long-term
observation and modeling leads to information about users' knowledge or preferences, those information that are more characterizing a user [81]. Ross enhanced
this classification with the attributes dynamic and static. A user model may reflect
users' information from the instance of time on which the data were gathered and
not provide the possibility to change or change during the interaction with the system [25]. Thus, users' abilities increases or changes during the working process
with a system, commonly dynamic user models are preferred. Another classification ways brought by Sleeman and applied beside others by Brusilovsky and Millàn
[16, 84]. Sleeman classified user models into the three main classes of *nature, structure*, and *user modeling approaches* [16]. *Nature* refers to user features and investigates the question *what is being modeled?* [16, p. 4]. The nature of user model
contains the users' characteristics and derived information. *Structure* refers to the
way how the acquired information is represented and investigates the different kinds
of maintaining the user model. For describing user models, we will use the procedure
of Brusilovsky and Millàn.

The *nature* of a user model refers according to Brusilovsky and Millàn to the following main user characteristics: *knowledge*, *interests*, *goals and tasks*, *background*, *individual traits*, and *context of work* [16]. Users' knowledge in particular in adaptive learning systems, is the most important feature that is modeled. Knowledge is commonly modeled as a scalar model of users' knowledge in a certain domain [16]. The scalar model consists of a single value on a qualitative or quantitative scale. A quantitative scale may be the range between 0 and 1, or 1 an 10, while the qualitative scale may describe the gathered values as good, average, excellent, or novice, intermediate, and expert [16]. Beside the scalar model, users' knowledge can be modeled as *structural models* [16]. The main purpose of structural models is to scale users' knowledge independently along sub-dimensions of the domain knowledge. Therefore *overlay models* can be used that reflect each knowledge fragment to experts' knowledge, either by a Boolean value or as scalar values. Users' knowledge about a certain domain is commonly dynamic and changes during the work with an adaptive system [16, 31]. *Users' interest* is the most important user feature in information retrieval and filtering systems, commonly applied in recommendation systems [16]. The most popular approach for representing the user interests are still weighted vector of keywords [16]. A vector model or set model is formed by independent, unrelated concepts without any internal structure. Another method for representing users' interest is the *concept model* that is similar to the overlay model of knowledge modeling [16]. These models allow a more accurate modeling of users' interest, thus they enable separately modeling different aspects of user interest based on given concepts. Users' *goals and tasks* are immediate purposes of users and the most changeable features. Goals and tasks may change within one session several times and refer to the current activity and task, which lead to identify goals [16]. Modeling and representation of users' goals and tasks are commonly performed with *goal catalogs*, overlay model with commonly predefined user goals and tasks. During users' interaction and work with the adaptive system the goals are recognized by users' interaction. The goal catalogs commonly include a small set of independent goals or tasks [16]. Further goals can be modeled in hierarchical structures with relatively stable high-level goals and lower stable low-level goals including even short-term goals as sub-set in the hierarchy. User or *task models* recognize the current activity and assign it to the related goal or task to provide adaptation in form of recommending pages, leading users' attention, or adapt content. The recognition of goals and tasks are still rated as a difficult and un-precise task in adaptive systems [16, 85]. *User's background* refers to a set of users' features in particular on previous experience that is not related to the domain of the adaptive system. While knowledge is modeled based on the users' knowledge in the specific domain of the adaptive system, the background of a user investigate the knowledge and experience that is not related to the core topic, but may have impact on the adaptation process [16]. In that context aspects like terminology, e.g. by using domain specific terms or common understandable terms, responsibility, e.g. access to certain information, or language ability, e.g. native or non-native speakers may be indicators of interest for adaptation [16]. Although, the background is similar to users knowledge the adaptive system commonly do not need the information in that granularity, therefore the representation of background is commonly performed by clustering various possible users into several groups (stereotypes) [16, 67, 81]. Users belonging to one stereotype are treated by the adaptation system in same way. A differentiation based on the user features is not performed. Users' background are stable features that commonly do not change during the work. User's *individual traits* are similar to user's background a stable set of user's features that commonly models a user as an individual [16]. These individual features refer for instance to personality traits, cognitive styles, preferences, or learning abilities. The representation of these feature are commonly modeled similar to backgrounds by stereotypes [16]. But there may exist adaptation effects that require a finer differentiation of the individual traits. In that cases a structural overlay model or vector model may be more adequate to represent the individual features. This category also embraces the visual preferences of users, aesthetic preferences, such as the look & feel of user interfaces, and representation preferences, e.g. visual graph-layout or list-based representation. Brusilovsky and Millàn introduced beside the user features, contextual features as *context of work* [16]. This aspect gains in recent adaptive application more and more relevance, thus the human's way of interaction with computer systems changed to more mobile and pervasive computing [9]. In general the context of work may include all environmental and contextual aspects that influence the work with a computer system. In that sense the location, device, interaction device, or situational aspect can be modeled to achieve an adaptation effect. In context of visual representation the screen size is an essential factor. The context is commonly not stable and changes during the work process depending on the observed features, e.g. mobile device in a train. The granularity of needed information is strongly depended to the adaptive application. Therewith the modeling approach can be performed by overlay structural model, stereotypes or vectors. The proposed *nature* of user models were used in various user-adaptive visualization systems [86–88], whereas the aspect of user preferences (as part of individual traits) plays an important role in modeling users' for adaptive visualizations.

In general Brusilovsky and Millàn subdivide the representation of knowledge for modeling users into *vector models*, *taxonomy models*, and *ontology models* [16]. While vector models represent the knowledge as independent concept with values, taxonomies consists of a parent-child relation that enables more accurate modeling [16]. Ontologies are the most powerful way of representation, thus not only parent child relationships can be modeled, but far more the entire structure of knowledge. The core element of all model types is the *knowledge model* and the *overlay model*. While the knowledge or domain model represents the domain knowledge of an adaptive system, the overlay model estimates the user's knowledge level in the given concepts. This level can be qualitative, simple numeric, and uncertainty-based [16]. Qualitative make use of qualitative value, e.g. poor, average, or good, while numeric values or quantitative values provide more precise information by using a continuous range of numbers. Uncertainty based values make use of predictive statistical models, e.g. Bayesian Networks to model the user [16].

Data Modeling
Commonly the modeling of data in adaptive systems refers to modeling the domain knowledge [16, 23, 31, 79, 80]. In the special case of visualization different aspects of the data may play an essential role for the visualization process. In this context the two data variables or features that were already introduced (see Sect. 2.5) play an essential role, the data structure and dimension [70, 89] and the order of data values [66, 90]. The structure and dimension can be theoretically used to determine the best suited visualization based on the data, whereas the order-ability of data values can be an indicator for choosing the right visual variables, e.g. color or size [24]. The representation of the required information can be performed by abstraction and information about the data [91]. Therefore a similar model as the user model can be generated in form of metadata. This data about data may include information about quantity, structure, dimension and order ability of the underlying data to be visualized.

4.4.3 Human Interface Adaptation

The acquired input data and generated representation of those enable the adaptation of certain aspects of a system to reduce the human effort in working with it. This section introduces in particular general adaptation effects that could be applied to information visualization systems. We focus on the work on visual variables and the adaptation of these based on the pioneering works of Bertin [92] and Mackinlay [24].

The fundamental statement to the definition and differentiation of visual variables was proposed by Bertin [92]. He differentiated between visual variables that use the two dimensions of a plane to encode information through graphical marks (*Implantation*) and those, which encode information through their relationship above the plane (*Imposition*).

A graphical mark is defined by basic geometrical elements of points, lines and areas. The position of a mark indicates a meaning between the values of the two dimensions. Marks could be changed through their size, saturation, texture, color, orientation, and shape. These visual variables of marks are classified further by their properties and perception abilities. Therefore Bertin introduced a classification of the visual presentation of marks on the plane, *selective*, *associative*, *ordinal*, and *quantitative* [92]. *Selective* variables can be arranged as groups (family), by isolating their visual relationships. *Associative* variables can be combined through their visual relationships to differentiate them from others. *Ordinal* variables can be ordered through their categories or levels in a generally accepted way. *Quantitative* variables can be ordered through their quantitative value. All introduced visual variables are associative and except shape all of them are selective. Bertin further proposed that only the variables size, saturation, and texture can be ordered and are therewith ordinal, whereas the only quantitative variable is size [92].

The second class of Bertin's visual variables (*Imposition*) encodes information through their relationships to each other above the plane. He differentiates based on how these relationships can be visually illustrated between *diagrams, networks, maps,* and *symbols*. A construction of the plane is a diagram, if the relationship of the plane-dimensions conveys all components of one dimension and all components of the other dimension [92, p. 58]. Networks are defined as a construct above the plane with relationships among all entities of a component. Maps are defined as networks of relationships between all the entities of a component and in relation to the position on earth (geographical position) [92, p. 59]. If the relationship is not between the entities or the components themselves, but the viewer associates a relationship (meaning) to graphical entity, this is not anymore investigated a graphical representation problem. Further he defined the relationship (meaning) between viewer and graphical representation as symbols. The viewer associates with the symbol a meaning as a result of habituality [92, p. 59].

The principle of differentiating between *Imposition* and *Implantation* was applied beside others by Mackinlay to generate automatic two-dimensional non-interactive presentation tools [24]. The foundation of his approach was the assumption that graphical presentation can be expressed as formal languages by semantic and syntax. The precise language definition leads to the expressiveness criterion, which can be seen as the Bertin's *Imposition* [24, 92]. With his introduced approach he formalized conventions that translated the Implantation theory to a practical and useable paradigm. Therefore he introduce the formal definition of a graphical sentence s as $s \subset \{< o, l > | o \in O \wedge l \in L\}$, where O is a set of graphical objects and L is a set of locations [24, p. 119]. The expressiveness of a graphical sentence is thereby equivalent to Bertin's Imposition theory and follows the paradigm of encoding information through their relationship to each other and on the plane (screen), which refers to the location. With this definition the expressiveness is defined by a well-formed graphical sentence consisting of a tuple o and L (x and y coordinates of a two-dimensional space) that have a finite non-zero height and width indicating the precise location of an object [24]. Based on the convention of position and object, syntax of the horizontal position was introduced as unary predicate (*HorzPos*) that consisted of a horizontal axis and a set of tuples placing objects at a constant height. The defined *HorzPos* language encoded for instance binary relations with an axis, a set of marks, and their position [24]. Mackinlay introduced a set of *Encodes* as triples $(s, facts, lang)$ to describe the semantic relations between objects and properties of a graphical sentence s, where $facts$ encodes the given semantic conventions of the language $lang$ [24].

Beside expressiveness that just depends on the formal syntax and semantics of a graphical language and refers to Bertin's Imposition and thereby to the placement of objects with their relations to each other and to the plane, Mackinlay addressed the *effectiveness* of graphical presentation by using the retinal variables [66, 92] as a component for presentation [24]. The effectiveness does not just depend on the syntax and semantic of the graphical language, far more it involves the capabilities of the perceiver (human) to present certain information accurately [24]. Therefore psychophysical results were used based on the observation of Cleveland and McGill

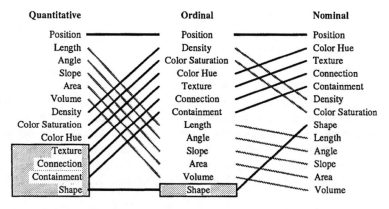

Fig. 4.3 Ranking of visual variables based on data value characteristics (from [24, p. 125], with kind permission of J. Mackinlay)

[93] that users solve visual tasks based on the interpretation with different degrees of accuracy of graphical presentations. The observed retinal variables of Cleveland and McGill [93] and Bertin [92] were enhanced in particular with the differentiation of color's attributes *hue* and *saturation* [24]. Mackinlay proposed based on psychophysical [93] and graphical vocabulary [92] works a ranking of retinal or visual variables for perceptual tasks that includes the data value characteristics of *quantitative*, *ordinal*, and *nominal* [24]. Figure 4.3 illustrates the proposed ranking. Mackinlay further extended the ranking by proposing the generation of lexicographic ordering with the *Principle of Importance* that "encode more important information more effectively" [24, p. 126]. Thereby the input of the expressiveness was a tuple of relations indicating the relative importance of the relations.

Mackinlay further introduced a *Composition Algebra* to provide alternatives to the generated design [24]. He proposed with his *Principle of Composition* to "compose two designs by merging parts that encode the same information" [24, p. 130]. This basis set of graphical languages derived from Bertin were classified into the six encoding techniques of *Single-position languages*, *Apposed-position language*, *Retinal-list language*, *Map languages*, *Connection languages*, and *Miscellaneous languages*. Single-position encodes information of a mark set *m* by the horizontal $h(m)$ or vertical position $v(h)$, whereas apposed-position encodes information positioned in two axes $vh(m)$. Retinal-list encodes information just with one of the visual variables *m* independent from the positioning and can therewith be moved for composing with other another encoding [24]. Map language encodes information on predefined position on a specific map $vh(m)$ in contrast to apposed-position the position on the map is predefined and provides commonly a meaning. Connection language encode information through the connection of two sets of visual variables $m_n(m_1)$, while Miscellaneous languages encode information through a variety of other graphical techniques $vh(m)$ [24]. Except Connection languages the entire composition propose sentences for visual presentation as conjunction of visual attributes and the position on the screen $m \cup v \cup h$ and lead to three operators of single-axis composition, double-axes composition and mark composition, which are formally well-defined [24, pp. 131].

The adaptation process in information visualization incorporates more adaptable variables that can support users in their information acquisition process as conventional list-based information systems. With the differentiation of Bertin [92] and the formalization of the grammar in particular for adaptive visual systems Mackinlay [24], we illustrated the foundation of what can be adapted. Although, the introduced model of Mackinlay just investigated the static presentation of information without involving the user as influencing factor in his adaptation rules, the differentiation between Imposition and the related rules and Implantation is a major foundation of this work.

4.5   Adaptive Visualizations

The acquisition, formal modeling, and representation, and computing uncertainty of various influencing factors plays an essential role in the adaptation process. The dominant influencing factor in context of adaptation was and is still the user with his various characteristics. While semantic and information visualizations investigate the user as a central element of design, adaptive visualizations goes one step beyond and provide a changing visual environment to fulfill users' demands. We already illustrated in this chapter that the main question regarding such an adaptation process is "what" can be adapted , "to what" can be adapted, "where" can adaptivity be applied, and "why" should be adapted [17]. In context of information visualization two main questions are of great interest, namely which adaptable variable are provided by visual systems and what are the influencing factors that changes those identified adaptation criteria.

This section aims to give a state of the art and technology in adaptive visualizations. To obtain a clear picture of existing systems and approaches, we will first define the term adaptive visualization. Thereafter we will introduce a classification for providing a comprehensible picture of the existing systems. The classification will be used to introduce the existing approaches and systems for adaptive visualizations.

4.5.1   Definition of Adaptive Visualizations

Adaptive visualization is a relatively young research area [94], whereas a disambiguated understanding of this field is not established yet. This section introduces the definition of adaptive visualizations aiming to have at least a common and disambiguated understanding of the term in this work. We outlined based on the work of Kobsa [11] that the adaptation process involves data acquisition for influencing certain adaptable parameters, the formal representation of these data, and the autonomous adaptation of content, structure, and presentation. Based on these criteria and the definition of Jameson [1] we introduced a definition for adaptation in computational systems in particular investigating the user interface adaptation (see Sect. 4.1). For investigating the information visualization variables the differentiation of Bertin [92] was introduced to illustrate in particular the difference between Imposition and Implantation [92]. This differentiation was used by Mackinlay to create a graphical language that addresses *expressiveness* by making use of Imposition and *effectiveness* by making use of Implantation and of the ability to order certain retinal or visual variables [24]. Further some common definition of information visualization and Visual Analytics were introduced that outline the main goal of information visualization, namely amplifying cognition for acquiring knowledge [66, 95]. An adaptive visualization should address all the mentioned issues, from the data-acquisition, over formal representation of the data to the autonomous adaptation of content, structure, and presentation, whereas the content, structure, presentation should investigate the *nature* of visualizations that are composed by different visual variables. Thereby the goal should still remain to amplify cognition for knowledge and information acquisition. Based on these premises we define adaptive visualization as follow:

> Adaptive visualizations are interactive systems that adapt autonomously the visual variables, visual structure, visualization method, or the composition of them by involving some form of learning, inference, or decision making based on one or many influencing factors like users' behavior or data characteristics to amplify cognition and enable a more efficient information acquisition.

4.5.2 Classification of Adaptive Visualizations

As information visualization gained during the last decade various classifications based on data, tasks, interaction techniques, or visualization techniques (see Chap. 2), the area of adaptive visualizations did not experienced that systematic classifications. In general there are two main aspects that could be used as classification criteria, the influencing factors that lead to the adaptation and the visual changes that are the result of the data processing. This thesis focuses on the visual appearance. We investigate the visual changes as classification criteria.

Ahn proposed based on the work of Bunt et al. [6] a classification for adaptive visualization with the four elementary categories of *Visualization Method Adaptation*, *Visual Structure Adaptation*, *Adaptive Annotations*, and *User Model Adaptation* [21]. The main differentiation was built upon two questions, "what to adapt" and "how to adapt" [21, p. 13]. The *Visualization Adaptation Method* refers to choosing and replacing the visualization technique, e.g. plot chart, pie chart or bar chart [21, p. 15]. The *Visual Structure Adaptation* refers according to Ahn to two main methods, first change of layout within a visualization and second providing methods for easily exploring visualizations [21, p. 16]. The change in layout is performed according the varying users' context, e.g. viewpoints, whereas the second method makes use of users' interactions to provide a kind of navigation support, e.g. by using different layouts for past and future (predicted) user interactions [21, p. 16]. *Adaptive Annotations* changes visual elements, e.g. color or icon to give more focus on certain data [21]. The last category of *User Model Visualization* does not refer to the adaptation process, it provides far more an insight into the derived user model, sometime with the ability to change and reduce noise [21]. Ahn further proposed that the structure of the first three categories can be seen as a stack [21]. Therewith the first class builds the foundation of the other two and the annotation is inherited from the structure, whereas the fourth category of *User Model Visualization* is completely independent [21].

The classification of Ahn was adopted from Bunt et al. [6], who used the already introduced (Sect. 4.1) classification of Kobsa [11] for categorizing adaptive systems in *content adaptation)* and *content presentation* [6]. Content adaptation refers to techniques that identify relevant content for user or context and process it based on the identified relevance [6, p. 410] by using for example *p*age and fragment variants [11], *c*ontent selection [96], or *c*ontent structuring [6, 97]. Content adaptation adapts by choosing or reducing content based on the derived relevance and addresses more the question "what is adapted" [6, 21]. In contrast to that, content presentation addresses the question of "How is adapted?". Bunt et al. introduced in this context methods for changing the view on certain content, whereas the content presentation premises that the relevance of content is already derived [6]. Methods for changing the view on contain could be for example *p*riority on focus [10] or *p*riority on context [98], in which the view on the relevant content is changed by color highlighting or size enlargement of content parts [6]. Another method for changing the presentation is *media adaptation*, which changes the way how content is presented by using different media channels, e.g. text, speech (audio), video, or graphical illustration [6, p. 422].

The introduced classifications provide a good foundation for classifying adaptive visualizations to provide a comprehensible view on the existing techniques. While the classification of Bunt et al. [6] contributed with two elementary questions of "what is adapted?" and "how is adapted?", their focus of investigation were not visualizations. Ahn applied and enhanced this classification and proposed a categorization into *Visualization Method Adaptation*, *Visual Structure Adaptation*, *Adaptive Annotations*, which fits better to the nature of visualizations. However, there are more variables provided by visualizations in general. First the differentiation of visual interfaces into Implantation and Imposition by Bertin [92] and the related differentiation of effectiveness and expressiveness by Mackinlay [24] are not clearly addressed. Another fact is that visual interface are more and more making use of juxtaposed visualization methods to use the entire screen and provide different perspectives, e.g. with brushing & linking techniques. Further the classification of Ahn just addresses the adaptation of visualization, but the content to be visualized and the related interaction techniques that reduce the visualized parts of data [70] are not considered. For classifying adaptive visualizations, we use both classifications and include the mentioned aspects for a more comprehensible view on visual adaptation from the information visualization perspective. Therewith we propose a differentiation of adaptive visualizations in *Visual Interface Adaptation*, *Visual Content Adaptation*, *Visual Layout*

Adaptation, and *Visual Variable Adaptation*. *Visual Interface Adaptation* can be seen as the adaptation of the entire user interface by placing or replacing different juxtaposed interactive visualizations. *Visualization Content Adaptation* addresses the changing amount of information entities or content that is visualized. *Visualization Layout Adaptation* addresses the changes in visualization layout or type. According to the works of Bertin [92], Mackinlay [24], and Card et al. [66] the layout addresses the placement of objects on the screen and their relation to each other, therewith a differentiation between type and layout is not appropriate. *Visual Variables Adaptation* refers to changes of retinal or visual variables, e.g. color, size, orientation, shape etc. to guide the attention to certain information or patterns.

4.5.3 Survey of Adaptive Visualization Techniques and Methods

This section will illustrate the state of art and technology for adaptive visualizations. For a clearer picture of the topic we will introduce the existing systems and approaches based on the above introduced classification. Systems that enable more adaptation abilities will be highlighted as those and assigned to their dominant adaptation characteristics. The survey will introduce in particular those systems that match with our introduced definition of adaptive visualizations. Systems that claim to be adaptive but understand under adaptivity aspects like parallel processing will not be focus of our investigation.

4.5.3.1 Visual Interface Adaptation

Information visualization systems may make use of the composition of different visual interfaces filling the entire screen space. The juxtaposed placement of visual interface are commonly designed to provide different perspectives on data, e.g. by different levels of detail or different aspects of the data. The *Visual Interface Adaptation* can thereby be seen as the adaptation of a user interface that primary makes use of visualizations. Here the general changes on the visual user interface, e.g. by replacing one of the given visualization layouts or changing the interaction modalities with the visualization, are the main focus of adaptation.

Wiza et al. proposed with *Persicope* a three-dimensional visual environment for illustrating search-results with an overview of the entire results and the contained relationships in an adaptive way [99]. *Periscope* uses the previously proposed "Adaptive Visualization Environments" (AVE) [100] as basic infrastructure for the adaptation process and provides beside an automatic selection of visualization types, the ability to choose visualization manually and highlight relevant attributes. The user interface consists of two main components, a 3D virtual scene for the visual representations and a (2D) area for selecting visualizations and refining the search process. The

composition of these two areas is the main component of the visualization environment, the *Interface Model* [99]. The user interface of *Periscope* is generated dynamically by a visualization engine based on the *Interface Models*, which is composed
by the fixed (on UI) two interface components. The system provides different types
of interface models, which are chosen based on certain *readability paradigms*. Wiza
et al. proposed that an interface model is the abstraction of interfaces and provide the
ability to create such interfaces dynamically and highlight different aspects in the
visualizations [101]. They introduced three types of interface model, *holistic interfaces*, *analytical interfaces*, and *hybrid interfaces* to support the exploration process
[99]. *Holistic interfaces* are used for voluminous results to prevent an overloading
detailed view. The results in this interface model are categorized based on one or
several criteria and visualized in groups of documents. *Analytical interfaces* provide
a detailed view on the results to enable solving analytical tasks, whereas the amount
of the result needs to be limited for representing each result-item with a graphical
metaphor representing documents' attributes by object properties [99]. *Hybrid interfaces* combines the visual potentials of analytical and holistic interfaces and may
contain categories and object properties (representing document attributes). Further
specialized interfaces are provided to support the focus on certain attributes, which
reduces the amount of data but provide a goal-directed interface [99].

The choice of the interface models is performed with the *interface readability
paradigms* [102], which describes prerequisites that have to be fulfilled to be considered as readable interfaces [99]. To rank the visualizations (interfaces) each interface
is formally described by *facets*, a set of interface properties. Thereby an interface
facet the conditions for a single dimension, e.g. color with (δ_i, c_i), where δ_i is a
visualization dimension and c_i is the capacity of the dimension and $i \in [1, n]$ is the
number of facets [99, p. 33]. Further the visualization dimensions are distinguished
in *classifying* and *presentation visualization dimensions* [99]. An *interface determi-
nant* formalizes as a pair the maximum number of objects that meets the readability
prerequisites. After a performed search, the amount of the results is compared to
the interface determinants of the interface models, if no visualization (or interface)
is able to visualize the amount, a holistic interface is chosen. In the next step the
interface model is selected based on the amount of the results and the facets [99].
Therefore the interface models are investigated, which provide for the given amount
of attributes possible best fitting attributes to meet the readability prerequisites. If the
amount of document attributes is higher than the visual dimensions, the attributes
are clustered to provide a more efficient readability [99]. Otherwise, if the attributes
can be resolved by more than one interface model, an interface model is selected
randomly or the user has to choose. The user is further able to force the use of a
certain interface model by deselecting the dynamic route of the system [99]. The
user has the ability to choose certain interface models, in particular for hybrid and
specialized interfaces the choice is performed completely by the user [99]. In these
conditions the user is able to map certain attributes to provided dimensions [99].

Periscope is a 3D-visualization environment that adapts the layout to the amount of data entities and the amount their attributes. Although the authors proposed that the interface is changing, the main change is performed by choosing a visualization layout from a set of given visualizations based on predefined rules or paradigms. Nonetheless we categorize their system as visual interface adaptation, thus this was the main goal of the work. It should be mentioned that the skeleton of the UI does not change. It always provides the mentioned two areas with different interaction abilities or visual interfaces. *Periscope* further does not make use of any enhanced adaptations, e.g. by adapting the visual variables or content. Further it just consider the amount of certain data entities and attributes, the data structure, data dimensions and all aspects regarding users are not investigated.

4.5.3.2   Visual Content Adaptation

Information visualization aims at providing a visual interactive environment for information (content) to amplify cognition [66], enable knowledge acquisition [95], or information provision and exploration [72, 103] for certain information related task like decision making [103]. *Visual Content Adaptation* addresses in particular the choice, reduction, or expansion of the data that are visualized. Thereby the human interface and the way of information presentation are performed mainly by graphical representations of the data, whereas the adaptation focuses on the content.

A prominent example for an adaptive visualization in particular to adapt content was brought by Ahn and Brusilovsky with their *Adaptive VIBE* [20, 21, 94, 104, 105]. *Adaptive VIBE* enhances the systems *TaskSieve* [20], an information-retrieval system for "intelligence-analysis" and *VIBE*, a *Visual Information Browsing environment* [106] with adaptation characteristics to enable a more sufficient search, based on retrieved user models [94]. The main idea is to provide two different types of points-of-interest (POI), one type represents the queried terms, whereas each term separated by a blank-space is considered as POI and the second type of POIs represent the task and user model of the user. For generating the user and task model, Ahn and Brusilovsky integrated two methods of gathering information about the user. They integrated the *sense-making* intelligence analysis method proposed by Pirolli and Card [107]: If a user selects a document, the entire document is presented to the user. The user is now asked and enabled to annotate text within the *Task Model Notes* environment [21]. The annotated text is the baseline for generating the user and task model respectively by extracting *named entities* with the assumption that relevant words are annotated more often than less relevant terms [21]. The named-entity extraction is performed with a statistical maximum-entropy model [108] that recognizes the top 300 named entities based on 32 types of nominal and pronominal named entities and 13 types of events [21, p. 52]. The main view of *Adaptive VIBE* is visual interface that places the two types of POIs as yellow and blue labeled dots (or circles) on the screen [94]. Yellow circles represents the queried term, whereas the blue circles are the extracted named-entities and represent the user model based on the current search [94]. The retrieved documents are visualized as squares on the visual interface, whereas the position of the squares indicates the similarity and relevance to each POI, either to the query or to the user model POI. To provide a more comprehensible spatial distinction of the documents, the POIs and perceive the distances in a better way three layouts are implemented. Beside the existing circular visualization in *VIBE* [106], where all POIs regardless, if they are generated by queries or by as a result of the user model, are placed in a circle, AHN and Brusilovsky enhanced the *Adaptive VIBE* with a hemisphere layout and a parallel layout [94]. The two developed layouts enabled a visual differentiation between the POI types and therewith an easier comprehension of the document similarity [21, 94]. The hemisphere layout places the POIs in two separated semicircles and the parallel layout places the different types of POIs in vertical lines [21, 94].

The user interface of *Adaptive VIBE* is static and provides a search field for entering search terms, the described visual interface with a manually changeable layout (circle, hemisphere, and parallel), that can be changed by the user, a view on the selected document with highlighted named-entities, a *Task Model Notes* interface on the right, where the user can annotate relevant data as described and a task or user model presented as a tag-cloud, where the relevancy of term is indicated by the presented size [94]. The visual interface of *Adaptive VIBE* was enhanced with various interaction features. It is for instance possible to move a certain POI and see the related changes in the layout. With this interaction users can arrange the POIs to comprehend the similarities. Users are further enabled to point a POI with the mouse. With the related mouse-over effect all the documents get the *Similarity Overlay Disc* that illustrates the similarity to the particular POI by an overlay disc consisting of color and a kind of blur-effect (disc). Furthermore the users are able to "dock" the POIs. With this interaction the POIS has no effect anymore on the placement of the documents [94].

Adaptive VIBE is an adaptive visualization system based for supporting users in the search process. The system supports users with a visual placement of documents based on the similarity of two different types of POIs. The task or user model in this context is generated by the user explicitly. Users have to annotate documents (by marking) actively. Based on this annotation such "user-POIs" are generated that enable a positioning of the documents with a certain distance to the POIs. The main aspect and outcome is to see the distance of documents to certain named-entities, which refers to the content of the document. Although, the system makes use of positioning the graphical elements that represent documents, we categorize it as content adaptation thus the layout itself is not affected. Only the documents relevance is indicated by the system. In other words *Adaptive VIBE* provides a kind of recommendation system that provides the results (content) visually. It does not support any kind of visual adaptation, e.g. by analyzing the users' past behavior. The mentioned changes on the visual variables based on the selection (by mouse over) are a traditional brushing and linking metaphor. Further each task requires a new annotation model and the user has to annotate again some documents for getting the POIs. Thus the positioning of the documents is only measured by the documents relevance, overlapping of the documents can appear and irritate the user. In general

Adaptive VIBE does not provide an adaptive behavior while using the system. The appeared changes are based on the results and the selected POIs.

Brusilovsky et al. introduced with their *ADaptive VISualization for Education* (ADVISE) an adaptive visualization for personalized access to educational documents as a suite including various versions: *ADVISE, ADVISE 2D, ADVISE 3D*, and *ADVISE VIBE* [109]. The main idea of ADVISE is to present a collection of educational documents based on user previous interactions and document investigation in spatial similarity-based visualizations. *ADVISE* is the visual enhancement of the previously proposed *NavEx* [110] to provide not only a navigation guide for learning material, but also face the problem of ordered lists, which do not offer help in selecting similar or dissimilar educational resources [109]. ADVISE combines a spatial similarity-based visualization (2d: ADVISE 2D or 3D ADVISE: 3D) with *adaptive annotation* to guide students in selecting the right documents for their learning goals [109]. The baseline for their visualization adaptation is the similarity-measurement that is performed with the cosine-algorithm based on document similarity. Two types of documents are investigated, if the document is a full-text, the terms are extracted and the weight or frequency is measured by TF or TF-IDF [109] and if the document contains program-code (thus the learning material is for learning programming languages), the language constructs are extracted. The extracted and weighted terms or language constructs are modeled in vectors in order to apply the Cosine-algorithm for similarity measurement. The visualization is performed with a simple spring model (Force-Directed Placement), where the measured similarity is used for the indicator of document distances. If the documents, or precisely the weighted terms and constructs in the document vectors, are similar they are placed next to each other and if they are dissimilar they are place with a higher distance [109]. The *adaptive annotations* are indicators for the user or learner, which document is recommended. Therefore two types of annotations are applied, a progress-based annotation illustrates the progress of users' interactions and investigations of documents. The progress-based annotation (green circles filling by percentage) is computed as the percentage of the explored documents in comparison to the entire corpus of one educational line. Further the adaptive annotations are illustrated as indicators for prerequisite "readiness" [109, p. 150] based on concept-level mode and indicate that the student is not ready for that course or document (annotated as a red X) [109].

ADVISE is a visual enhancement of the adaptive *NavEx* [110], *ADVISE 2D* and *ADVISE 3D* are more generic visualizations that can be applied to various document collections, and *ADVISE VIBE* is the visual enhancement of the POI-based information retrieval system *VIBE* [106]. The main purpose of *ADVISE* is to visualize similar documents adaptively and provide with the document annotations a guidance for the learner. The system is categorized as content adaptation, thus the entire visual appearance considers the content in terms of their similarity. The visualization does not make use of any visual variable, the visual interface is static and the layout is fixed to a precomputed document similarity. The approach is interesting, thus it uses a similarity algorithm for the adaptation, but users' similarities are not investigated.

Shi et al. introduced with *HiMap* an adaptive visualization approach for visualizing masses of social network connection in a hierarchical and comprehensible manner

[111]. They introduce their adaptive visualization approach based on a pipeline with three main steps, *Offline Data Manipulation*, *Adaptive Data Loading/Summarization*, and *Clustered Graph Visualization*. They make use of a single force-directed algorithm [112] for visualizing the nodes and links of the social networks. In the first step of *Offline Data Manipulation* the individuals of the social networked are clustered in a hierarchical structure by using the *Fast algorithm for detecting community structures in networks* [113] to detect and cluster the strongest relationships within the graph structure [111]. The second step of *Adaptive Data Loading/Summarization* consists of two sub-steps. In the first sub-step a ranking algorithm is applied to order the nodes based on two weight-degrees, first based on the overall amount of the outgoing links from the cluster and second based on the amount of connections between the clusters. The adaptive data loading loads the related data based on these two criteria until a quantification criteria of the screen-size is indicating that more nodes are appropriate to be placed [111]. The visualization of the clustered graph (third step) is performed with an enhanced Kamada-Kawai force-directed algorithm [112] (Stable Kamada-Kawai (SKK-C)), which enhances and generalizes the algorithm to work recursively for the clustered graph and includes an adaptive stabilization term [111]. The enhanced graph-layout algorithm limits the visualized levels of cluster-hierarchy to not overload the user with information. To compensate this limitation two enhanced zooming abilities are included beside a geometric zoom, *hierarchical zoom* and *semantic zoom*. *Hierarchical zoom* enables the zooming through the different hierarchies by zooming in and out of a cluster-hierarchy. The *semantic zoom* provides more detailed information by loading more data to fill the screen-place, instead of simply magnifying the graphical representations [111].

HiMap provides an interesting approach for providing a comprehensible view on complex network data. The main adaptive feature of the system is the adaptive data loading functionality that incorporates various aspects to reduce or expand the amount of the visualized entities. However, *HiMap* just provides one visual layout algorithm and does not make use of the composing and adapting the user interface, visual variables, or the layout. The reduction and expansion of loaded data is the only adaptive functionality. Further the user and the relevance based on any user characteristics is not considered at all.

4.5.3.3   Visual Layout Adaptation

The positioning of graphical marks in correlation to the screen or to other marks is assigned as a visual Layout (Imposition) [92] and addresses commonly the expressiveness of visualizations [24]. The adaptation of Layout changes both the visualization technique that is commonly a composition of the placement of graphical objects on the screen and any correlations of the marks to each other, e.g. bar charts, hierarchical graph visualization, or multidimensional point-based visualizations, and the changes in the positions of the marks that is addressed by Ahn (and Brusilovsky) [20, 21, 94, 104, 105] as structure adaptation, e.g. the change of graph visualization in placing the objects in another order (hierarchical vs. concentric).

Gotz and Zhou observed and evaluated the interaction behavior of users to find out, if there exist any repeated interactions that can be used as an input for adaptation in visual analysis environments [114]. Their evaluation incorporated different commercial Visual Analytics tools and a self-developed visual environment. The result of their observation and study was applied to two features (*Pattern* and *Trail*) and five recommendations. They proposed that 96 % of their participants reflected perceived or real limitation that leads to the widespread of behavioral structures as repeated interactions [114]. In their work they identified two prevalent patterns: the scan and the flip pattern. Their second recommendation based on patterns proposes to recognize the performed users' patterns and provide proactively assistance [114]. *Trails* as logical paths of discovering insights were evidenced in their study by 100 % of the participants, their first recommendation regarding these trails was to record and preserve the trails. Further they recommended to re-use and by "capturing and exposing previous trails and their associated insights" [114, p. 9]. The last recommendation on trails was bookmarking capability of visualization and note-taking within visual environments. In second step they enhanced the outcomes of their study to model the user analytic behavior in a multi-level granularity of *Tasks*, *Sub-Tasks*, *Actions*, and *Events* [115]. Their model assigned *tasks* as the highest level and stands for the analytic goal of the user. *Sub-tasks* refer to more concrete sub-steps of the overall goal [115]. *Actions* are more generic and represent an atomic analytic step, e.g., query, split, or filter. *Actions* can be represented as a triple of $Action = < Type, Intent, Paramters >$, whereas $Type$ is a unique (semantic) signature for the action, $Intent$ is the users' intention, and $Parameters$ define the functional scope [115, p. 126].

Gotz and Wehn applied the outcomes of their study in an adaptive information visualization tool [116]. They followed their own recommendation to capture the interaction patterns and provide, if sufficient an alternative visualization layout. Their system recognizes the repeated interactions of users as four main patterns, *Scan Pattern*, *Flip Pattern*, *Swap-Pattern*, and *Drill-Down Pattern* [116]. The order of these interaction and the related patterns are stored in an action model. Therewith the defined behavioral patterns are repeated actions as defined in the multi-level model that was previously proposed [115, 116]. The *Scan Pattern* is the repeated action of *Inspect*, where the user interacts with object to get further information [116]. If the *Inspect* pattern is occurred repeatedly, it can be assumed that a comparative view on the data would be more appropriate. The *Flip Pattern* is recognized by repeated interactions with the *filtering* functions of the visualization. This behavioral pattern leads to the same assumption, namely that a comparative view on data would be more appropriate. The *Swap-Pattern* occurs, if the user changes repeatedly the position of objects on the screen and leads to the assumption that the user wants to compare the correlation of attributes. The *Drill-Down Pattern* occurs, if the user repeatedly filters down along orthogonal dimensions and leads to the assumption that the user wants to focus on a targeted subset of data. The user interface of their visualization is composed by a windows where the user is able to select a data set and filter some data dimensions [116]. The main window is dominated by a single visualization layout, with an action-based history at the bottom and a visualization recommendation at the right side.

Gotz et al. enhanced their adaptive visualization system in particular to meet the requirements of Visual Analytics for insight provenance [8]. Their *Harvest* visualization was designed to support non-experts in complex and exploratory visual analytic processes. Based on the previous work [114–116] *Harvest* contained three main components, *Smart Visual Analytics Widgets*, *Semantics-based capture of insight provenance*, and *Dynamic visualization recommendation* [8]. The *smart visual analytics widgets* provide a set of reusable visualizations that support semantic based user interaction and handle dynamic data. The *Semantics-based capture of insight provenance* is the implementation of the proposed layer model [115] of visual analytic tasks and identifies user analytic trails and behavioral patterns based on sequential user interactions in combination with predefined library of interaction sequences that include predefined rules for the patterns [8]. Based on the patterns a visualization recommendation is performed. The *dynamic visualization recommendation* recommends based on the identified pattern an alternative visualization that fits better to the identified pattern. *Harvest* provides eight different visualization widgets, whereas the recommendation chooses five and the most appropriate one is listed at the top of the recommendation toolbar.

Gotz et al. introduced an interesting and easy to implement approach for supporting users in complex and exploratory visual analytical tasks. The core of their technology can be identified as the pattern library and the related rules that lead to change or recommend alternative visualizations. The main shortcoming of their approach is that new visualizations are patterns have to be observed, evaluated, and trained by experts with predefined rules. There exist no learning or inferencing from past users behavior, beside the already identified simple rules. The heterogeneity of users is not considered in their approach at all. The entire approach just focuses on repeated patterns that lead to changes in layout as visualization widgets. Further a combination of two or more visualization is not provided to enable different perspective on the data. Another main shortcoming from our point of view is that the authors neither made use of changeable visual interface (always a single visualization) nor of visual variables. Although they propose that the visualization widgets are dynamic, the visual presentation is always the same. *Harvest* is an adaptive visualization that makes use of predefined patterns to change the layout and is limited to this functionality.

Mackinlay's work on the graphical language for providing effective and expressive composed visual presentation was applied several years later to the commercial visualization system *Tableau* [117]. The *Tableau* enhancement for visual analysis *Show Me* provides a set of different visualization layouts and enables the design of visual representation of data with the design principles proposed in [24] and [92] in particular for those data aspects that are intended to be presented [117]. One main goal is to enable the design of visualizations or small multiple displays for enhancing the *user experience* in the visual analysis flow for non-visualization experts, whereas the aspect of user experience was not investigated in the previous attempt of *APT* [24].

Therefore the *VizQL* [118] specification language is used that describes the structure of visualizations and the applied data-base queries and is based on the proposed *APT* Algebra [24] for generating automatic presentations [117]. The automatic presentation and enhancements in *Show Me* contains four main components *Automatic Marks, Add to Sheet, Show Me*, and *Show Me Alternatives* that categorize data fields based on integrated rules and heuristics for providing visual presentations considering user experience issues. Therefore the data are categorized in *data type*, containing text, data, date & time, and numeric and Boolean values, *data role*, consisting of data dimensions and measure, e.g. quantitative data fields, and *data interpretation* that distinguishes between discrete and continuous data. The categorizations are used to generate the mentioned *small multiple displays* for multi-dimensional data, whereas the widgets are connected visualizations of the same type and provide therewith contextualized visualizations for the user. The *small multiple displays* are generated, if various data fields are used for defining the column and row axes [117]. Further the used data fields are highlighted by different colors for differentiating between dimensions and measures of data fields. *Automatic Marks* decide based on predefined rules and the categorization of data fields an appropriate layout for the given data, e.g. for temporal quantitative data a line-chart is automatically chosen. With the *add to sheet* command the user is enabled to place a field to a view [117]. An integrated *affinity* heuristic supports the generation of effective visual displays by adding related fields to the visualization, e.g. the heuristic determines for instance in a bar view if the new added field should be assigned as a column or a row [117, p. 1139]. This heuristic can be applied to hierarchical data too [117]. Further the definition of some visual variables is a tasks of the *add to sheet* functionality [117]. Therefore an integrated set of various colors and shapes are available that are assigned to the data based on the number of categories (up to ten shape and up to twenty color). The *show me* functionality is a core component of the visualizations system. It decides based on the data dimensions and values the most appropriate layout based on predefined rules that considers the categorization of the data fields and the number of each category. Therefore the visualizations in *Tableau* contain a predefined ranking for the various data fields and their attributes. *Show me alternatives* enables the choice of an alternating visualization. This functionality does not rank the visualization and provides all possible alternatives.

The *Show Me* functionality of *Tableau* is a very interesting and nice environment for designing effective visual interfaces based on the graphical language proposed by Mackinlay [24]. It provides various functions in particular to design the underlying data in more effective and expressive way and make use of studies of user experience. The system itself cannot be categorized as adaptive or intelligent, although it makes use of some useful features. There are not recommendations, e.g. for further data dimension, based on users' choice or behavior. Further the visualization system does not change automatically. The interactions and commands of users are always required. The adaptation of visual variables is limited to two different variables (color and shape), whereas only the amount of attributes (shape ≤ 10 and color ≤ 20) is an indicator for an initial choice. In general the system is a non-adaptive visualization design environment that makes use of data characteristics recommending layouts, whereas the layout is not changing without users' explicit choice or action (show me alternatives).

Golemati et al. proposed a context-based adaptive visualization system for digital libraries and historical archives [88] and enhanced their approach later [119]. Their proposed visualization environment were based on a visualization library with a set of different visual layouts and selected the most appropriate one based on combined contextual information [119]. Their environment consisted of two main components, *Context Modeling* and *Visualization Method Selection*. *Context Modeling* included *user context*, which was enhanced later to *implicit* and *explicit user context* [119], *system context*, and *document collection context* [88, 119]. The *user context* is the history of a user building the individuality and strongly influencing the choice of a visualization [88], whereas *explicit user context* is an initial user profile, e.g. with demographic data, educational level, cognitive abilities, and experience in computer-use [119]. The *explicit context* of user is acquired through interviews and explicit questions. *Implicit user context* builds additional information extracted while using the visual environment, e.g. likes, dislikes. This information is used to enhance the user context that is built based on interviews. *System context* contains information about the available software and hardware and the capabilities of those [88, 119]. *Document collection context* refers to the information relative to the portion to be visualized, e.g. document formats like text or scanned images, or metadata.

Visualization methods and their selection is the second main component of their approach and consists of two sub-components, *Visualization Method Properties* and *Visualization Method Selection* [88, 119]. Golemati et al. used for their *visualization method properties* the classification proposed by Katifori et al. [120] (see Sect. 3.5.2) and divided each of those categories in 2D and 3D visualizations according to Katifori et al. Further they investigate a number of basic visualization features according to data properties, e.g. number of dimension (2, $2\frac{1}{2}$, and $3D$), metaphor, interactive browsing in document types, e.g. article, book, or graphs, user-defined grouping, e.g. articles or books, color coding, e.g. file system navigator, and term frequency, e.g. tile bars [119, p. 194]. The *visualization method adaptation* is performed through a rule-based engine with rules defined as triples in form of *(context-property, vis-method, score)* [119, p. 194]. The *context-property* is either a property of user, system or collection. The *vis-method* is a visualization system property and the *score* is a range between $[-10, 10]$ and indicates how appropriate a visualization method is for the expressed context [119]. The rule-based approach contains rules, such as *(sysctx-display-3D, vismeth-noDimensions-3, 6)* [119, p. 194] to define that visualization methods employing three dimension are appropriate for systems with 3D displays. Beside the visualization method selection their approach provides *adaptive features in method selection* that investigates in particular users' specific preferences regarding whether a user has considered a visualization as suitable or not in a specific context and if the user like or dislikes a visualization method [119, p. 195]. The user specific information is collected when a visualization is closed by explicitly asking the user for his preferences [119]. If a user dislike a visualization method, e.g. for a certain data-set and he or she is selecting "dislike" a dynamic user profile is augmented with *dislike, viz-method* [119, p. 196]. The dynamic user profile is considered in the visualization selection method by a set of rules that increments, decrements, and measures similarity for the incrementation and decrementation the score of the visualization method. Another functionality of the visualization method selection enables the choice visualization by user in descending order of their score.

Golemati et al. introduced with their approach an adaptive visualization environment for selecting an appropriate single view on data based on different contextual factors, whereas always one factor is considered for the visualization method selection. The proposed approach premises a rigorous modeling of the different contextual factors for each visualization type and requires therewith expert knowledge for modeling the scores and appropriateness of visualization methods. Further the do not consider the adaptation of certain visual variables, content, or interface composition. The visualization method selection is performed once at the beginning of a task and does not change during the work, which cannot be assigned as real adaptation. All contextual models are predefined and response to some predefined rule. The only dynamic factor is the user profile that makes use of explicit user feedback to score a visualization up or down.

Bai et al. proposed with *CAVE* a contextual adaptive visualization in particular for knowledge worker to support the detection of contextual changes based on users' qualification, knowledge, and experience [86, 87, 121]. Their focus was on developing visualizations that enable in particular the knowledge acquisition process. Therefore they identified four elementary "requirements of developing contextual adaptive knowledge visualization" [86, p. 189], the *visual solution creation, visual solution modification/customization/enhancement, visual solution integration*, and *visual solution transformation* to support the knowledge acquisition process in changing and transforming contexts. The *visual solution creation* assumes that the changes of context let a visualization become irrelevant in a new context with new requirements and propose to build new visualization in flexible way. The development of new visualization can be performed based on existing visualization or from scratch. They proposed that this problem can be faced by their *CAVE* system through selecting and mapping new visual components [87, 121]. The *visual solution modification/customization/enhancement* addresses changes in user and problem context by providing the user the ability to change the visualization component, the used visual variables, or adjusting transformation parameters [121]. The *visual solution integration* (or visualization integration) addresses the aspect that changes on context visualization either require the visualization of different data or reflect different features of the data. Their main purpose is that there does not exist a single visualization that is able to visualize all types of data for all visualization purposes (tasks). Therefore Bai et al. propose to integrate various visualization methods (layouts) for the different and changing contexts. The *visual solution transformation* (or visualization transformation [121]) proposes the necessity of transforming visualizations form one type to another in a seamless way to provide different views on the same data.

Beside the requirements Bai et al. introduced a *contextual adaptive knowledge visualization model* [86], which was renamed to the *contextual adaptive visualization environment framework* [121] and included three main components, the *problem context, purpose context*, and *knowledge worker context*. The *problem context* refers to the given problem (task) and possible solutions, which contain beside other problem situation, knowledge types, visualization and knowledge tasks, time and so forth. In the application layer the problem context is described with problem data, problem model, problem solver, and problem scenarios that enable processing and managing the identified problems [86]. Thereby the problem solver is responsible for the data and the problem model responsible for changes in the visualizations and enables users to react to identified changes. The purpose context refers to the contextual information about the goals of a knowledge worker by applying visualizations, e.g. domain related purposes, knowledge worker purposes, task related purposes, and so forth [121]. In the application layer the *visualization context and component manager* refers to the purpose context and is responsible for the creation of visualization with the components visualization data, visualization models, visualization solvers, and visualization scenarios [86]. The knowledge worker context refers to the user (knowledge worker) and the related attributes, such as type, profile, or ability [121]. In the application layer the knowledge worker context and adaptation manager is responsible for the communication between the two introduced components and further enables the interfacing between user and the three components [86].

CAVE proposes some interesting approaches in particular for describing the contextual problems. The implementation of the system [86, S.192f] demonstrated the adaptivity of the system by the ability of the user to perceive a problem form different perspectives with various visualization layouts. An adaptive character could not be identified, although the system claims to be adaptive. It is more an adaptable system, where the user (or stakeholder) is able to choose based on different contextual aspects different visualizations types, whereas each variance has to be modeled by experts.

Conati [122, 123], Toker [124, 125], Steichen [126–128] and colleagues investigated and evaluated different aspects of users' cognitive abilities and tasks in mainly two different visualizations with equivalent information provision but diverse visual layouts (bar chart and radar graph) [122, 127]. Their studies conducted in particular eye-tracking methods to predict users' tasks, cognitive abilities, perceptual speed, visual working memory, and verbal working memory [127]. Therefore they developed an eye-tracking architecture that enables calculating numerous summative statistics, e.g. user's fixation-rate, mean absolute saccade angles, or proportionate amount of *Areas of Interest* [127, 128]. They identified five *Areas of Interest*, *high area* contains relevant data values and covers the upper half of a visualization, *low areas* covers the lower part of visualizations, *labels* are the data labels, *questions* describe the tasks to be solved, and *legend* covers the legend [127]. Further they defined based on tasks and AOIs a set of eye-tracking features, consisting of task-level features, e.g. fixation, fixation duration, and saccade lengths and angels, and AOI-level features consisting of features with regards on fixation on AOIs, e.g. number, sum & mean, time to first fixation, and so forth [127]. Their evaluations and studies conducted in general two levels of tasks, simple tasks compared by the values of two vectors and complex tasks, included values of three vectors [124]. With their evaluation they could evidence beside others that the eye gaze behavior provides evidence about tasks and cognitive abilities [127] further they could evidence a correlation between perception speed and task completion time [124]. Another main finding was that the gaze-behavior-based predictions are significantly better than a baseline classifier and the accuracy at the beginning is higher ate each task [127].

Although, the studies and evaluations do not really provide an adaptive visualization system, the work on these aspects is mentionable, due to future appliances in adaptive visual environments. The results provide promising evidences that eye gaze interaction analysis would be an adequate influencing factor driving adaptive visualizations and systems [127]. Thus the evaluations focused on the visual layout and the exemplary use case [128] and some other studies were applied on mainly two visual layouts (bar chart and radar graph); we categorize the results in visual layout adaptation. The studies are very promising and a visualization system that uses the eye fixations beside other influencing factor might be useful, if the adaptation is enhanced to other criteria, e.g. content, interface, or variables too.

Godehardt introduced a contextualized visualization environment that adapts the visualization layout to different sensor information in particular for knowledge worker [35]. The proposed approach investigate various sources of sensor data, e.g. operation system data with the information about open or closed documents and the relation of this documents, to generate a knowledge worker's context. The generated context information is used as input to derive the current task of a user and provide autonomously a visual layout [35]. A rule-based engine creates from the sensor information a context that chooses between two existing visualization layouts, a process-visualization, and a graph visualization. Godehardt proposes that the approach is generic and could involve more sensor information or include more visualization layouts [35], but we could not find any other works that includes these aspects. Figure 4.4 illustrates the two visual layouts.

The proposed approach of provides theoretically the ability to consider various influencing factors as sensor information to model a context. However, the focus of his work seemed to be more the investigation of sensor information to model a context rather than the visual adaptation itself. The proposed system selects one visual layout from existing two layouts to support knowledge workers. An adaptation on visual interface, content, or visual variables is not considered in that work.

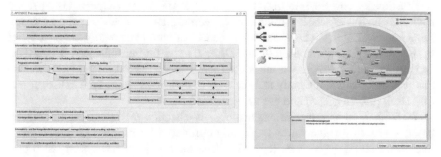

Fig. 4.4 Contextualized visualization by Godehardt: two visual layouts for supporting the contextual task (from [35, pp. 71–74], with author's permission)

Voigt et al. proposed with *VizBoard* an adaptive visualization that includes the collaborative filtering method in selecting different types of visualization [82, 129]. For selecting the visualization type and layout respectively they propose a system that makes use of implicit and explicit user information [129]. The information about user is stored in an ontological knowledge-base (VISO) [82]. The implicit user information relies on the work three main user actions, *repeated use*, *glimpse*, and *related rate*. *Repeated use* registers the user interactions with the same visualization in a certain time interval and if it is used more than three times, the visualization is upgraded. *Glimpse* refers to the situation that the user does not reach his goals within a predefined time-interval or interaction counts, which leads to a downgrade of the visualization layout. *Related rate* refers to a previously explicit rating of the user in another context (with other data) and leads to an upgrade in the new situation. Beside the implicit analysis they introduced explicit ratings according to *factual visualization knowledge*, *quality of domain assessment*, *context knowledge*, and *user rating for collaborative filtering*, whereas they premise a direct cooperation of two or more users for the collaboration filtering factor [129]. The overall rating of visualizations is performed by measuring the arithmetic mean of all factors, which leads to a visualization selection.

VizBoard provides an interesting approach by combining the collaborative filtering aspect in the adaptation process of visualization. However, the system premises a direct collaboration and affects the visual layout, whereas the adaptation process makes not use of the possible visual granularities of visualizations.

4.5.3.4 Visual Variables Adaptation

According to various works and studies in information visualization [24, 66, 70, 92] the visual (or retinal) variables, e.g. color, size, saturation, brightness, shape etc. are information carrier. Further these variables are perceived by human in a field of distractors commonly parallel [130–133] (see Sect. 2.2. The adaptation of these variables is one main issue to direct the attention of users to certain parts of visualizations.

Brusilovsky and Su introduced with their *WADEIn* an adaptive visualization environment for expression execution as learning environment for the programming language *C* [5]. This system was enhanced with explanatory characteristics by Brusilovsky and Loboda [69] (*WADEIn II*) and further enhanced (*cWADEIn*) and evaluated [134, 135]. We introduce their approach based on the latest version that was described sufficiently and the core of their work, namely *WADEIn II* and *cWADEIn* [69, 134, 135] and use the term *WADEIn II*.

WADEIn II is an adaptive explanatory visualization that enhances the model-based approach for generating explanation proposed by Kumar [136] with adaptive visualization approaches [69, 134]. The main goal of the system is to provide students an adaptive learning environment for the expressions and operators of the programming-language *C* [134]. Therefore a static user interface is subdivided in four main areas, *Goals and Progress* (Fig. 4.5a) (a), *Settings* (Fig. 4.5a) (b),

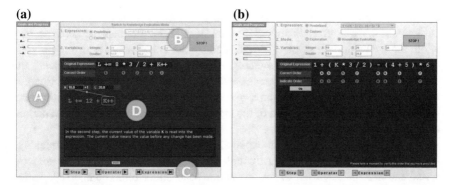

Fig. 4.5 The *WADIn II* adaptive explanatory visualization. **a** illustrates the application in exploration mode with the four introduced areas and **b** illustrates the application in evaluation mode
(from [135, p. 195] with permission)

Navigation (Fig. 4.5a) (c), and *Blackboard* (Fig. 4.5a) (d) [134, p. 252]. *Goals and
Progress* illustrates the learning goals as a list of concepts and the students' progress
on these goals as "skillometers" [134, p. 252]. The *settings* enable the selection of
certain expressions to be evaluated and provide the setting of initial values for variables [134, 135]. Further the *settings* area provides the choice between two main
modes, the exploration mode (Fig. 4.5a) and the evaluation mode (Fig. 4.5b) [69].
The *navigation* area enables users to go through the different process step-by-step
or on operator-by-operator basis. These steps can be performed by the user forward
and backward.

The main area of their system is the *Blackboard* that makes use of a real blackboard metaphor and illustrates the operations (called visualizations in the work),
explanations, and controls that indicate more complicated concepts are involved
in the operation [69, 134]. *WADEIn II*'s blackboard contained three main aspect,
Visualizations, *Explanations*, and *Adaptation*. This version included in their visualization different expression that could be solved and explored step-by-step, whereas a
color-coding indicated a specific context and animation were used for showing some
further aspects. They differentiated in their "visualization" five different steps, *reading variable*, *producing value*, *writing variables*, and *pre- and post-incrementation
and decremention*. The color encoding were used in the different steps to illustrate a
context, e.g. in *variable read* the name of the variable is highlighted in red, an animation shows the insertion of the value (red too), and after the value is inserted the
color changes to green. Further they identified some more difficult concepts that are
highlighted with orange flags. *Explanation* is one main functionality of the system
that has more the characteristics of an intelligent tutoring system [135]. Based on
the degree of knowledge the student gets either a short or long explanation of the
step to perform [69]. With this procedure novice students are supported with detailed
information, while experienced may be discouraged by long explanations [134]. The
adaptation in *WADEIn* is adapting the tasks and provides for each concept an illustration in terms of exploration knowledge and evaluation knowledge. The progress

of students' knowledge in illustrated in the *Goals and Progress* area, whereas the exploration knowledge is illustrated by the length of the progress bar (*skillometer*), while the evaluation knowledge is indicated by the color intensity of the progress bar for each concept [69]. These factors are computed based on the concepts and the evaluation of the results, whereas each user is starting with the assumption that no knowledge about the expression is given. The adaptation functionality influences beside the explanations the speed of the animations or the appearance of animations at all.

WADIn II (or *cWADEIn*) is an explanatory learning environment for students to learn a programming language. Although the application claims to be an adaptive visualization it makes no use of any adaptation of visualization. The main adaptation is based on students' knowledge level and the explanation. The only visual adaptation that we could identify was the change of the progress bar in color intensity and length. For this reason, we categorize this system as adaptive visual variables, but the real use of the variables in visualization context is not given.

de Jongh et al. introduced an adaptive platform for creating interactive visualizations by considering user preferences on the community structure [137]. The user preferences are collected by allowing users to define *cliques* from their perspectives. The user defined cliques are combined with the original data that are processed of the DBLP digital library with the *Jaccard Index* [137, 138]. The clustering of the data is performed with an enhanced *Label Propagation Algorithm* (LPA) [139] with the ability to run on weighted graphs and the frequencies are modified by edges [137, p. 4]. Based on the two retrieved metrics the data are visualized with a derivative of the force-direct algorithm [140], where the node size is based on degree centrality and the nodes illustrates based on pie charts the overlapping degree. The edges refer to co-authorship, where the thickness indicates the similarity based on the Jaccard Index measurement [137]. The user is able to select based on the computed LPA correlations and the pie charts a group of edges. The visualization changes the visual variables of the edges from pie charts to rhombus, whereas the color refers to the machine-derived clusters and the change of icons (to rhombus) to user-defined clusters [137]. Finally the user-defined cliques are automatically passed to the clustering algorithm and their opacity is changed to enable a differentiation between the original data and the user-defined cliques (Fig. 4.6).

(a)   **(b)**   **(c)**   **(d)**

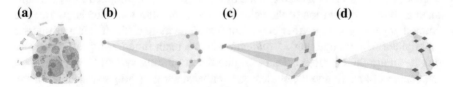

Fig. 4.6 Adaptive visualization of research communities: **a** illustrates an overview of the clusters, **b** illustrates the initial pie charts and the correlation that enables users to group cliques, **c** illustrates the user-defined cluster, where the pie charts are replaced by rhombus, and **d** the final adapted graph (opacity and color) (adapted from [137, pp. 2–4] with permission)

The adaptive visualization by de Jongh et al. makes in particular use of the visual variables, e.g. color, opacity, thickness, and icon (pie-chart icon vs. rhombus). Both, the measured similarities on original data and the user-defined cliques just use visual variables for distinguishing the several aspects and provide information with those variables. The approach enables users to define their own cliques or groups of researcher, whereas an adaptation of the interface, layout or content is not performed. But they combine the similarity of original data with the similarities proposed explicitly by users, which is for sure an interesting approach.

4.6 Summary and Findings

Computer-based information and knowledge acquisition is a promising approach with various facets and variables that should be investigated by developing such systems. Adaptive systems provide a useful and promising way to face in particular the heterogeneity of users, tasks, context, and data with adaptive methods that reduce human effort in complex information acquisition processes. This chapter introduced in particular adaptive visualizations as an approach for bridging the gap between the heterogeneous influencing factors, above all the human with his variety of attributes, and information visualization. We introduced first the term adaptation in context of human-centered system and could show that beside the different terms, e.g. personalization, customization, or intelligent, the term adaptation gained various and partly differing definitions. Therefore, we defined adaptive systems by generalizing two common definitions and delimited it from other related terms. Thereafter we introduced the main idea of adaptive systems by investigating various models and processes. Based on a general model of adaptation, we outlined the main idea of adaptive systems that can be summarized with helping users to achieve their intended tasks in faster, easier, or with better results [25]. For that purpose we introduced different models and outlined three main steps in adaptation, *acquisition*, *representation*, and *production* [11]. The general process can be summarized by the acquisition of relevant information for adaptation, the formal representation of this information, and the production of certain changes of the system behavior. Further we could show that one main goal of adaptive system is to support the information acquisition process [1], which is in line with our purposes in this thesis.

After the general view on adaptive systems, we focused on the adaptation process in information visualization and categorized the process in *influencing factors, knowledge modeling*, and *human interface adaptation*. We defined as influencing factors, those information that may influence the behavior, appearance, or view of information visualization systems. Here we identified three main factors: human, data, and task based on our work on the foundations on information visualization in Chap. 2. Knowledge Modeling investigated the representation and relevant factors that can and should be modeled in context of information visualization. Here a prominent model was introduced, which investigated the two main questions "what is being modeled?" and "how". We could identify based on the model that tasks and goals are derived commonly from user behavior and introduced mainly user modeling and data modeling. The human interface adaptation focused on adaptation criteria of information visualization. Here we introduced the pioneering work of Bertin [92] and its appliance to a graphical language by Mackinlay [24]. We could identify the most common visual parameters that can be adapted in terms if adaptive visualizations. This identification and the related differentiation is the baseline for our conceptual work.

The main goal of this chapter was to give a comprehensive and comprehensible state of art and technology for adaptive visualizations. For this purpose we first defined adaptive visualization based on the definition of adaptive systems and the definition of information visualization. The clear definition and the worked out models and classifications enabled us to classify adaptive visualizations into the four main classes of *Visual Interface Adaptation*, which refers to the adaptation of the user interface and placement of visual layouts, *Visual Content Adaptation* that refers to content reduction or filtering, *Visual Layout Adaptation* that investigated the visualization layout or type, and *Visual Variables Adaptation* that refers to the retinal variables. Our review on the systems was performed based on the introduced categories with the assumption that not all systems can be exclusively assigned to one category. They may include a variety of adaptation possibilities of the visualization systems. In that cases the most dominant adaptation criteria was used for the categorization.

Our review on the existing systems covered the last decade. The goal was to find a system or approach that make use of all the introduced adaptation criteria, but at least combine some of them to provide a real benefit out of the visual structures. Our review clearly signals that the emerging area of adaptive visualizations did not investigate the "human interface adaptation" in depth, yet. The most systems are replacing visualization types and layouts respectively based on some users' implicit or explicit demands. The focus of today's systems is more "to what" should be adapted rather than "what can be adapted". None of the introduced systems adapted the entire range of possible visual adaptations. Just few systems combined two visualization criteria, whereas commonly just one was dynamically adapted. In general none of today's system makes real use of the potentials of information visualization in the adaptation process.

References

1. A. Jameson, in *The Human-Computer Interaction Handbook: Fundamentals, Evolving Technologies and Emerging Applications*, 2nd edn., ed. by A. Sears, J.A. Jacko (CRC Press, Boca Raton, 2008), pp. 433–458
2. C. Stephanidis, A. Paramythis, M. Sfyrakis, A. Stergiou, N. Maou, A. Leventis, G. Paparoulis, C. Karagiannidis, in *Proceedings of the 5th International Conference on Intelligence and Services in Networks: Technology for Ubiquitous Telecom Services, IS&N'98* (Springer-Verlag, London, UK, 1998), pp. 153–166. http://dl.acm.org/citation.cfm?id=648010.741686
3. P. Brusilovsky, User Model. User Adapt. Interact. **11**(1–2), 87 (2001). doi:10.1023/A:1011143116306, http://dx.doi.org/10.1023/A:1011143116306

4. P. Brusilovsky, M.T. Maybury, Communications of the ACM 2002 (2002)
5. P. Brusilovsky, H. Su, in *Proceedings of 6th International Conference on Intelligent Tutoring Systems (ITS'2002)* (Springer Verlag, 2002), pp. 229–238
6. A. Bunt, G. Carenini, C. Conati, in *The Adaptive Web*, vol. 4321, Lecture Notes Computer Science, ed. by P. Brusilovsky, A. Kobsa, W. Nejdl (Springer, Berlin, 2007), pp. 409–432
7. V. Alvarez-Cortes, B. Zayas-Perez, V. Zarate-Silva, J. Uresti, in *Electronics, Robotics and Automotive Mechanics Conference, 2007. CERMA 2007* (2007), pp. 312–317. doi:10.1109/CERMA.2007.4367705
8. D. Gotz, Z. When, J. Lu, P. Kissa, N. Cao, W.H. Qian, S.X. Liu, M.X. Zhou, in *Proceedings of the first international workshop on Intelligent visual interfaces for text analysis, IVITA'10* (ACM, New York, USA, 2010), pp. 1–4. doi:10.1145/2002353.2002355, http://doi.acm.org/10.1145/2002353.2002355
9. M. Hartmann, Context-aware intelligent user interfaces for supporting system use. Ph.D. thesis, Technische Universität Darmstadt (2010)
10. K. Höök, Interact. Comput. **12**, 409 (2000). doi:10.1016/S0953-5438(99)00006-5
11. A. Kobsa, J. Koenemann, W. Pohl, Knowl. Eng. Rev. **16**, 111 (2001)
12. M. Baldoni, C. Baroglio, N. Henze, *Reasoning Web*, LNCS Tutorial (Springer, Berlin, 2005), pp. 173–212
13. R. Oppermann (ed.), *Adaptive User Support: Ergonomic Design of Manually and Automatically Adaptable Software* (Lawrence Erlbaum Associates Inc., Hillsdale, 1994)
14. A. Micarelli, F. Gasparetti, F. Sciarrone, S. Gauch, Personalized search on the world wide web, in *The Adaptive Web*, ed. by P. Brusilovsky, A. Kobsa, W. Nejdl (Springer, Berlin, 2007), pp. 195–230. http://dl.acm.org/citation.cfm?id=1768197.1768205
15. S. Marathe, S.S. Sundar, in *Proceedings of the SIGCHI Conference on Human Factors in Computing Systems CHI'11* (ACM, New York, USA, 2011), pp. 781–790. doi:10.1145/1978942.1979056, http://doi.acm.org/10.1145/1978942.1979056
16. P. Brusilovsky, E. Millán, User models for adaptive hypermedia and adaptive educational systems, in *The Adaptive Web*, ed. by P. Brusilovsky, A. Kobsa, W. Nejdl (Springer, Berlin, 2007), pp. 3–53. http://dl.acm.org/citation.cfm?id=1768197.1768199
17. P. Brusilovsky, User modeling and user-adapted interaction **6**(2), 87 (1996). http://www.springerlink.com/index/X33Q23N15373K164.pdf
18. J. Fink, A. Kobsa, User Modeling and User-Adapted Interaction **10**(2–3), 209 (2000). doi:10.1023/A:1026597308943, http://dx.doi.org/10.1023/A%3A1026597308943
19. A. Goy, L. Ardissono, G. Petrone, in *The Adaptive Web*, vol. 4321, Lecture Notes in Computer Science, ed. by P. Brusilovsky, A. Kobsa, W. Nejdl (Springer, Berlin, 2007), pp. 485–520. doi:10.1007/978-3-540-72079-9_16, http://dx.doi.org/10.1007/978-3-540-72079-9_16
20. J.W. Ahn, P. Brusilovsky, D. He, J. Grady, Q. Li, in *Proceedings of the 17th International Conference on World Wide Web WWW'08* (ACM, New York, NY, USA, 2008), pp. 1–10. doi:10.1145/1367497.1367499, http://doi.acm.org/10.1145/1367497.1367499
21. J.W. Ahn, Adaptive visualization for focused personalized information retrieval. Ph.D. thesis, School of Information Sciences, University of Pittsburgh (2010). http://etd.library.pitt.edu/ETD/available/etd-08262010-150850/
22. H. Maeda, K. Haro, N. Ikoma, in *Fuzzy Systems Conference Proceedings, 1999. FUZZ-IEEE'99. 1999 IEEE International* vol. 2 (1999), pp. 1153–1158. doi:10.1109/FUZZY.1999.793118
23. A. Kobsa, in Workshop im Rahmen der 17. Fachtagung für Künstliche Intelligenz (KI 93), Humboldt-Universität zu Berlin, ed. by 17. Fachtagung KI. (Springer, Berlin, 1993)
24. J. Mackinlay, ACM Trans. Graph. **5**, 110 (1986). doi:http://doi.acm.org/10.1145/22949.2295
25. E. Ross, Intelligent user interfaces survey and research directions. Technical Report, (University of Bristol, Bristol UK, 2000)
26. E. Knutov, P. De Bra, M. Pechenizkiy, New Rev. Hypermedia Multimed. **15**(1), 5 (2009). doi:10.1080/13614560902801608. http://dx.doi.org/10.1080/13614560902801608
27. A. Paramythis, Adaptive systems: Development, evaluation and evolution. Ph.D. thesis, Johannes Kepler Universität Linz (2009)

28. K.M. Feigh, M.C. Dorneich, C.C. Hayes, Hum. Factors J. Hum. Factors Ergon. Soc. **54**(6), 1008 (2012). doi:10.1177/0018720812443983, http://hfs.sagepub.com/content/54/6/1008. abstract

29. K. Nazemi, N. Bhatti, E. Godehardt, C. Hornung, Montgomerie, C., Association for the Advancement of Computing in Education–AACE-: ED-Media 2007, in *Proceedings : World Conference on Educational Multimedia, Hypermedia & Telecommunications* (2007)

30. P. Langley, in *UM'99: Proceedings of the seventh international conference on User modeling* (Springer-Verlag New York Inc, Secaucus, USA, 1999), pp. 357–370

31. A. Kobsa, in *Grundlagen der Information und Dokumentation*, ed. by R. Kuhlen, T. Seeger, D. Strauch, 5th edn. (K.G. Saur, München, 2004)

32. A. Neumann, Recommender systems for scientific and technical information providers. Ph.D. thesis, University of Karlsruhe (2008)

33. D. Bouneffouf, CoRR, (2013). arXiv:1305.1114

34. D. Sonntag, in *Human Interface and the Management of Information. Interacting in Information Environments*, vol. 4558, Lecture Notes in Computer Science, ed. by M. Smith, G. Salvendy (Springer, Berlin, 2007), pp. 645–654. doi:10.1007/978-3-540-73354-6_71, http://dx.doi.org/10.1007/978-3-540-73354-6_71

35. E. Godehardt, Kontextualisierte visualisierung am wissensintensiven arbeitsplatz. Ph.D. thesis, Darmstadt, TU, Diss., 2009 (2009). http://bibcd.igd.fraunhofer.de/bibcd/INI_Science/dissertations/diss_godehardt.pdf.181 S

36. N. Reithinger, G. Herzog, A. Blocher, KI—Künstliche Intelligenz 2/07, 30 (2007)

37. E. Frias-Martinez, S. Chen, X. Liu, IEEE Trans. Syst. Man Cybern. Part C Appl. Rev. **36**(6), 734 (2006). doi:10.1109/TSMCC.2006.879391

38. T. Mitchell, *Machine Learning* (McGraw-Hill, New York, 1997)

39. L. Fausett, *Fundamentals of Neural Networks* (Prentice-Hall, Englewood Cliffs, 1994)

40. B.E. Boser, I.M. Guyon, V.N. Vapnik, in *Proceedings of the fifth annual workshop on Computational learning theory COLT'92* (ACM, New York, USA, 1992), pp. 144–152. doi:10.1145/130385.130401, http://doi.acm.org/10.1145/130385.130401

41. I. Zukerman, D.W. Albrecht, in *User Modeling and User-Adapted Interaction* (Kluwer Academic Publisher, 2001), pp. 5–18

42. C. Stab, K. Nazemi, (Supervisor), D.W. Fellner (Supervisor), Interaktionsanalyse für adaptive benutzerschnittstellen. Diploma thesis at the Technische Universität Darmstat. Supervised by K. Nazemi (2009). http://bibcd.igd.fraunhofer.de/bibcd/INI_Science/theses/2009/Stab.pdf. 161 S

43. K. Nazemi, C. Stab, D.W. Fellner, in *Advanced Intelligent Computing Theories and Applications*, vol. 6215, Lecture Notes in Computer Science, ed. by D. Huang, et al. (Springer, Berlin, 2010), pp. 362–371. doi:10.1007/978-3-642-14922-1_45, http://dx.doi.org/10.1007/978-3-642-14922-1_45

44. E. Manavoglu, D. Pavlov, C.L. Giles, in *ICDM'03: Proceedings of the Third IEEE International Conference on Data Mining* (IEEE Computer Society, Washington, DC, USA, 2003), p. 203

45. J. Noguez, L.E. Sucar, in *ENC'05: Proceedings of the Sixth Mexican International Conference on Computer Science* (IEEE Computer Society, Washington, DC, USA, 2005), pp. 2–9. doi:10.1109/ENC.2005.7

46. R. Burns, S. Carberry, S. Schwartz, in *User Modeling, Adaptation, and Personalization*, vol. 7899, Lecture Notes in Computer Science, ed. by S. Carberry, S. Weibelzahl, A. Micarelli, G. Semeraro (Springer, Berlin, 2013), pp. 114–126. doi:10.1007/978-3-642-38844-6_10, http://dx.doi.org/10.1007/978-3-642-38844-6_10

47. K. Nazemi, C. Stab, D.W. Fellner, in *IEEE International Conference on Intelligent Computing and Intelligent Systems. Proceedings* (IEEE Press, New York, 2010), pp. 607–612. doi:10.1109/ICICISYS.2010.5658514, http://bibcd.igd.fraunhofer.de/bibcd/INI_Science/papers/2010/10p127.pdf

48. V. Mihajlovic, M. Petkovic, Dynamic bayesian networks: A state of the art. Technical Report TR-CTIT-01-34, University of Twente, Enschede (2001)

49. T.A. Stephenson, An introduction to bayesian network theory and usage. IDIAP-RR 03, IDIAP (2000)

50. D.W. Albrecht, I. Zukerman, A.E. Nicholson, User Model. User Adapt. Interact. **8**(1–2), 5 (1998)

51. K.P. Murphy, An introduction to graphical models. Technical Report, Faculty of Science. University of British Columbia (UBC) (2001)

52. K. Murphy, S. Mian, Modelling gene expression data using dynamic bayesian networks, Technical Report, Computer Science Division, University of California (1999)

53. F.V. Jensen, *Introduction to Bayesian Networks* (Springer, Secaucus, 1996)

54. T. Grenager, A quick romp through probabilistic relational models. http://nlp.stanford.edu/grenager/papers/prm_tutorial_2003_02_20.ppt. Accessed Oct 2013

55. L. Getoor, D. Koller, N. Friedman, A. Pfeffer, B. Taskar, in *Introduction to Statistical Relational Learning*, ed. by L. Getoor, B. Taskar (The MIT Press, Cambridge, 2007)

56. N. Friedman, K. Murphy, S. Russell, in *15th Annual Conference on Uncertainty in Artificial Intelligence* (Morgan Kaufmann, San Francisco, 1999), pp. 139–147

57. L. Getoor, M. Sahami, in *Working Notes of the KDD Workshop on Web Usage Analysis and User Profiling* (1999)

58. L. Getoor, N. Friedman, D. Koller, B. Taskar, J. Mach. Learn. Res. **3**, 679 (2003)

59. C.R. Anderson, P. Domingos, D.S. Weld, in *KDD'02: Proceedings of the eighth ACM SIGKDD International Conference on Knowledge Discovery and Data Mining* (ACM, New York, USA, 2002), pp. 143–152. http://doi.acm.org/10.1145/775047.775068

60. F. Brosy, Lprm—learning probabilistic relational models (2007). http://www2.informatik.hu-berlin.de/wm/seminar2006w/ProbabilisticRelationalModels_FranziskaBrosy.pdf

61. C.M. Grinstead, L.J. Snell, Introduction to Probability (American Mathematical Society, 1997). http://www.amazon.ca/exec/obidos/redirect?tag=citeulike09-20&path=ASIN/0821807498

62. B. Rabiner, L. Juang, A.S.S.P. Mag., IEEE **3**, 4 (1986)

63. L.R. Rabiner, Proc. IEEE **77**(2), 257 (1989). doi:10.1109/5.18626, http://dx.doi.org/10.1109/5.18626

64. A. Künzer, F. Ohmann, L. Schmidt, MMI-Interaktiv **7**, 61 (2004)

65. V.I. Levenshtein, Sov. Phys. Dokl. **10**(8), 707 (1966)

66. S.K. Card, J.D. Mackinlay, B. Shneiderman, *Readings in Information Visualization: Using Vision to Think*, 1st edn. (Morgan Kaufmann, 1999). http://www.amazon.com/exec/obidos/redirect?tag=citeulike07-20&path=ASIN/1558605339

67. E. Rich, Cogn. Sci. A Multidiscip. J. **3**(4) 329–354 (1979)

68. W.J. McGuire, A. Padawer-Singer, J. Personal. Soc. Psychol. **33**, 743 (1976)

69. P. Brusilovsky, T.D. Loboda, in *Proceedings of the 11th Annual SIGCSE Conference on Innovation and Technology in Computer Science Education, ITICSE'06* (ACM, New York, USA, 2006), pp. 48–52. doi:10.1145/1140124.1140140, http://doi.acm.org/10.1145/1140124.1140140

70. B. Shneiderman, in VL (1996), pp. 336–343

71. J.S. Yi, Y. ah Kang, J. Stasko, J. Jacko, IEEE Trans. Vis. Comput. Graph. **13**(6), 1224 (2007). doi:10.1109/TVCG.2007.70515

72. G. Marchionini, Commun. ACM **49**(4), 41 (2006). doi:10.1145/1121949.1121979, http://doi.acm.org/10.1145/1121949.1121979

73. C. Fluit, M. Sabou, F. van Harmelen, in *Visualizing the Semantic Web. XML-Based Internet and Information Visualization*, Second edn., ed. by V. Geroimenko, C. Chen (Springer, London, 2006)

74. R.W. White, R.A. Roth, in *Exploratory Search: Beyond the Query-Response Paradigm*, vol. 1, Synthesis Lectures on Information Concepts, Retrieval, and Services, ed. by G. Marchionini (Morgan & Claypool Publishers, 2009). doi:10.2200/s00174ed1v01y200901icr003, http://dx.doi.org/10.2200/s00174ed1v01y200901icr003

75. E. Horvitz, J. Breese, D. Heckerman, D. Hovel, K. Rommelse, in *Proceedings of the Fourteenth Conference on Uncertainty in Artificial Intelligence* (Morgan Kaufmann, 1998), pp. 256–265

76. M.S. Hancock, A bayesian network model of a collaborative interactive tabletop display, Technical Report, University of British Columbia (2003)
77. E. Horvitz, in *Proceedings of the SIGCHI conference on Human Factors in Computing Systems CHI'99* (ACM, New York, USA, 1999), pp. 159–166. doi:10.1145/302979.303030, http://doi.acm.org/10.1145/302979.303030
78. E. Horvitz, in *Intelligent Systems, September 1999, IEEE Computer Society*, pp. 17–20 (1999). ftp://ftp.research.microsoft.com/pub/ejh/mixedin.pdf
79. W. Wahlster, A. Kobsa, in *Proceedings of the IEEE, Special Issue on Natural Language Processing* (1986), pp. 948–960
80. W. Wahlster, A. Kobsa, in *User Models in Dialog Systems*, Symbolic Computation, ed. by A. Kobsa, W. Wahlster (Springer, Berlin, 1989), pp. 4–34. doi:10.1007/978-3-642-83230-7_1, http://dx.doi.org/10.1007/978-3-642-83230-7_1
81. E. Rich, Int. J. Man Mach. Stud. **18**, 199 (1983)
82. M. Voigt, S. Pietschmann, K. Meissner, in *Semantic Models for Adaptive Interactive Systems*, Human Computer Interaction Series, ed. by T. Hussein, H. Paulheim, S. Lukosch, J. Ziegler, G. Calvary (Springer, London, 2013), pp. 83–107. doi:10.1007/978-1-4471-5301-6_5, http://dx.doi.org/10.1007/978-1-4471-5301-6_7
83. C. Seeberg, Life Long Learning; Modulare Wissensbasen für elektronische Lernumgebungen (Springer, 2003)
84. D. Sleeman, Int. J. Man-Mach. Stud. **23**(1), 71 (1985). doi:10.1016/S0020-7373(85)80025-0, http://dx.doi.org/10.1016/S0020-7373(85)80025-0
85. L. Bossi, S. Braghin, A. Datta, A. Trombetta, in *User Modeling, Adaptation, and Personalization*, vol. 7899, Lecture Notes in Computer Science, ed. by S. Carberry, S. Weibelzahl, A. Micarelli, G. Semeraro (Springer, Berlin, 2013), pp. 38–50. doi:10.1007/978-3-642-38844-6_4, http://dx.doi.org/10.1007/978-3-642-38844-6_4
86. X. Bai, D. White, D. Sundaram, IJEEEE **1**(3), 193 (2011)
87. X. Bai, D. White, D. Sundaram, A. Auckland, N. Zealand, in *Proceedings of the 12th European Conference of Knowledge Management* (Academic Conferences Limited, 2011), pp. 56–64
88. M. Golemati, C. Halatsis, C. Vassilakis, A. Katifori, G. Lepouras, in *Proceedings of the conference on Information Visualization, IV'06* (IEEE Computer Society, Washington, DC, USA, 2006), pp. 62–67. doi:10.1109/IV.2006.5, http://dx.doi.org/10.1109/IV.2006.5
89. D.A. Keim, C. Panse, M. Sips, in *Exploring Geovisualization*, ed. by J. Dykes, A. MacEachren, M.J. Kraak (Oxford: Elsevier, 2003)
90. S.K. Card, J.D. Mackinlay, in INFOVIS (1997), pp. 92–99
91. E.H. Chi, in *INFOVIS'00: Proceedings of the IEEE Symposium on Information Vizualization 2000* (IEEE Computer Society, Washington, DC, USA, 2000), p. 69
92. J. Bertin, *Semiology of Graphics* (University of Wisconsin Press, Madison, 1983)
93. W.S. Cleveland, R. McGill, J. Am. Stat. Assoc. **79**(387), pp. 531 (1984). http://www.jstor.org/stable/2288400
94. J.W. Ahn, P. Brusilovsky, Inf. Process. Manag. **49**(5), 1139 (2013). doi:10.1016/j.ipm.2013.01.007, http://www.sciencedirect.com/science/article/pii/S0306457313000137
95. D. Keim, G. Andrienko, J.D. Fekete, C. Görg, J. Kohlhammer, G. Melançon, in *Information Visualization*, vol. 4950, Lecture Notes Computer Science, ed. by A. Kerren, J. Stasko, J.D. Fekete, C. North (Springer, Berlin, 2008), pp. 154–175
96. E. Reiter, R. Dale, *Building Natural Language Generation Systems* (Cambridge University Press, New York, 2000)
97. A. Knott, R. Dale, in *Selected papers from the Fourth European Workshop on Trends in Natural Language Generation, An Artificial Intelligence Perspective* (Springer, London, UK, 1996), pp. 47–67. http://portal.acm.org/citation.cfm?id=646188.683402
98. K. Höök, Knowl. Based Syst. **10**(5), 311 (1998). doi:10.1016/S0950-7051(97)00034-8, http://www.sciencedirect.com/science/article/pii/S0950705197000348, Intelligent User Interfaces
99. W. Wiza, K. Walczak, W. Cellary, in *Proceedings of the ninth International Conference on 3D Web Technology, Web3D'04* (ACM, New York, USA, 2004), pp. 29–40. doi:10.1145/985040.985045, http://doi.acm.org/10.1145/985040.985045

100. W. Wiza, K. Walczak, W. Cellary, in *Web Engineering*, vol. 2722, Lecture Notes in Computer Science (Springer, Berlin Heidelberg, 2003), pp. 204–207. doi:10.1007/3-540-45068-8_36, http://dx.doi.org/10.1007/3-540-45068-8_36

101. K. Walczak, W. Wiza, in *Proceedings of the IFIP TC6/WG6.4 Workshop on Internet Technologies, Applications and Social Impact, WITASI'02* (Kluwer, B.V., Deventer, The Netherlands, 2002), pp. 45–60. http://dl.acm.org/citation.cfm?id=647071.713757

102. W. Wiza, K. Walczak, W. Cellary, in *IADIS International Conference e-Society 2003* (Lisbon, Portugal, 2003), pp. 365–372

103. J. Kohlhammer, Knowledge representation for decision-centered visualization. Ph.D. thesis, Technische Universität Darmstadt (2005)

104. J.W. Ahn, P. Brusilovsky, Inf. Vis. **8**(3), 180 (2009). http://ivi.sagepub.com/cgi/content/short/8/3/167

105. J.W. Ahn, P. Brusilovsky, in *User Modeling, Adaptation, and Personalization*. Lecture Notes in Computer Science, vol. 6075, ed. by P. Bra, A. Kobsa, D. Chin (Springer, Berlin Heidelberg, 2010), pp. 4–15. doi:10.1007/978-3-642-13470-8_3, http://dx.doi.org/10.1007/978-3-642-13470-8_3

106. K.A. Olsen, R.R. Korfhage, K.M. Sochats, M.B. Spring, J.G. Williams, Inf. Process. Manag. **29**(1), 69 (1993). doi:10.1016/0306-4573(93)90024-8, http://dx.doi.org/10.1016/0306-4573(93)90024-8

107. P. Pirolli, S. Card, in *Proceedings of International Conference on Intelligence Analysis* (2005), pp. 2–4. https://analysis.mitre.org/proceedings/Final_Papers_Files/206_Camera_Ready_Paper.pdf

108. R. Florian, H. Hassan, A. Ittycheriah, H. Jing, N. Kambhatla, X. Luo, N. Nicolov, S. Roukos, in NAACL/HLT (2004), pp. 1–8

109. P. Brusilovsky, J.W. Ahn, T. Dumitriu, M. Yudelson, in *Tenth International Conference on Information Visualization, IV 2006* (2006), pp. 142–150. doi:10.1109/IV.2006.16

110. P. Brusilovsky, in *Proceeding of WebNet'2001*, ed. by W. Fowler, J. Hasebrook, World Conference of the WWW and Internet (2001), pp. 124–129

111. L. Shi, N. Cao, S. Liu, W. Qian, L. Tan, G. Wang, J. Sun, C.Y. Lin, in *Visualization Symposium 2009, PacificVis'09. IEEE Pacific* (2009), pp. 41–48. doi:10.1109/PACIFICVIS.2009.4906836

112. T. Kamada, S. Kawai, Inf. Process. Lett. **31**(1), 7 (1989). doi:10.1016/0020-0190(89)90102-6, http://dx.doi.org/10.1016/0020-0190(89)90102-6

113. M.E.J. Newman, Phys. Rev. E **69**(6) (2004)

114. D. Gotz, M.X. Zhou, An empirical study of user interaction behavior during visual analysis, Technical Report, IBM Research Division, NY (2008)

115. D. Gotz, M. Zhou, in *IEEE Symposium on Visual Analytics Science and Technology, 2008. VAST'08* (2008), pp. 123–130. doi:10.1109/VAST.2008.4677365

116. D. Gotz, Z. Wen, in *Proceedings of the 14th international conference on Intelligent user interfaces, IUI'09* (ACM, New York, USA, 2009), pp. 315–324. doi:10.1145/1502650.1502695, http://doi.acm.org/10.1145/1502650.1502695

117. J. Mackinlay, P. Hanrahan, C. Stolte, IEEE Trans. Vis. Comput. Graph. **13**(6), 1137 (2007). doi:10.1109/TVCG.2007.70594, http://dx.doi.org/10.1109/TVCG.2007.70594

118. C. Stolte, D. Tang, P. Hanrahan, Commun. ACM **51**(11), 75 (2008). doi:10.1145/1400214.1400234, http://doi.acm.org/10.1145/1400214.1400234

119. M. Golemati, C. Vassilakis, A. Katifori, G. Lepouras, C. Halatsis, in *Intelligent User Interfaces: Adaptation and Personalization Systems and Technologies*, ed. by C. Mourlas, P. Germanakos (IGI Global, 2009), pp. 188–204

120. A. Katifori, C. Halatsis, G. Lepouras, C. Vassilakis, E. Giannopoulou, ACM Comput. Surv. **39** (2007)

121. X. Bai, D. White, D. Sundaram, Electron. J. Knowl. Manag. **10**(1), 01 (2012)

122. C. Conati, H. Maclaren, in *Proceedings of the Working Conference on Advanced Visual Interfaces, AVI'08* (ACM, New York, USA, 2008), pp. 199–206. doi:10.1145/1385569.1385602, http://doi.acm.org/10.1145/1385569.1385602

123. C. Conati, G. Carenini, M. Harati, D. Tocker, N. Fitzgerald, A. Flagg, in *AAAI Workshops* (2011). http://www.aaai.org/ocs/index.php/WS/AAAIW11/paper/view/3944

124. D. Toker, C. Conati, G. Carenini, M. Haraty, in *User Modeling, Adaptation, and Personalization*, vol. 7379, Lecture Notes in Computer Science, ed. by J. Masthoff, B. Mobasher, M. Desmarais, R. Nkambou (Springer, Berlin Heidelberg, 2012), pp. 274–285. doi:10.1007/978-3-642-31454-4_23, http://dx.doi.org/10.1007/978-3-642-31454-4_23

125. D. Toker, C. Conati, B. Steichen, G. Carenini, in *Proceedings of the SIGCHI Conference on Human Factors in Computing Systems, CHI'13* (ACM, New York, USA, 2013), pp. 295–304. doi:10.1145/2470654.2470696, http://doi.acm.org/10.1145/2470654.2470696

126. B. Steichen, G. Carenini, C. Conati, in *UMAP Workshops, CEUR Workshop Proceedings*, vol. 872, ed. by E. Herder, K. Yacef, L. Chen, S. Weibelzahl (CEUR-WS.org, 2012). http://dblp.uni-trier.de/db/conf/um/umap2012w.html#SteichenCC12

127. B. Steichen, G. Carenini, C. Conati, in *Proceedings of the 2013 International Conference on Intelligent User Interfaces, IUI'13* (ACM, New York, USA, 2013), pp. 317–328. doi:10.1145/2449396.2449439, http://doi.acm.org/10.1145/2449396.2449439

128. B. Steichen, O. Schmid, C. Conati, G. Carenini, in *UMAP 2013 Extended Proceedings. First International Workshop on User-Adaptive Visualizations (WUAV 2013)* (2013)

129. M. Voigt, M. Franke, K. Meißner, in *UMAP 2013 Extended Proceedings: First International Workshop on User-Adaptive Visualization (WUAV 2013)* (2013)

130. A.M. Treisman, J. Souther, J. Exp. Psychol. Hum. Percept. Perform. **12**, 107 (1986)

131. A.M. Treisman, S. Gormican, Psychol. Rev. **95**(1), 15 (1988)

132. J.M. Wolfe, W. Gray (ed.), *Integrated Models of Cognitive Systems* (2007), pp. 99–119

133. R.A. Rensink, Annu. Rev. Psychol. **53**, 245 (2002)

134. T.D. Loboda, P. Brusilovsky, in *Times of Convergence. Technologies Across Learning Contexts*, vol. 5192, Lecture Notes in Computer Science, ed. by P. Dillenbourg, M. Specht (Springer, Berlin, 2008), pp. 250–261. doi:10.1007/978-3-540-87605-2_28, http://dx.doi.org/10.1007/978-3-540-87605-2_28

135. T.D. Loboda, P. Brusilovsky, User Model. User Adapt. Interact. **20**, 191 (2010). doi:10.1007/s11257-010-9077-1, http://dx.doi.org/10.1007/s11257-010-9077-1

136. A.N. Kumar, in *Supplementary Proceedings of the 11th International Conference on Artificial Intelligence in Education (AI-ED 2003)* (IOS Press, Amsterdam, 2003), pp. 425–432

137. M. de Jongh, P.M. Dudas, P. Brusilovsky, in Berkovsky et al. (eds): *UMAP 2013 Extenden Proceedings. First Internation Workshop on User-Adaptive Visualizations (WUAV)* (2013)

138. P. Jaccard, Bulletin del la Société Vaudoise des Sciences Naturelles **37**, 547 (1901)

139. U.N. Raghavan, R. Albert, S. Kumara, Phys. Rev. E **76**(3) (2007)

140. S.G. Kobourov, CoRR (2012). arXiv:1201.3011

Part II
Model for Adaptive Semantics Visualization

Chapter 5
The Methodological Approach of Adaptive Semantics Visualization

The literature review and the state of the art analysis in information visualization, semantics visualization, and adaptive visualization illustrated studies on human visual perception, existing approaches, and existing systems and methods. We identified in context of our review on semantics and adaptive visualizations various potentials that do not exist yet. The goal of work is not only to identify those potentials but also to provide methods that face the existing problems and make use of the existing approaches for a more appropriate visual adaptation of semantics. This chapter will summarize some of the main outcomes of our literature review and propose based on the identified potentials, requirements that should be fulfilled to provide scientific and technological advancements in adaptive visualizations. Therefore, we will first identify the requirements that build the foundation on the conceptual work. Thereafter a high-level design of our conceptual model will be presented. The high-level design aims at giving a short and comprehensible overview of our main intentions and related contributions that will be described in Chap. 6 more detailed. The main contribution of this chapter is the high-level design of our approach that can be seen as a roadmap for the detailed description of algorithms, methods, and models of our conceptual model in the following Chap. 6.

5.1 Analysis and Derivation of Requirements

The rigorous investigation of literature, approaches, models, and systems of information visualization, semantics visualization, and adaptive visualization enabled on the one hand the identification of valuable models for improving the human interaction with visual applications and on the other hand the identification of gaps in existing systems and approaches. To provide a beneficial design on adaptive semantics visualization, the identified gaps and beneficial values should be outlined in order to define requirements on an improved high-level design for adaptive visualizations. To perform this elementary work, we investigate the entire spectrum of the investi-

© Springer International Publishing Switzerland 2016
K. Nazemi, *Adaptive Semantics Visualization*, Studies in Computational
Intelligence 646, DOI 10.1007/978-3-319-30816-6_5

gated literature and start with the main goal of information visualization that aims at amplifying cognition by interactive visual representation of abstract data [1, p. 6].

Amplifying cognition can be supported, as outlined in Sect. 2.2, mainly by investigating how human perceive visual information. In a preattentive stage human's attention can be directed through certain visual variables within a field of distractors to visual information. We refer in this work with the term visual variables to all the retinal variables that are used in the differentiation of Bertin [2] as Implantation and include those visual features that encode information on plane, e.g. color, size, shape. For the adaptation of visualization the *Guided Search Theory* is of interest, thus the study results evidenced that the preattentive visual perception can be used to guide the user too her expected visual data entities [3]. The active attention of users on a visual representation leads according to Ware [4] to goal directed processing that is supported by a non-volatile representation of objects on the screen and refers to the *Layout* of a visualization [5]. While visual variables enable a fast direction of attention to certain visual entities, layout provides a goal directed and attentive processing of visual information. We defined in this work *Layout* according to Bertin [2] as Imposition and therefore as the positioning of graphical objects above the screen with their relations to each other. This definition matches with the proposed model of Rensink and leads to acquire meaning (gist) from the provided visual layout [5]. Beside these two elementary visual representation ways, we could identify that the content (data) is of interest for the adaptation process (Sect. 4.5.3.2). The reduction or filtering of content reduces the complexity of visual representations and provide a more comprehensible view on the underlying data. To provide a coherent model of adaptive visualization it is necessary to investigate the content to be visualized and adapt it to the demands of users. Another important aspect in this context is the investigation of multiple-layouts for the same (or different) data. A single layout visualizes the data with one perspective and may lead to complex representations of information. The reduction of the complexity can be performed by using multiple-view visualizations and arrange them into a user interface. Consequently, an adaptive visualization should provide at least the ability to orchestrate different visualizations in a juxtaposed manner for providing different and in best case complimentary visual representations. The investigation of existing systems of adaptive visualizations illustrated clearly that there exist no adaptive visualization approach that investigates the entire range of the mentioned adaptation criteria. Based on these findings, we define the first requirement (R1) as follow:

R1 (Variables in adaptive visualizations): An ideal adaptive visualization should cover the entire range of adaptation capabilities of information visualization to amplify cognition and support the human visual perception. This includes the adaptation of content, visual layout, visual variables, and the visual user interface.

The process of adaptation in visualizations is commonly dominated by two main influencing factors: data and user (see Sect. 4.4.1). Data and data properties are valuable information sources for the choice of visualization types or visual layouts, ordering of data values [2, 6], and an important indicator for the amount and therefore the complexity of visual representations. Even in user-adaptive visualization the aspect of data should be considered and investigated as influencing factor, thus the right design of visual representations depends strongly on the underlying data. Beside data, the user and her aptitudes plays an increasing role for the adaptation process. The heterogeneity of users requires the investigation of their behavior, knowledge, and further visualization related characteristics. The investigation of systems and literature illustrated that existing systems either consider the user as influencing factor for the adaptation ("to what") or the underlying data [6, 7]. An adaptive visualization that investigates both, user characteristics and data characteristics does not exist. This main gap leads to our second requirement (R2):

R2 (Influencing factors in adaptive visualizations): An ideal adaptive visualization should perform the adaptation to both, user and data characteristics. These two influencing factors should be seen as mandatory but not exclusive. Further influencing factors may complement the adaptive behavior of visualizations.

The review of existing works in adaptive visualization illustrated further that the user model is commonly and limited based on the underlying knowledge domain of a certain adaptive system. Although, the ways of modeling knowledge may vary, the inferred users' behavior and the determined user models are commonly limited to the knowledge domain in which the adaptive systems are applied. Visualizations may be designed to visualize just one knowledge domain or data-base. In this case the proposed existing approaches on user modeling would be valuable and reflect the domain knowledge of a visualization. But commonly visualizations, in particular our approach, is not designed for a certain predefined knowledge domain or data-base. Far more they enable the visualization of a variety of data bases with different domain of and knowledge and handle different data sources. A predefined knowledge model would not meet the requirements of visualizing different data from various data-bases. It is therefore necessary to find a solution that investigate the given data and combine the user interaction behavior to infer a user model that can be applied to different data and knowledge bases. The user model should further contain in best case the interaction context consisting of users' interaction behavior and the data characteristics. Further the unobtrusive way of gathering users' information is of great importance, which can be performed by implicit user interaction analysis. This leads to our next requirements (R3) and (R4):

R3 (User model generation for different data): The varying data of visualization environments require the design of user models that are able to handle different and varying data-bases or knowledge domains. The user model should anyhow be transferable to different data-bases and support the adaptation process.

R4 (User model generation by implicit interaction analysis): The most unobtrusive way to gather information about users and their behavior is the analysis of their interaction with the system as the natural consequence of their system behavior. An ideal adaptive visualization should primarily make use of user's interactions for analyzing users' behavior.

User interaction analysis is commonly performed to gather information about the behavior of an individual user. Just one known system [8] is making use of the interaction behavior for recommending or adapting visualization layouts to "canonical" users. Canonical user is the sum of all users working with a certain system. Even this one system just considers the interactions as repeated patterns that are predefined and stored in interaction pattern libraries to adapt or recommend visualizations. Each interaction pattern is modeled by experts. Their approach does not consider data characteristics or changing user characteristics. It assumes that the repeated patterns lead in any case to the definition of a certain behavior. Although, this approach without the use of a user model evidenced promising improvements [8–10], the approach can or should be enhanced by considering the data context and the use of real canonical user models that trains the system based on the average usage behavior without the necessity of experts to model certain patterns. This leads to the use of a "canonical user model" that enables a visualization adaptation based on the general user behavior considering the data context, as we could find out a preliminary user study [11]. The general user behavior by considering the data context leads to find the most used combination of visualizations and provides a self-learning system that considers the average usage behavior of users to adapt and recommend visual layouts. The use of canonical user models provides a promising foundation for improving the visualization system in a general manner. Based on the outcomes of the mentioned studies and identified limitations, we define our next requirement [R4] as follow:

R5 (Appliance of a canonical user model in adaptive visualizations): The user interaction analysis should be enhanced to cover the usage behavior of all users with certain data for improving the general behavior of the visualization system and obviate the need for an expert training or modeling of the user model.

The interaction analysis for the generation of a canonical user model and individual user models enables measuring the similarity of users' behavior and the deviation of individual's user behavior compared to the canonical user. These techniques are in particular used in different recommendation systems [12] for recommending similar items to similar users. In context of visualization this aspect would improve the general adaptation behavior thus similar users are recognized in a particular data context and similar views can be recommended. The view on literature could outline that this aspect is rarely considered for adaptive visualizations [13, 14] and the proposed approaches does not support the similarity-measurement on a visual level. They use it far more for recommending the content rather than the visualizations. The inclusion of similarity-based adaptation would enhance the adaptation process and reduce the "cold start" problem of such systems. Another aspect that is in particular interesting in context of adaptive visualizations is the analysis of users' deviation compared to the canonical user. By analyzing the deviation, certain individual anomalies in interaction behavior would be revealed and can be used to adjust the level of adaptation based on the canonical user model. Therefore we define a further requirement [R6] as follow:

R6 (Analysis of user-similarity and deviations in adaptive visualizations): The analysis of user-similarities enhances adaptive visualizations by reducing the "new user" problem. Users' deviation analysis leads to identifying anomalies in user behavior and consequently to the adjustment of the adaptation effects. An adaptive visualization with these features enhances the adaptation capabilities.

Beside the requirements on adaptive visualization, we outlined in our review on semantics visualizations that the process of exploratory search is not sufficiently supported by existing systems. Although, the visualization of semantics provide a good foundation for supporting the exploratory search process, today's systems focus more on visualizing the formalization of knowledge rather than providing tools for information acquisition. And those systems, which support the process of information acquisition, are commonly question-answering systems that make use of the semantic formalisms. We further outlined that formulating questions requires knowledge about a certain topic or domain. If the user wants to know a specific fact to enhance his previous knowledge, those question-answering systems are appropriate. But commonly the process of search involves some kind of learning and exploration to reveal the informational context. In simple search engines this process is performed by querying a known item, acquiring information that enclose the context and enable a kind of learning, and reformulating the query until the entire intended informational space is revealed. In contrast to that process, information visualization follow more a top down principle, by overviewing the entire information space, zooming into certain parts of interest (intention), and gathering the intended details on demand [15]. Exploratory search should include both principles with semantics visualizations to support the exploratory search process in the increasingly growing semantic and linked data bases. To illustrate different views on exploratory search, we introduced two models of exploratory search that provide a good foundation for supporting the entire exploration process. Further we introduced various definitions and classifications of tasks in context of information visualization (see Sect. 2.4) and classified them to abstract high-level tasks (see Sect. 2.4.2). Our review on existing semantics and ontology visualizations revealed that there exists no system that supports the exploratory search process. Therefore one main requirement on our system is the support of the exploratory search process including, such sub-tasks like comparing different data or viewing the same data with different perspectives (overview and detailed view). The semantics visualization system should enable the process of exploration ate least by providing different views on the same data (R7) and visualizing different data-bases on the same visual interface (R8). Further the top-down and bottom-up views on information support the exploration process as described in Sect. 3.4.2 (R9).

R7 (Support of exploratory search by different visual perspectives on the same data): The exploratory search process in semantics visualization can be supported by different views or perspectives on the same data by the orchestration of a multiple-visualization user interface. Ideal semantics visualizations for search and exploration purposes should enable the different view on the same at the same time.

R8 (Support of exploratory search by juxtaposed visualizations on different data-bases): The exploratory search process in semantics visualization can be supported with investigation and comparison tasks on different data ideally with the same visualizations that are juxtaposed arranged. An ideal semantics visualization for search and exploration purposes should enable the visualization on different data, in terms of parts of data or totally different data-bases to support the comparison and investigation tasks.

R9 (Support of exploratory search by bottom- and top-down views and interactions): The exploratory search process in semantics visualization is characterized by learning and investigating information and data. The process of learning that leads to refining and precising the user queries can be supported by the top-down principle (overview to details) and complemented by the bottom-up (detail to overview) approach. An ideal semantics visualization for search and exploration purposes should provide both principles to support the entire exploratory search process.

One further aspect in the process of exploratory search is the ability to verbalize and reflect the process of information finding. The verbalization or expression ability leads to recapitulate the search process and as an indicator for learning [16]. Beside the analysis of the study outcomes, analysis of models, and analysis of existing systems, we conducted a preliminary user study to find out, if the prior knowledge of a user has implications on his verbalization abilities [17]. The evaluation evidenced that one factor for enabling the verbalization in subjects is the higher self-assurance and thereby the prior knowledge. The efficiency of a search task is strongly related to users' abilities in verbalization and recognition. The verbalization ability is in turn related to users' knowledge. This covers the knowledge in a specific domain, in a certain language or in usage with computer systems [17]. Table 5.1 summarizes the deduced requirements for our conceptual model.

The review on literature evidenced that a system or approach that meets the introduced requirements **does not exist at all**. Different studies, models and systems evidenced that **there is a necessity to adapt visual environments** and support the exploratory search process through visualizations. Table 5.1 gives an overview of the derived and analyzed requirements on such a model.

Table 5.1 Requirements for adaptive semantics visualization

	Requirement
R1	Variables in adaptive visualizations
R2	Influencing factors in adaptive visualizations
R3	User model generation for different data
R4	User model generation by implicit interaction analysis
R5	Appliance of a canonical user model in adaptive visualizations
R6	Analysis of user-similarity and deviations in adaptive visualizations
R7	Support of exploratory search by different visual perspectives on the same data
R8	Support of exploratory search by juxtaposed visualizations on different data-bases
R9	Support of exploratory search by bottom- and top-down views and interactions

5.2 High-Level Design for Visualization Adaptation

The derived requirements on adaptive semantics visualization are the foundations of providing a high-level design of the proposed solution. This part of the section introduces the general proposed model on a high but comprehensible level to ensure that the derived requirements are met with our conceptual model. Figure 5.1 illustrates the design of our model on that abstract level. It consists of four main layers, *Influencing Factors*, *Knowledge Model*, *Adaptation Process*, and *Visual Adaptation* and meets the structure of our review on literature to enable a clear comparison with the state-of-art and the identification of added values in a comprehensible way.

The high-level design illustrates in particular the procedure of adaptation in a cyclic-way and should be seen more as a schematic illustration. The general process can be described as follow: the user interacts with the system by entering a search term, interacting with a visual representation, or interacting with the user interface that may contain static elements. This interaction is performed with a dedicated *user interaction* component that decides based on the interaction type, if a data-query should be sent. In case of querying data, one or more queries are sent to one or more data-bases that are commonly spread in the Web and may contain heterogeneous data formats. The retrieved data are then transformed to one or more data models that can be handled by the visualization system. Regardless if the system adaptivity is turned

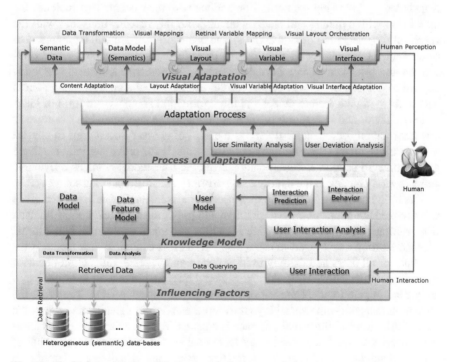

Fig. 5.1 The high-level design of the methodological approach for adaptive visualizations

on or off, the data model is sent to the visualization pipeline that starts with semantics. In cases of turned-off adaptivity or not-logged-in user, the information is sent just to a canonical user model. In case of logged-in user the information are sent beside the canonical user model to the logged-in individual user model. The transformation of the retrieved data is an important aspect, thus the varying data may have various forms but have to be visualized in a similar way. The data model itself does not contain any further information about the data and just represents the semantic correlations of the data. According to (R2) data characterizations and information about the data are necessary to provide a sufficient visual adaptation. This process is performed by analyzing the retrieved data and modeling the analysis results in *data features model*. For each data model, a data feature model is provided that includes information about the data, their amount, and some further computed contextual information. The data feature model is a part of the data model and enhances the data model with information about the retrieved data. The retrieved information about the data characteristics are used in the adaptation process.

The second influencing factor according to (R2) is the user. The users' behavior with a system is observed and gathered by analyzing their interactions according to (R4). Each interaction of the user is first formalized in a way that the interaction context can be derived and different levels of abstractions can be determined. The formalization of the user interaction is a fundamental step for the following processes of behavior derivation and interaction prediction. The behavioral pattern of users' is gathered by following the entire interactions with the system. This includes the interaction with dynamic and static elements with the visualizations and with the graphical representation of data entities. The behavioral pattern will be represented as normalized probability distribution and provides information about users' preferences and prior-knowledge. Further an enhanced predictive statistical method is applied to determine the next actions of the user. The gathered information about data in form of data feature model and the information about user in form of inter-action behavior and interaction prediction flow into the user model that represents this information for a canonical user as the sum of all user interactions in context of the retrieved data (R5) and individual user models for providing individualized adaptations. To support different data-sources the user model contains a dynamic part as overlay model on the data and supports therefore different data sources (R3).

The individual user models are then used to measure similarities between given users for reducing the "new user" problem and providing a sufficient view on data. If no similarities with other users are found the canonical user model is the initial user model for a new logged-in user (R6). Further this canonical user model is the baseline for all users and thereby used to derive anomalies in interaction behavior by measuring deviations (R6). All the gathered information is used in a rule-based adaptation engine to adapt the various layers of visualization (R1). Therefore the content or semantics, the visual layout as representation of visualization types, the visual variables, and the visual interface is adapted (R1). The dynamic character of the general user interface (visual interface) enables the juxtaposed positioning of various visual layouts (R7) for the same data model and visualizing different data model with same or different visualizations (R8). The choose of the visualization layout is further enabling both a top-down view on the data, according to the *visual information seeking mantra* [15] or a bottom-up view to expand the visual context based on a query [18] (R9).

The following sections describe the general idea and approach on an abstract level to obtain the main contributions of this work. In the following Chap. 6 a more detailed and reconstructable description of the approach and its components will be given.

5.3 Influencing Factors

Influencing factors can be defined as any kind of information that are used or can be used to perform adaptation. According to the four elementary questions of Brusilovsky [19], the influencing factors respond to the question *To What* can be adapted. Various aspects can be used as influencing factors in adaptive systems and visualizations. Commonly the identification and gathering of available information that leads to adaptation, is the first step of the entire adaptation process and may include data about users, usage behavior or usage environment [20] (see Sect. 4.2).

The most common influencing factor in adaptive systems is the individual behavior of users. The acquisition of users' information are either gathered explicitly, e.g. by asking users for their preferences or knowledge, or implicitly by analyzing the user or usage behavior. With the emerging growth of mobile computing devices further environmental aspects are considered in the adaptation process (e.g. [21]). In this work, the investigation of those influencing factors is important that affect the adaptation of information and semantics visualization. The assumption that all influencing factors, e.g. usage environment or location, can be used for adapting visualizations cannot be neglected. It is far more important to identify those factors that strongly affect the visual parameters in semantics and information visualizations.

The investigation of literature in information visualization, semantics visualization, and adaptive visualization enabled us to identify the most valuable influencing factors for adapting visualizations. The early and still ongoing process on visualizing information makes commonly use of data characteristics for visualization. The data characteristics include beside other the structure and dimension of data and the ability to order data-values. Both data characteristics are important in the adaptation process. Thus information and semantics visualizations build an information bridge between users and data, the second valuable and important influencing factor is the user. In this context the visual information processing of users plays an important role. An ideal adaptive visualization should therefore investigate in the adaptation process at least the two influencing factors: data and user. This reflects our identified and formulated requirement (R2), where *an ideal adaptive visualization should perform the adaptation to both, user and data characteristics. These two influencing factors should be seen as mandatory but not exclusive. Further influencing factors may complement the adaptive behavior of visualizations.*

The investigation of literature in adaptive visualizations in Sect. 4.5 could clearly outline that existing systems and approaches do not make use of both influencing factors for adapting visualizations. Most of the existing approaches for adapting visualizations are limited to the investigation or consideration of users as influencing factor. This is due to the origin of adaptive systems that are more focused on user-centered design. In contrast to that, approaches and systems from information visualization or Visual Analytics focus on the data and data characteristics. Although, adaptive visualization gained recently enormous interest in research, a combined investigation of both, data and users is not yet proposed in the literature.

Our main contribution related to influencing factors is the combined investigation of data and user characteristics in the visual adaptation process to enable a more effective adaptation of visualizations. We mean with a combined investigation of both influencing factors, a real combination of both in terms of which data characteristics correlates with which user attributes.

The acquisition, transformation, and representation of influencing factors for the adaptation process will be described in Sect. 6.1 *Knowledge Model*. This is to ensure a comprehensible way of illustrating the transformation and representation of the influencing factors as models that affect the adaptation process. Beside acquisition of influencing factor information, various algorithms for measuring different relevance values, and formal representations, in particular the Sect. 6.1.3.4 *Modeling Users* will illustrate the combined investigation of data and users in one model.

5.4   Knowledge Model

The knowledge model can be described as the entire set of information that is used for the adaptation process. Our model includes beside the information about adaptation, the data to be visualized too. This is due the investigation of data as one influencing factor for the adaptation process. However, the visualization of data can be performed without the knowledge model, whereas the adaptation effects are then not applied. Knowledge model can be assigned according to Kobsa et al. as *representation* or *secondary inferencing* [20] (see Sect. 4.2). Thereby knowledge model is responsible for expressing the acquired information in a formal and machine-processable way. Although, the model of Kobsa et al. proposes a bisection or separation of information acquisition and knowledge modeling in terms of representations, the strict bisection is not adequate in all adaptive systems. Our model for adaptive semantics visualizations for instance makes use of the visualized data. As we will describe in the conceptual chapter, the process is investigating not only external information as representations for adapting the visual environment. The combination of persistent and volatile models that is related directly to a data query is investigated too. Although, the high-level design proposes a similar bisection of information acquisition of the influencing factors and the formal representation of the acquired representation, Sect. 6.1 *Knowledge Model* will investigate both aspects to enable a more comprehensible illustration of the entire process. The section will introduce the way how this information is gathered, analyzed, weighted, and structured as knowledge model for the adaptation process.

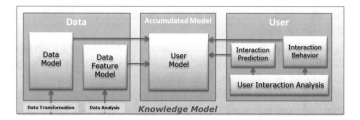

Fig. 5.2 Simplified view on the knowledge model layer of our high-level Design

Figure 5.2 illustrates a simplified view on the knowledge model as layer of our high-level design. It includes three main parts that outline the already mentioned combined investigation of the two different influencing factors of data and user in the adaptation process. The three parts are the core components of the knowledge model and describes the entire steps from gathering information to a formal representation.

The 'data' part investigates the entire process of querying data from distributed sources, transforming data, including semantics even from not semantically formalized data, and representing data in a processable way. The data model consists of three main components that will be described in Sect. 6.1.1 in a detailed way: *inclusion of semantics by iterative querying*, *inclusion of semantics from metadata*, and *internal data representation*. The approaches for including semantics in terms of contextual information are used to represent data in a machine readable way.

With our approach for iterative querying heterogeneous semantic data bases, we enable a semantic contextual visualization for search terms. The main contribution of the iterative querying for including semantics is the creation of a semantic context that enables a more sufficient visualization of contextual information of search results from different data bases. We further apply this approach to non-semantic data to illustrate the sustainability of the proposed approach. Therefore a method for including semantics from metadata will illustrate the way how the approach can be applied to further domains. Our contribution in this part is the transfer of our iterative semantic inclusion method for non-semantic data. The last component is the formal representation of semantics for visualization purposes. We introduce in this context a model that contain all semantic information and referenced resources for gathering structural information about the underlying data. It should be outlined that distributed semantic data, e.g. Linked-Open Data, do not necessarily provide a schema that enables gathering information about the data structure. This is in particular not given for the result of a search query. The resulted data are commonly provided as single instances or entities with semantic references. Our main contribution in this context is to provide a data model that contains information about data structure, dimension, and even order abilities of data values. The model enables consequently a proper semantics visualization of search results. Our investigation of existing systems and approaches illustrated that the above mentioned aspects in processing semantics for search and visualization purposes are not investigated yet. Our data model and the related approaches for gathering semantic information and representing them with structural information are not only of interest for the adaptive visualization, but also for semantics visualization.

The data feature model enhances the data model with further information about data characteristics. This model consists of two main components, the quantitative analysis of data and weight-analysis of semantic relations that are described more detailed in Sect. 6.1.2. The quantitative analysis of data makes use of the iterative querying approach and enhances this with numerical values of each data type. This process enables to determine which structural aspect of the semantics is more dominant. With this analysis step, the structure of the data as result of a search query according to our introduced classifications can be gathered to provide sufficient visual layouts. The main goal is to quantify instances, concepts, incoming and outgoing relations, and the entire set of data for providing adequate visual layouts. In our literature review, a quantification of semantic elements for visualization purposes could not be found. Our contribution in this part is to enable a more sufficient choice of visual layout and further relevance measurements by quantifying each semantic element. Based on the quantitative measurement, we propose weighting-analysis of semantic relations for measuring the relevance of semantic neighbors and thereby the semantic context. To measure the semantic context of a particular selected instance, we introduce two algorithms: the *Inverse Instance Frequency* (iIf) and the *Direct Relation Frequency inverse Relations Frequency* (dR-iRf). These two algorithms enable to measure the relevance of semantically related instances and concepts and enable a volatile visual adaptation based on the currently focused instance. The review on adaptive visualization could outline that the aspect of semantic context and the visual adaptation was not considered in existing works. We provide with these algorithms first attempts for measuring the current contextual relevance for visual adaptation.

The 'user' part of the knowledge model investigates the entire process from gathering user information, analyzing the gathered information, predicting users' actions, and determining users' behavior to a formal representation of users' information. The model consists of three main components that lead, together with the formalized information about data, to the 'user model' as the main part of the knowledge model. The entire process including the representation of users' model will be described more detailed in Sect. 6.1.3.

Thereby users' interactions with the visual environment are implicitly gathered as a natural consequence of the interactions with the system. The users are not confronted with explicit questions from the system for analyzing their behavior according to our requirement (R4). To ensure the analysis of an unambiguous interaction with the system, we introduce the formal representation of users' interactions as a relation of leaf values according to *Relational Markov Models*. This relation is in our context constrained to three main domains of *interaction type, visual layout*, and *data* (see Sect. 6.1.3.1) that enables various measurements and derivation of behavioral patterns in visualizations. The identification of the three domains and the formal representation build the main contribution of this part, which is used for further computations, e.g. the analysis of users' behavior. Therefore the three domains are set in correlation to model the behavior of users with particular data and visual layouts as a vector that includes weighted measures for each correlation (see Sect. 6.1.3.2). We contribute here with the formal definition and model of this particular correlation of data and visual layouts that enable a more efficient adaptation and selection of appropriate visual layouts. The formal representation is further used to determine the next action of users. We evaluated in this context various unsupervised predictive statistical methods with the reference files from Greenberg [22] to find the best suited algorithm for predicting interactions (see Sect. 6.1.3.3). In this context, we contribute with our *KO*/19-Algorithm* that calculates a probability distribution for predicting the next possible interaction based on an observed sequence of n-interaction events. Our algorithm is an enhancement and modification of the *KO3/19-Algorithm* proposed by Künzer [23] that is in contrast to our enhancement limited to three interaction events.

The core component of our knowledge model is the user models that combine the gathered information about data characteristics, usage behavior with data and visual layouts, prediction results, and interaction analysis results in one consistent and reusable model. In this context, we first introduce a model that enables a user modeling beyond the domain borders of one data-source according to our requirement (R3). We make use of the semantic structure of data to gather an abstract term that is given by the structure of semantics to model the behavior of users' with a vector of weighted data and visual layout correlations. The main contribution is the formal description of the user model generation with normalized weighted concepts in correlation to visual layouts. The user model further considers transitions and users' interactions that go beyond the interactions with visual layouts. Based on the formalized user model generation and the introduced measurements on interaction analysis and data characteristics, we introduce our concept of the canonical user model according to our requirement (R5). The canonical user model represents the interaction behavior of all users that interacted in anyway with the visualization environment and thereby the average usage behavior with the system. The canonical user model enables a general adaptation and consequently improvement of the entire system for all users and in particular for unknown or new users. It further trains the system by real users and obviates the need of experts for training or modeling the system. Further it enables to measure similar behavior to define groups of users' and thereby average group user models and the measurement of certain behavioral deviations for identifying behavioral anomalies. Thereby one of our main contributions is the canonical user model and the formal description of the model to enable the replication for other adaptive visualization systems.

5.5 Process of Adaptation

The gathered information about users' behavior and data characteristics enables the adaptation of visualization in various ways. As we could outline in our review of existing attempts of adaptive visualization commonly just one factor, either information about users or data characteristics is considered in the adaptation process. Our contribution, to combine both influencing factors, leads not only to a different

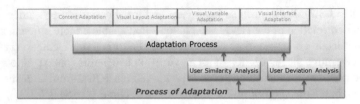

Fig. 5.3 Simplified view on the process of adaptation

modeling of the required knowledge about users and data. It influences the entire process of adaptation. As illustrated in Fig. 5.3 the process of adaptation contains of three main components, whereas the main component of *adaptation process* adjusts and applies the gathered information to adapt the different layers of visualization. Further the study on existing systems could illustrate clearly that approaches for analyzing user similarities and deviations are not applied in adaptive visualizations. The use of a canonical user model, instead of training the system by experts, as it is commonly performed, requires the analysis of both users' similarity and users' deviations. Thereby the three main components build the adaptation processing in a high level manner.

The user similarity component is responsible to gather throughout the usage of the visualization environment, similarity measures between the individual user and the canonical user and between the individual user and certain user groups according to our requirement (R6). The goal of measuring the similarity is two-fold: first, the similarity between an individual user to the canonical user model indicates the similar behavior of the individual user to the average usage behavior. It provides thereby the possibility to apply the canonical user model for the individual user. This is important and useful for new users, new domains of knowledge, and new application scenarios. The similar behavior is adopted in those cases, where a lack of information about the certain user exists. Further the similarity analysis leads to group users and measure the average usage behavior of a certain user group. The average behavior of the user group may differ from the canonical modeled behavior with the visualization. The information of the canonical user model would lead to adaptation effects that are not appropriate for that individual user. Instead of filling the information gap from the canonical user model, the average group user model (canonical user model of the group) is used to fill the gaps. Our main contribution is the appliance of similarity measurements on individual, canonical, and group-level in adaptive visualizations to provide a more efficient adaptation to individual users.

In contrast to user similarity analysis, the deviation analysis of users aims at adjusting the level of adaptation according to the deviated behavior of individual users according to the defined requirement (R6). As the canonical user model or the group user model may fill the information gap for new user, new knowledge domains, or even new application scenarios, an individual user may interact with the visualization in a way that is not only not similar to those canonical model, but differs strongly from the modeled behavior. In these cases, the application of a canonical user model would lead to confusing and frustrating the user. Thus an individual user may expect a certain system behavior that is not given in those models at all. The user deviation analysis is the contrary part to the similarity analysis and determines in particular behavioral anomalies to the average usage. The more a certain interaction behavior of an individual user differs from the canonical user model; the lower is the adaptation effect of the system based on the model. Instead the adaptation effect is reduced first to observe the interaction behavior of the user and provide a more specific adaptation effect based on the individual user model. Our main contribution is the appliance of similarity algorithms in a contrary way. We propose an algorithm that measures the anomalies in interaction behavior and reduces the adaptation effect based on the canonical user model. So the degree of adaptation is dependent to the similar or deviated user behavior.

The third and main component is the adaptation process that describes in detail the way how the various influencing factors that were modeled in a machine-readable way are applied in the adaptation process. We applied therefore an adaptation process that includes all models. Due to the different influencing factors, their representations, and the various layers of adaptation, it is necessary to define at least a process model that describes if, when, and how the models are applied and affect which visual layer. We introduce first the Semantics Visualization Markup Language (SVML) that describes the visual layouts and their visual capabilities. This can be enhanced for each visual layout is the baseline of our process model. The information of visual layouts and their capabilities regarding data, data amount, data structure, adaptable visual variables and similar layouts are stored in the SVML. The SVML is thereby the model for all integrated visual layouts that can be adapted. Based on this, we introduce the general process model of adaptation that includes the entire adaptation process based on various influencing factors for adapting different layers of adaptation. The general process model guides through the entire adaptation process step-by-step and illustrates how the different models are applied. To enable a comprehensible view on the adaptation process, each sub-process is illustrated separately. Our main contribution in this part is two-fold: the SVML as a reusable markup language for visual layouts and later for an entire visual environment and the process model for adaptation that enables a transparent and comprehensible adaptation process. The here described process of adaptation, including the similarity and deviation analysis are described in detail in Sect. 6.2 *Process of Adaptation*.

5.6 Visual Adaptation

The last layer of our high-level design is visual adaptation. The main task of visual adaptation is the transformation of the processed information about users and data to a differentiated model of visual layers to enable an efficient adaptation according to our main requirement (R1). The transformation process is two-fold: a horizontal transformation process that includes the steps of *Data Transformation*, *Visual Mapping*, *Retinal Mapping*, and *Visual Layout Orchestration* and a vertical process that

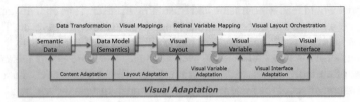

Fig. 5.4 Simplified view on the visual adaptation

manipulates each visual layer based on the adaptation process. The transformation processes can be considered separately, thus the horizontal transformation pipeline that is an enhancement of the reference model proposed by Card et al. [1], can be used in non-adaptive visualization scenarios too. The visual adaptation layer consists of five main components as illustrated in Fig. 5.4. The horizontal transformation pipeline starts (in our case) with semantic data that are transformed in a formal data model. Based on this data model, which includes already information about data structure and characteristics, a visual mapping to one or more visual layouts is performed. The visual layouts should be seen as placements of objects and their relations to each other and to the screen (scene) according to Bertin's *Imposition* [2]. After the placement of objects on screen the retinal variable mapping is performed. This step includes retinal or visual variables to the computed and selected visual layout, e.g. color, size, transparency etc. With this step the visual layouts contain beside an appropriate placement of the objects on screen, appropriate visual variables that may be illustrated as colored icons or glyphs. The horizontal transformation process ends with the orchestration of visual layouts on screen. This step determines if one or more visual layouts are necessary to visualize the underlying data and which of those provide complementary information. The result is a juxtaposed placement of one or more visual layouts that builds the user interface (visual interface).

While the horizontal transformation enhances the established reference model of Card et al. (see Sect. 2.1 and [1]), the main goal still remains an adaptation of visualizations in different levels of visual appearance. We therefore investigate various models and outcomes of studies on human visual perception to refine the reference model for adaptation purposes. Further the main tasks as identified in Sect. 2.4.2 are investigated. The adaptation of visualizations does not refer just to one criterion that is affected by the adaptation. Visualizations provide more variables that can be adapted. The adaptation of these variables may have different effects on the usage of a visual environment and the acquisition of information. We could outline in our literature review that **none of the existing adaptive visualizations is investigating the entire visual adaptation capabilities** (see Sect. 4.6). The existing approaches commonly focus on one factor of the visualization in their adaptation process, e.g. changing the visualization type. But as we discussed in the chapters that investigate the existing literature on information visualization, human perception, and adaptation, the human visual information processing differs strongly in the way how information are visually presented. There exists a kind of bisection that could be used to guide the human attention to certain graphical representations of data or to lead to a serial processing of information. We introduce with our **layer-based reference model of adaptation**, the main contribution of our work in Sect. 6.3. Our reference model considers the outcomes on studies of visual perception and the related models. Based on the models of Rensink [5], Ware [24], TREISMAN [25–27], and Wolfe [3, 28] we separate in our reference model the layers of visual layout in visualizations from visual variables. The separation should lead to more efficient visual adaptations, due to the different way how human perceive the related visual information. This differentiation is further argued with the proposed differentiation of Bertin [2] that proposes the differentiation of *Imposition* and *Implantation*. Further our reference model enhances the visual adaptation by including the layer of visual interface. We have observed in the literature and in many existing systems and approaches that the juxtaposed visual arrangement on screen plays an important role for conveying information. We further observed that none of the existing adaptive approaches considers this aspect in the adaptation process.

This enhanced reference model is our core contribution and enables a fine granular adaptation of visualization in different levels. As illustrated in Fig. 5.4 the gathered information about users and data are influencing each of the illustrated layers. An efficient adaptation is thereby performed by adapting the content that is visualized. This affects in particular the layers of data and data model and the related transformation process of data. The layout adaptation refers to the visual placement of objects on screen and their relations to each other. The adaptation of visual layout leads to a different view on the data on layout level (see Sect. 6.3.3) and manipulates the visual mapping transformation. The visual variable adaptation affects the visual or retinal variables and enables a guiding and viewing of relevance. This adaptation step manipulates the retinal variable mapping and the related variable, such as color, size, shape, and so on. The visual interface adaptation refers to the arrangement of visual layouts on screen and manipulates the visual layout orchestration. With this layer one or more visual layouts including the visual variable layer are placed on the screen based on user' behavioral information and information about data characteristics. The result of the entire visual adaptation is a user and data adapted visual interface that is influenced by user and data and adapts each layer of our reference model based on the underlying information.

5.7 Support of Exploratory Search

The high-level design of our conceptual model illustrates the adaptation process in an abstract way. The highest layer of visual adaptation concludes the model, but provides one main further aspect that is part of our conceptual model and the contributions of this thesis: the visual interface adaptation. We have outlined on our review of existing systems and approaches for semantics visualization (see Sect. 3.5) that the search process and in particular the exploratory search process according to the introduced models of exploratory search (see Sect. 3.4.2 and [16, 29, 30]) are not supported. The main goal of semantics visualization seems still to be the visualization of the ontological structure and schema. Although, semantics enables an enhanced search process, the systems that are focusing on information seeking are more question answering systems rather than enabling enhanced search capabilities by adopting visual interfaces. One of our goals is to provide a visual interface that supports the entire process of exploratory search by our model of visual interface adaptation. We therefore investigate in Sect. 6.4 first the main differences of conventional search processes and the search process with visual environments. We will illustrate in this context two main approaches that follow a bottom-up and a top-down information seeking process. This aspect will enlighten the complementary information seeking procedures. Based on the illustrated differentiation, we introduce our **visualization cockpit model**. The visualization cockpit model provides by the placement of juxtaposed visual layouts in combination of different data sets and data bases, a model that supports the entire process of exploratory search. Beside a top-down and bottom-up search support according to the requirement (R9), the model supports following visual perspectives on data:

- [Perspective view]: Visualization of the same data with different visual layouts (R7).
- [Perspective-comparative view]: Visualization of different sub-set of data from the same data-base with different visual layouts.
- [Comparative view on level-of-details]: Visualization of the same data using the same visual layouts with different parameters.
- [Comparative view on data sub-sets]: Visualization of different data sub-sets from the same data-base with the same visual layouts.
- [Comparative view on data]: Visualization of different data-bases with the same visual layouts (R8).
- [Non-linked view]: Visualization of different data-bases with different visual layouts.

Our visualization cockpit model enables the linking and dislinking of visual layouts for visualizing the same data, different data-sets from one data-base, and even various data-bases. With the customizable character of a user interface that allows users to select, dismiss, and rearrange visual layouts on screen and connecting and disconnecting the visual layouts from data-sets or data-bases, the visualization cockpit model provides enhanced exploratory search capabilities. Our contribution in this context is a model that enable enhanced visual search by combining different visual layouts and link them to data-sets, data-bases, or to other visual layouts. The model will be described in Sect. 6.4 *support of Exploratory Search*.

5.8   Chapter Summary

This chapter introduced the high-level design of our conceptual model. We first summarized some important outcomes of the literature review and the review of existing system and approaches to identify requirements. The identification and formulation of the requirements was the baseline of the high-level design. In summary nine requirements were identified that should lead to a more appropriate conceptual model for adaptive visualizations. Thereafter, we introduced our high-level conceptual design with the goal to provide a comprehensible view on the complex conceptual models that will be introduced in the following chapter. Our high-level design used the terms and methods that were already outlined in the literature review to enable a mapping to the existing approaches and enlighten the contributions of our work. Our high-level design with its five layers of *influencing factors*, *knowledge model*, *process of adaptation*, and *visual adaptation* was then set in relation to the identified requirements. Each of these layers was then described in a very short and comprehensible way with two main intentions: to illustrate their relation to our requirements and to enlighten our own contributions in this work. The methodological view on our conceptual model aimed at illustrated the general idea and the related contributions.

References

1. S.K. Card, J.D. Mackinlay, B. Shneiderman, *Readings in Information Visualization: Using Vision to Think*, 1st edn. (Morgan Kaufmann, 1999). http://www.amazon.com/exec/obidos/redirect?tag=citeulike07-20&path=ASIN/1558605339
2. J. Bertin, *Semiology of Graphics* (University of Wisconsin Press, 1983)
3. J.M. Wolfe, W. Gray (ed.), *Integrated Models of Cognitive Systems* (2007), pp. 99–119
4. C. Ware, *Information Visualization Perception for Design* (Morgan Kaufmann Publishers, 2004)
5. R.A. Rensink, Annu. Rev. Psychol. **53**, 245 (2002)
6. J. Mackinlay, ACM Trans. Graph. **5**, 110 (1986). http://doi.acm.org/10.1145/22949.2295
7. J. Mackinlay, P. Hanrahan, C. Stolte, IEEE Trans. Vis. Comput. Graph. **13**(6), 1137 (2007). doi:10.1109/TVCG.2007.70594. http://dx.doi.org/10.1109/TVCG.2007.70594
8. D. Gotz, Z. When, J. Lu, P. Kissa, N. Cao, W.H. Qian, S.X. Liu, M.X. Zhou, in *Proceedings of the first international workshop on Intelligent visual interfaces for text analysis IVITA '10* (ACM, New York, USA, 2010), pp. 1–4. doi:10.1145/2002353.2002355. http://doi.acm.org/10.1145/2002353.2002355
9. D. Gotz, M.X. Zhou, An empirical study of user interaction behavior during visual analysis. Technical report, IBM Research Division, NY, 2008)
10. D. Gotz, M. Zhou, in *IEEE Symposium on Visual Analytics Science and Technology VAST '08*, (2008), pp. 123–130. doi:10.1109/VAST.2008.4677365
11. K. Nazemi, R. Retz, J. Bernard, J. Kohlhammer, D. Fellner, in *Advances in Visual Computing*, vol. 8034, Lecture Notes in Computer Science, ed. by G. Bebis, R. Boyle, B. Parvin, D. Koracin, B. Li, F. Porikli, V. Zordan, J. Klosowski, S. Coquillart, X. Luo, M. Chen, D. Gotz (Springer, Berlin, 2013), pp. 13–24. doi:10.1007/978-3-642-41939-32. http://dx.doi.org/10.1007/978-3-642-41939-3_2
12. A. Neumann, Recommender systems for scientific and technical information providers, Ph.D. thesis, University of Karlsruhe (2008)
13. P. Brusilovsky, J. wook Ahn, T. Dumitriu, M. Yudelson, in T*enth International Conference on Information Visualization, IV 2006*. (2006), pp. 142–150. doi:10.1109/IV.2006.16
14. M. Voigt, S. Pietschmann, K. Meissner, in *Semantic Models for Adaptive Interactive Systems*, ed. by T. Hussein, H. Paulheim, S. Lukosch, J. Ziegler, G. Calvary, Human (at) Computer Interaction Series (Springer, London, 2013), pp. 83–107. doi:10.1007/978-1-4471-5301-65. http://dx.doi.org/10.1007/978-1-4471-5301-6_5

15. B. Shneiderman, in *VL* (1996), pp. 336–343
16. R.W. White, R.A. Roth, Exploratory Search: *Beyond the Query-Response Paradigm*, vol. 1, Synthesis Lectures on Information Concepts, Retrieval, and Services, ed. by G. Marchionini (Morgan & Claypool Publishers, 2009). doi:10.2200/s00174ed1v01y200901icr003. http://dx. doi.org/10.2200/s00174ed1v01y200901icr003
17. K. Nazemi, O. Christ, in *Advances in Affective and Pleasurable Design*, Advances in Human Factors and Ergonomics Series (Taylor & Francis, 2012). http://books.google.de/books?id= WHtwWU7C_vYC
18. F. van Ham, A. Perer, IEEE Trans. Vis. Comput. Graph. **15**, 953 (2009)
19. P. Brusilovsky, User modeling and user-adapted interaction, J. Pers. Res. **6**(2), 87 (1996). http:// www.springerlink.com/index/X33Q23N15373K164.pdf
20. A. Kobsa, J. Koenemann, W. Pohl, Knowl. Eng. Rev. **16**, 111 (2001)
21. M. Hartmann, Context-aware intelligent user interfaces for supporting system use. Ph.D. thesis, Technische Universität Darmstadt (2010)
22. S. Greenberg, Using unix: collected traces of 168 users. Research report 88/333/45. Technical report, Department of Computer Science, University of Calgary, Calgary, Canada (1988)
23. A. Kzer, F. Ohmann, L. Schmidt, MMI-Interaktiv **7**, 61 (2004)
24. C. Ware, *Information Visualization Perception for Design* (Morgan Kaufmann (Elsevier), 2013)
25. A.M. Treisman, G. Gelade, Cognit. Psychol. **12**(1), 97 (1980)
26. A.M. Treisman, Comput. Vis. Graph. Image Process. **31**(2), 156 (1985)
27. A.M. Treisman, S. Gormican, Psychol. Rev. **95**(1), 15 (1988)
28. J.M. Wolfe, G.G. Gancarz, Lakshminarayanan, V. (eds.) B*asic and Clinical Applications of Vision Science* (1999), pp. 189–192
29. B.S. Bloom, *Taxonomy of Educational Objectives* (David McKay Co., Inc., NY, New York, 1956)
30. G. Marchionini, Commun. ACM **49**(4), 41 (2006). doi:10.1145/1121949.1121979. http://doi. acm.org/10.1145/1121949.1121979

Chapter 6
Conceptual Model of Adaptive Semantics Visualization

The high-level design introduced in the previous chapter gave already a short and comprehensible view on a methodological level of our conceptual model. The main goal of this chapter is to introduce the conceptual model that was introduced in an abstract way in detailed and replicable way. The detailed description should not only enable to review the value of our contributions and advancements in visual adaptation, it should far more enable interested audience to apply the methods, algorithms, and models for enhancing the idea of adaptive visualization. To provide such a replicable description some parts of the following chapter may be detailed for the general audience. To enable a comprehensible illustration of the proposed model, despite their detailed description, we will renounce the description of the relations to the identified requirements and the description of outlining our contribution to the scientific community within this chapter. For retrieving this information, we refer to the previous chapter that already gave a short summary of our conceptual model. However, we will keep the structure of this chapter according to the introduced high-level design to enable a better comparison as illustrated in Fig. 6.1.

This chapter introduces first the knowledge model subdivided in three main sections of *data model*, *data feature model*, and *user model*. Data model will describe the way semantic information is gathered from Web-sources and from non-semantic metadata. Here the approaches of iterative querying will be described that lead to a formal representation of data as data model. Data feature model will illustrate the retrieving of quantitative measure of the underlying data with the same iterative querying approach. Further we will introduce two weighting-algorithms (iIf and dRf-iRf) that measure the relevance of semantic neighbors of focused instances. The core part of this section is describing the relation of data and user behavior as user model. We will first introduce a formal representation of users' interactions based on Relational Markov Models. Thereby, we constrain the domains to those that are really needed in context of our adaptation purposes. Thereafter, we will introduce an approach for determining and weighting the user behavior. Thereby the user behaviors is measured based on both, related data and the interaction with the visual layouts. Based on the measurement of the interaction behavior, we introduce the prediction algorithm that determines the next possible action of users based on

© Springer International Publishing Switzerland 2016

K. Nazemi, *Adaptive Semantics Visualization*, Studies in Computational Intelligence 646, DOI 10.1007/978-3-319-30816-6_6

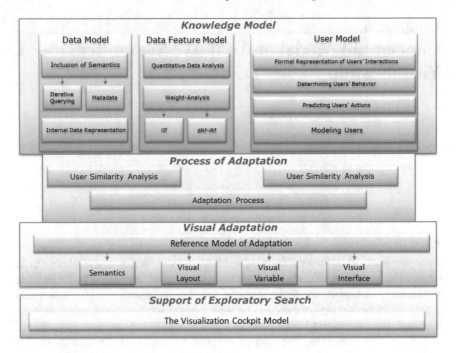

Fig. 6.1   Structure of the chapter conceptual model

the observed previous interactions. Thereafter the formal representation of the user model including the relations between data and visual layouts will be described. The formal description of the canonical user model and the group definition will conclude this chapter.

Section *Process of Adaptation* starts with the formal description of the similarity algorithm based on the canonical user model. The user similarity analysis further illustrates how similar users can be clustered in groups and measures average group user models. Thereafter the deviation analysis will be presented based on the canonical user model to measure users' behavioral anomalies that lead to reducing the system's adaptive behavior. Then the general adaption process will be described that guides through the entire process of adaptation and illustrates when and how the measured values and modeled knowledge is applied. The adaptation process will be described step-by-step to enable the replication of the process. In this context a formal representation of visual layout capabilities as markup language will be introduced too. Section *Visual Adaptation* will introduce our layer based reference model of adaptation. Therefore, we first introduce an abstracted task model for information visualization and the different models of visual perception. Based on these models, we describe our reference model, its different layers and the adaptation capabilities of the identified layers. In this context the set of visual layouts that are part of our conceptual model will be presented too. The last section in this chapter introduces our visualization cockpit model and illustrates how this model can be applied to support the exploratory search with juxtaposed visual layouts. We will identify a set of different visual perspectives on data and illustrate how the model supports solving even complex exploratory and comparative tasks.

6.1 Knowledge Model

The adaptation effect of adaptive visualizations is commonly based on further information that goes beyond the information that should be visualized. We introduced various approaches that make use of certain information like contextual data, user data, or other data that effect the visual environment. These systems commonly make use of one influencing factor that is modeled in a processable way. The adaptation of visualizations provides more attributes for adaptation, but they should investigate at least data and user for the adaptation process. Our approach makes use of data and data characteristics, and of users' interaction behavior to adapt the visualization. This section introduces the way how this information is gathered, analyzed, weighted, and structured as knowledge model for the adaptation process. We will first introduce the data model with our iterative approach for including semantics from semantic Web-sources. Thereafter the internal representation of data in various levels of abstraction will be illustrated. Based on the retrieved information to be visualized, we introduce our approach for data feature modeling. This will include beside quantitative information, semantic weightings of related objects. The formal description of these weighting-algorithms should provide a more comprehensible way to measure the relevance of semantic neighbors of focused instances. This section concludes with the formal and processable representation of users. Therefore we first illustrate how and which information of users is retrieved and describe it in a formal way. Based on this information the user behavior and prediction modeling are described. In this context we will introduce our canonical user model and models for individual users. These are used in the following Sect. 6.2 to measure similarities between users and deviations from the canonical user.

6.1.1 Data Model

The visualization of data from different resources and data bases and in particular with different levels of semantic annotations or even no explicit semantics requires the transformation of those data in a uniform representation. The gathering and transformation of the data is a main step in the visualization process [1]. This part will illustrate how data are queried from different data-bases, where the schema of the data is unknown and represented in a uniform and processable way. We start with our iterative querying approach for semantic data bases, thus commonly semantic data are processed and visualized. Thereafter we introduce how well-structured metadata can be processed and introduce the approach for semantic enrichment of those data exemplary. The gathered data are then represented in a uniform way including aggregated information about the data-set and information about the structure of data. This internal representation of data will give us the foundation for measuring relevance of certain contextual information, which will be introduced in the following parts of this section.

6.1.1.1 Inclusion of Semantics by Iterative Querying

The increasing role of Linked-Data and Linked Open Data (LOD) [2–5] in Web, where masses of data are daily added and annotated with semantics provides excellent conditions for information search. LOD has experienced great growth in the open internet and became an established way of data representation for conceptualizing knowledge entities and describing semantic relationships between knowledge entities and domains. The Linked Data format is not only used to model a specific domain by a small set of knowledge engineers, it is more a reflection of knowledge interpretation of an entire community, which models domain-comprehensive knowledge for structure and disseminate it to a diversified audience. A single Linked Data data-base gains millions of knowledge entities per day and grows faster than expected [5].

Although, the data in the LOD data-bases are semantically well-defined, the amount of data is more than sufficient and their structure provides the opportunity for the usage of alternative knowledge-acquisition and interaction with semantics, today's user interfaces of Linked Open Data do not really evince an added value to existing visualization for exploratory search, as already outlined in Chap. 3. Existing semantics visualization techniques do not consider the surpluses of the Linked Open Data structures, where the semantics structure has to be built-up with a routine of query requests. They focus on various but specific ontology characteristics. The complex structure of the Linked-Data varies, based on the data-base and the way how these data are queried. The heterogeneity of the requested data should be exploited for the visualization and enable a more efficient interaction with the underlying semantics.

The semantic structure of Linked-Data is that of a light-weight ontology, which consists of concepts, sub-concepts and instances, commonly available through RDF-Triples. This structure provides useful information for grouping data into categories and subcategories and supports the process of exploratory search. For instance if a user query results in thousands or millions of result entities, it is necessary to provide a way to refine the search. This refinement can be performed by using different semantic relationships to illustrate the context of the searched query and enable a visual refinement on the data. Although the semantic and LOD data-bases contain such a semantic, users' queries are commonly responded as a set of instances that matches the query. The resulted set contains in the first step no semantic information and is commonly a "list" of instances corresponding to the users' search query. Therefore, it necessary to re-build the semantic of the data in a way that supports the search process and helps users to refine their query either visually or by reformulation.

For information search, where the schema of the semantic structure is not known before querying or searching, the semantic rebuilding or inclusion can be performed by iterative querying on the data-base as we proposed in [5].

One method to gather relevant structural information is the categorization of the instances to their related concepts. The categorized view on the data enables a more efficient differentiation between the resulted data entities in form of instances [5]. Let us take the term "Merkel" as an example for the proposed categorized view on data: If the user searches for *Merkel* in a semantic data-base, a set of instances that are "labeled" with *Merkel* results. Without a categorization of the instances in relation to their concepts, the user has to inspect [6] the instances and build knowledge about the term to refine it, e.g. with "Merkel city" or "Merkel film actor". A conventional search engine would provide, based on the general rankings, for the term *Merkel* in the first pages information about the politician and chancellor "Angela Merkel" and would not support explicitly the exploration process. Although the most results on a conventional search engines would refer to *Angela Merkel*, there are other data entities, named *Merkel*, e.g. a city in Texas (United States), a film actor, or a medical disease. A user, who searches for the term *Merkel*, would not get this information on a high-level of abstraction without the categorization to the concepts of the instances. A categorized view would further help to refine the search, e.g. if the user is interested in the city, she would choose the category of cities or locations (depending on the schema). But another main aspect is that the user gets the categories of all instances of her search in a single and comprehensible graphical view [7] that enables a comprehension of the entire searched term and leads to a learning effect [8].

The inclusion of categories can be performed by using the inheritance-relation (e.g. *instanceOf*) combined with an iterative querying. The iterative querying can be performed bottom-up or top-down. In both cases, first the number of the resulted instances is queried to ensure that a categorization makes sense. In case of low amount of instances, e.g. three resulted instances, the categorization of the instances would not make really sense, thus these can be visualized at once. The information from which concepts the resulted instances inherit, is important regardless of the way how they are visualized, thus this information are further used to model the users abstract level of interest and knowledge. The iterative top-down querying model starts with the search-term of the user and queries all instances for that particular term. The semantic data-base returns a set of instances that contain the searched term. Each instance contains an URI for a unique identification of the instance and commonly some properties, e.g. geographical or temporal information. After that the top-down query method queries for each instance the highest concept from which the particular instance inherits. The direct querying of the highest concept (commonly called *domain*) is supported by some LOD data-bases, e.g. *Freebase* [9, 10]. At this stage, for each queried instance the highest concept is given and each of the high-level concepts contains a set of the queried instances, whereas this is a subset of the entire queried instances. The iterative querying starts with requesting the sub-concepts of each high-level concept, where the result is a set of concepts that inherits from the high-level concept and each sub-concept contains a set of queried instances with at least one resulted instance. This approach enables to consider the amount of sub-concepts and if there exists just one sub-concept it can be ignored for the data model to reduce the interaction costs. Let us take our example of the term *Merkel* and let us assume there is an instance with the domain (highest-level) concept *Persons* and there is a sub-concept *Film Actor* and the only one instance is an instance of the sub-concept *Film Actor*, this sub-concept can be ignored thus both concepts *Persons* and *Film Actor* contain the same instance. But on the other hand this information is not provided for the user that the instance is a film actor. The procedure of iterative querying the sub-concepts is a routine that ends with concepts from which the instance is inheriting (instanceOf). The top-down procedure is illustrated schematically in Fig. 6.2.

The bottom-up iterative querying of the concepts is similar to the introduced approach with the main difference that the super-concepts are requested instead of the sub-concepts. The procedure starts with querying the instances for a search term. For each instance the related concept is queried. Further for each concept the super-concepts are queried until there are no more super-concepts. The main advantage of this procedure is that it can be applied on any data-base that provide in some way an abstraction level in form of concepts or categories. Disadvantages are that aspects like multiple-inheritance or concepts and super-concepts with one instance cannot be differentiated. This may result in more interactions of users through the concept hierarchy to get the instance and therewith the information entity. Figure 6.3 illustrates the results of our example *Merkel* with both procedures.

Our approach improves users' search by making use of the existing categorization on schema level by helping to refine the query based on a category and having an overview of the entire abstract concepts that are related to the searched term. As search results are commonly illustrated as a lists sorted by relevance, with semantic data, we can improve the visualization of search results to help the user to further refine his query and gather the conceptual context. The existing categorization of the

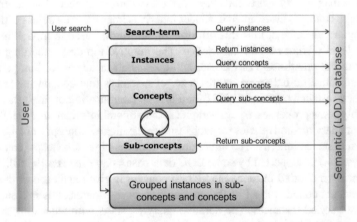

Fig. 6.2 Schematic illustration of the top-down iterative querying of concept relations [5, p. 245], [11]

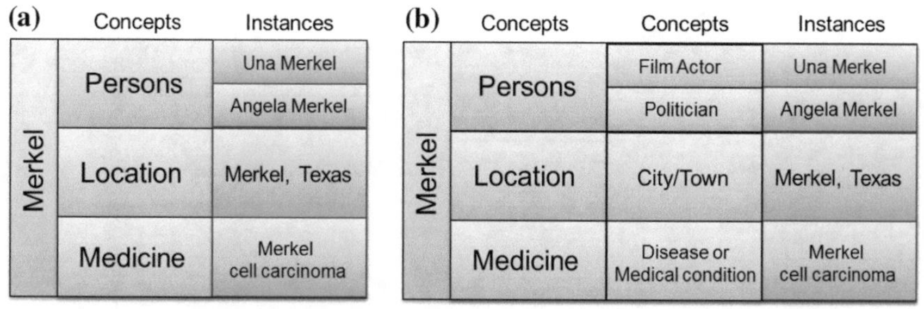

(a)

	Concepts	Instances
Merkel	Persons	Una Merkel
		Angela Merkel
	Location	Merkel, Texas
	Medicine	Merkel cell carcinoma

(b)

	Concepts	Concepts	Instances
Merkel	Persons	Film Actor	Una Merkel
		Politician	Angela Merkel
	Location	City/Town	Merkel, Texas
	Medicine	Disease or Medical condition	Merkel cell carcinoma

Fig. 6.3 Example for the semantic inclusion of concepts in search: **a** illustrates the top-down inclusion of concepts with ignored sub-concepts and **b** illustrates the bottom-up approach with deeper concept hierarchy (adapted from [11])

data can be used to structure the search results within these categories. Users can choose high order categories to see the conceptual context or lower level categories to decrease the number of relevant search results. As this procedure enables to refine the search and illustrates a limited context of the search term results, the contextual schema of the search results is not generated. This includes the semantic relations of instances (SR) in terms of $SR = \langle F, R_o, T \rangle$ where $F = \{f_1, f_2, \ldots, f_n\}$ is a set of instances that were received based on the users' search, and $R_o = \{r_1, r_2, \ldots, r_m\}$ is a set of instance-relations that describes a direct relation to T and is a subset of all relations $R_c \subseteq R$, and $T = \{t_1, t_2, \ldots, t_k\}$ is a set of instances. As the first iterative querying provided us with the concepts the semantic structure to enable a categorized view on data, the semantic relations to other concepts were not investigated. The reason to split this both querying mechanism is quite simple. If we would query all related instances of the search, a semantic context would be provided but will confuse users as there would be results that were not part of our search. For example, if we would search for the term "Merkel", we would get the categories *Cities*, *Persons*, and *Religion*. The category *Persons* would have results like *Angela Merkel, Una Merkel*, but also persons who have direct relation to the queried terms, e.g. *Joachim Sauer* (thus married with Angel Merkel). Further the category *Religion* would include the instance *Protestantism* (thus Angela Merkel's is protestant). This will be applied to all instances that were the result of the users' query. The diversity of information leads to confusion and reduces the precision of the search.

To face this problem but provide the semantic context of a data entity of interest, we designed an iterative querying on instance-level that loads the related instances too but does not visualize them. The visualization of the instances is performed on-demand. Further relations, e.g. instances that are related to instances are not queried. These kind of instances can be queried completely on-demand, if the user is interested in that information. To query the instances that are directly related to the results of the search term, we applied a similar procedure as those for gathering the categorical information. The iterative querying model on instance-level starts with the search-term of the user and queries all instances for that particular term. The semantic data-base returns a set of instances that contain the searched term. Each instance

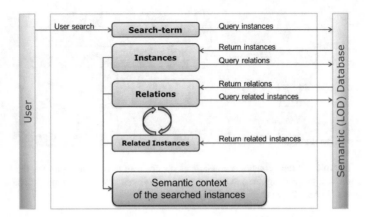

Fig. 6.4 Schematic illustration of the iterative querying of instance relations [5, 11]

contains again a URI that is used for querying the direct relation of that particular instance. For each relation the direct instance is queried, whereas the instances remain until the user selects that particular instance invisible. The iterative querying ends with the last related instance of the last searched instance that is queried by the user. The procedure of iterative querying on instance-level is illustrated schematically in Fig. 6.4.

The iterative querying approaches introduced in this section enable a semantic inclusion of various Linked Data bases that gained popularity and widespread in recent years. The categorization of resulted search entities enables in particular to refine the search and have an insight into the different areas, a searched term appears. The semantic relations enable to see the first level correlations to other entities and gather a contextual meaning from a graphical picture. This further enables to support the exploratory search process by navigating through the semantic relations and acquire knowledge through the navigation process. In general the inclusion of semantics from semantic data-bases is the first step to provide semantics visualizations. But semantics visualization is not limited to visualize only formal or semi-formal semantic data. As described in Sect. 3.5 a meaningful relationship can be gathered from Metadata too and visualized as semantic relations. The next section will therefore introduce the way how semantic information can be gathered from Metadata.

6.1.1.2   Inclusion of Semantics from Metadata

Semantic or Linked data already provides a simple access to the structure of the data and enables therewith an easy visual representation. With the above described iterative querying knowledge about the schema of the underlying semantics is not necessary anymore. Although, the amount of semantic data in particular as Linked-Data increased enormously, there still exist the necessity to visualize structured

metadata that do not contain explicit semantic notations. This section will introduce an approach for generating light-weight semantics from well-structured metadata. It should be clarified that although the approach can be applied to various
forms of metadata, there will still be the need of some manual configurations. The
manual configuration on generating light-weight semantics from metadata are performed through a XML-based configuration file that we named *Semantics Visualization Markup Language* (SVML). This part of the section will introduce based on
a common data-structure for bibliographic entries, namely BIBTEX the approach of
semantics inclusion exemplary. It should be outlined that other metadata structures
may need the mentioned configuration as a prerequisite for a proper visualization of
the underlying data.

Databases containing metadata are returning, similar to semantic data-bases, a set
of entities that match the search. Each entity of those structured data can be identified
by an opening and closing tag. In our example, Listing 6.1, the tag *@article* opens
the content of one entity. The contained content is further labeled with several tags
that all belong to the opening tag, which provides in most cases the type of the entity,
in our example an article. The contained tags provide for human already a semantic
correlation of the information in that article. A human might see that the journal in
which the article appeared is Computer Graphics Forum and there are two authors,
who wrote the article. To visualize this information a similar routine was provided
as the iterative querying for semantic data-bases with the main difference that the
iterative inclusion was made based on the enclosed tags of the returned entities.

```
@article{CGF28-3:751-758:2009
journal = {Computer Graphics Forum},
author = {Harald Sanftmann and Daniel Weiskopf},
title = {Illuminated 3D Scatterplots},
pages = {751-758},
url = {http://diglib.eg.org/EG/CGF/volum28/issue3/
        v28i3pp0751.pdf},
volume= {28},
number= {3},
year = {2009},
abstract = {In contrast to 2D scatterplots, the existing 3D
    variants have the advantage of showing one additional data
    dimension, but suffer from inadequate spatial and shape
    perception and therefore are not well suited to display
    structures of the underlying data. \ldots },
note = {Categories and Subject Descriptors (according to
        ACM CCS): I.3.7 [Computer Graphics]:Three-Dimensional
        Graphics and Realism},
}
```

Listing 6.1 Example of a standard BIBTEX entry

After a set of such metadata are received by the data model, a routine starts for each entity that first identifies the opening and closing tag. After that each contained tag is assigned as relation or properties. A rule-based engine assigns commonly all short values in particular numerical values, such as pages, volume, and year as properties of the instance. Further we defined some predefined names such as title, author, person, book, institution, etc. that are used in most metadata. The names are saved in a bag of word with a XML-based description on how to handle them. For example the tag author leads to a relation "author of", and uses the words "and", ";", or "&" as delimiters for the content. In an enhanced version weighting algorithms are used to differentiate between authors and disambiguate the name [12]. Therefore the frequency distribution of the co-authorship of not unique names is used to identify for example if an author "D. James" is the same "D. James" of another article and if "Daniel James" is the same person [12]. For all unknown tags the name of the relation is generated by the tag's name. For example let us assume that "author" is not in our bag of words. The approach will draw a direct relation between title and the author and name the relation "author". We have experienced that this simple semantic enrichment already leads to a better understanding of the entire topic. Further those tags that contain more than 100 characters are handled as content. In the case of our example the abstract would be assigned as content and will be displayed on demand. Our investigation on metadata showed that some text documents that are full articles are returned with the metadata. Figure 6.5 illustrates schematically the described procedure.

The introduced approach works best, if all the metadata tags are known and rules are defined on how to handle them. Thus, the most metadata structures do not contain a complex structure, this can easily be performed for the metadata. However, the introduced approach provides a way to generate a light-weight semantic on-the-fly without the need of any configurations on data level. Every enhancement of our *bag of words* makes it more probable that in future cases the entities are recognized and the need for a manual configuration decreases.

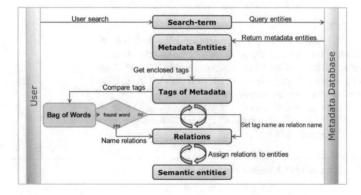

Fig. 6.5 Schematic illustration of the iterative inclusion of semantics from metadata [12]

6.1.1.3 Internal Data Representation

The introduced methods and routines enable to gather the semantic structure of data in a search query. It should be outlined that the iterative querying methods are not necessary to perform for ontologies that should be visualized from files. In each case, regardless if the semantic structure is gathered through the search query or is predefined in a schema, the data to be visualized have to be transformed in an internal data model to perform various measurements and analysis. For this purpose, we defined an internal data representation that contains the most important data values and characteristics. This internal representation of data aims at providing a fast visualization of the underlying data and fast measurements and computations of several weightings and frequencies that will be introduced in the next sections.

The internal data representation contains of three main components, *objects representation*, *structure representation*, and *informational representation*. In the first step the entire gathered semantic information are stored in one data-set-object. Thereby each data-set is represented by one data-set-object that contains in an informal way the entire data, including those that are not visualized. The data-set object is the informal representation of the entire set of data from a certain data-source. The visualization of this data requires a formalization of the data. Therefore data-objects are generated that includes the main values of data if this should be visualized. This instantiation step generates from the data-set-objects, data-objects that include all the required information and values of each data-entity. The data-objects represent therewith a data entity with the following attributes:

- *Semantic Entity*: This attribute contains the entity to be visualized with a unique resource identifier. The entity is an element of the result set of users' query and is the main attributes of our data-object.
- *Semantic Label*: The label of the data-object is the label that is visualized next or on the data-entity. In many cases the data-bases provide a dedicated label for each entry. If a data-base provides the URI of an entity without a label, the label is generated through the URI by using the last string separated by "/" and discarding the previous URI parameters.
- *Semantic Relations*: This attribute contains all direct relations of the data-entity. Thereby both, incoming and outgoing relations are stored in this attributes together with their labels and the target or source URI or entity.
- *Semantic InstanceOfRelation*: This attribute contains information about concepts from which the data entities inherit. Thus commonly data-source allow multiple-inheritance, this attribute contains a list of all concepts with a direct inheritance relation.
- *Semantic SubClassOfRelation*: This attribute contains all concepts from which the concepts of the data entity inherit. This information is gathered commonly through the iterative querying process as described in the previous parts.
- *Semantic Property*: In some cases the data-entity contains one or a set of certain properties. This attribute contains the values of all existing properties.

- *Semantic propertyType*: If the data-entity contains properties, the type of the properties is stored in this attributes. The data-object differentiates between nominal, ordinal, quantitative, temporal, and geographical properties.
- *Semantic SourceValue*: To enable the visualization of the content the source of the URI is gathered and stored in this attribute. In some cases there exist no source and the URI is just used to ensure a unique identification. Further there are cases, in which the URI refers to another source, e.g. the Wikipedia. In these cases the reference is used to get the sources and enable a visualization of the content.
- *Semantic SourceType*: Similar to the mentioned property type, the source type is determined, if a sources is given. Again the data-object differentiates between nominal, ordinal, quantitative, temporal, and geographical source types.

The object representation enables already to visualize the requested data and would be sophisticated for many proprietary visualizations that just aim to visualize the semantics. It contains all information of the requested data-entities with their relations to concepts and other entities. Figure 6.6 illustrates the object representation of our approach.

The *structure representation* goes one step beyond the data and represents the structure of data in form of hierarchical and arbitrary relations and their related entities. This enables a more detailed view on the data structure and the choice of adequate visual layouts. In general the structure representation is divided into two major parts, the hierarchical and arbitrary relations. The hierarchical part contains all instances in form of children of certain sub-concepts that are parent of the instances in the data model. Further the relations, their number, the instances, and the overall

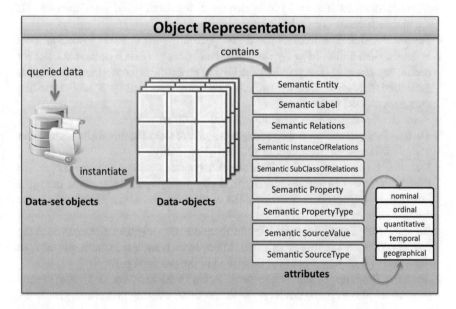

Fig. 6.6 Schematic illustration of the object representation

number of instances that inherit from sub-concepts are stored in the children object. The second object contains all sub-concepts of the concepts in the data-set objects. It contains all concepts that are sub-concepts of another concept in the data domain, their numbers, the inheritance relations on concept-level, and the numbers of all those relations. The third object of the hierarchical relations, the super-concept-object provides the inverse information for faster data processing and contains all super-concepts, their number, the super-concept relations, and the number of those relations.

The arbitrary relations contain all structural information on instance-level and are divided into the objects, *incoming relations* and *outgoing relations*. The object incoming relations provides information about the source of a relation that is commonly another instance with a reference (URI), the relations that comes with their type, and the number of the instances and relations that are commonly the same number. With the incoming relation object, instances with many references can be determined. The second object, the outgoing relations object, contains all information about outgoing relations, the target instance, and their numbers. These two objects enable to identify instances with huge or less amount of incoming and outgoing relations and lead to assume and calculate the relevance of the instances. The relevance measurement will be described in the following sections. Figure 6.7 illustrates schematically the described objects and the contained information.

With the structure representation the general structure of the data is determined and builds the foundation for determining sophisticated visual layouts. Based on the structure representation and the object representation, we defined a third object

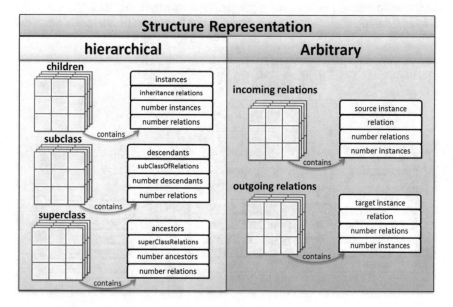

Fig. 6.7 Schematic illustration of the structure representation

that provides general information about the underlying data to be visualized. The information object aims at giving general information about the data and enable to invalidate parts of data. We mean with invalidate the inverse process of instantiate as described for the object representation. As the data are requested from certain data-bases, more data are stored in the internal data-set object than visualized. By instantiating, this data are transformed in objects and are then visualized. It is necessary to reduce the information objects during the users' interaction with data. Thus users' interaction may lead to further instantiation of object or rather to new queries, the visualized information are then not of interest anymore and would lead to confusing results. The invalidation of the data objects, enable to remove those data-objects from the visualization. Further the information representation summarizes and aggregates all information of the data at once. It contains three main objects, the instances-object, the relations-object, and the concepts-object. The instances-object contains all instances in the data set, information about the root concept of the instances, the parent concepts of the instances, the relations, and the source of the instances. The source enables to illustrate the content of a certain instance. The relations-object contains all relations in the data set, all related instances and concepts, and the URI of the relation as source. The concepts-object contains all concepts in the data set, the inherited instances, and all types of relations to other concepts. Further the top or root concept of the entire data-set is stored to identify the data-set itself. Our approach is able to visualize more than one data-set at once, so this information is necessary to identify the data-set. Figure 6.8 illustrates schematically the described objects and the contained information.

The internal data representation is stored in three main components that contain the semantic data and various and aggregated information about data. This structure of representation allows not only a sufficient visualization of the underlying data but

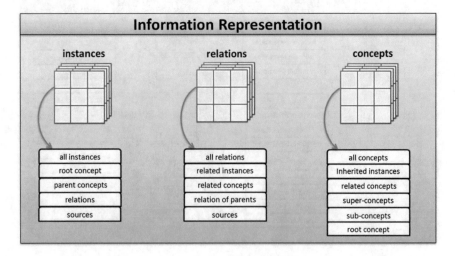

Fig. 6.8 Schematic illustration of the information representation

far more the computation of several further aspects that refers, e.g. to the relevance of contextual information. The next section will introduce the main relevance measurements of "focused objects". Further we describe how the iterative querying is used to gather the stored quantitative information.

6.1.2 Data Feature Model

The data model consists of an internal representation of the underlying data that can be visualized. It includes the data structure in form of semantic relations, instances or entities, and concepts that build the categorical structure of data. This information enables a more efficient visualization of the data. However, the data model does not contain at this stage any information about the *data characteristics* or *data features*. We have illustrated in Chap. 2 that one of the main factors for visualizing information are the data characteristics in form of dimensions or data-values and their ability to be ordered. Based on these features the process of information visualization is able to perform the visualization and provide efficient visual layouts and visual variables. A data-set for instance with numerical values of longitudes and latitudes can be visualized by its quantitative or ordered characteristic, which may help some experts to extract the needed information or it can be visualizing by using a geographical map that makes the information acquisition process more efficient. Another example could be a data-set with strong hierarchical correlations of the data, which can be visualized in different graph structures or even as scatterplots. But the most efficient way would be to enlighten the hierarchical structure by choosing an adequate visual layout for the underlying hierarchical information.

The here addressed visualization *design* is of great importance for the adaptation of visual environments. Only if a system has the information about the features of the underlying data, an efficient adaptation is possible. As the two mentioned examples could outline different kind of information leads to different level of adaptation. As data values in form of longitudes or latitudes may change the visual layout, numerical values that cannot be assigned to any predefined content may lead to change of the use of different visual variables (e.g. color or size) [1, 13, 14]. This part of the thesis illustrates two main aspects, first the gathering of quantitative information of semantic data to determine the main characteristics of the underlying data and afterward two algorithms for computing the relevance of semantic neighbored instances and concepts by new approaches of weighted frequency distribution. This section was partially published in [15].

6.1.2.1 Quantitative Analysis of Data

Semantic data are in general structured data that can be described best as graphs. As we illustrated in the previous sections the type of relations can be abstracted to three main relation types: relations between instances of semantic data describe the contextual "meaning" of instances and are instantiated and valid for the related instances. They express in best case a meaningful sentence in terms to disambiguate the term and provide a context. An example for such a sentence could be "Obama was born in the USA". In this sentence there are two instances "Obama" and "USA", whereas the relation is built with the predicate "was born". The second type of relations, the inheritance relations, enable the formulations of similar meaningful sentences, whereas the main goal is to disambiguate a term and categorize it into a concept of entities with the same attributes. Examples for inheritance relations may be "Obama is a person" or "Obama is a city". The main difference is that inheritance relations connect an instance to concepts. The concepts are commonly abstracted information and provide a high-level view on data. The third category, the relations between concepts or concept-relations (class-relations) provide on the abstract level information that are valid for all entities that inherit from the concepts. Examples for such kind of relations could be "Country has a capital" or "Person has a gender". Concept relations are abstract formulation of the schema of data. The two example sentences say that each element of the set "Country" has an element of the set "Capital" and every person has a gender.

This introduction should outline that although relations in semantic data may be considered as very similar in terms of linguistics, they contain different types that are important for their visualization. Relations between instances and concept-relations can be considered as arbitrary graphs. Their structure does not provide any kind of hierarchical information, whereas the inheritance-relations provide hierarchical structures that can be used to outline or highlight the hierarchical structure of the data in a visual manner. Furthermore, the data that we use to visualize are queried on demand. Therewith various information about the number of returned entities and their properties, number of concepts, strength of the hierarchical structure and so forth are not available before the search process. Thus the data bases commonly change in course of time; a preprocessed analysis of the data-base does not make sense. Therefore it reasonable to use the iterative querying approach introduced in Sect. 6.1.1.1 for gathering quantitative statements of the data and determine the data characteristics from the gathered quantitative measures.

In each of the iterations of our iterative querying, we apply a quantification of the data elements, their properties, their concepts, their relations and in particular their relation types. As illustrated the in Fig. 6.9, the common starting point of visualizing semantics with our approach is a user's search. The search term returns a set of instances, where each instance may have further attributes, such as temporal or geographical attributes. The instances are stored, together with their attributes, as the entities to be visualized. In the next step, the iterative concept querying starts: We start with the concepts that any of the visualized instances is related to. For each of these concepts, we query the sub-concepts and add them to the set of all active concepts. At the end of this iteration, we have all concepts that build the schema of the underlying semantic data that was requested by the user, and the set of inheritance relations, which give use information about the hierarchical structure of the data. In order to have information about all instances that are related to any of the active concepts, the iterative querying of instance relations is applied: For each of

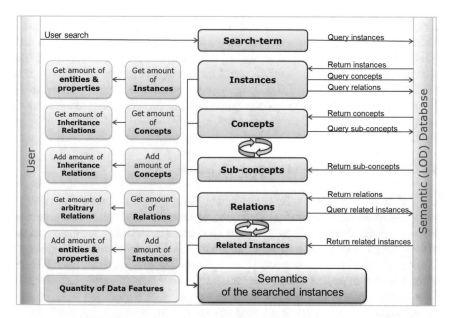

Fig. 6.9   Approach for gathered quantitative information of the semantic data

the active concepts, we query all relations that this concept is involved in, and the instances that are related to the concept via any of these relations. The entire set of quantitative information of the data is stored as the "Quantity of Data Features", as illustrated in Fig. 6.9.

The retrieved quantitative information enables to choose an appropriate set of visual layouts for the results the users' search. The number of entities is important to choose a visualization that is capable to illustrate that amount of data if necessary, or to categorize the entities by their concepts to support the exploration process. The categorization ability can be determined by the number of concepts that are given in a data-set. However, the number of relations enables to determine the structure of the data, the correlation between arbitrary relations and inheritance relations is an indicator for the choice of the right graph-layout to visualize the structural information and support therewith the exploration process. As we will illustrate in the following Sect. 6.1.2.2 the relations and in particular the type of relation can be used to determine the weight of neighbored entities to a selected instance and derive the contextual relevance. The retrieved number and type of entities' properties enables to recommend alternating visualizations. In many semantic data-bases temporal and geographical information are assigned as properties of entities. By retrieving the existence and the number of such properties, the entities can be visualized by their temporal or geographical attributes through maps or temporal visual layouts.

The here gathered quantitative information are essential to provide adaptive visualizations. The number of data entities, the structure that is indicated by the type and number of relations, and the properties of the entities enables to assume the complexity of the data to be visualized. The introduce approach just counts the different characteristics of the semantic data and gives a picture of the entire search results. These numbers gives no contextual information about selected entities in the semantic data. The next section will introduce two algorithms that enable to determine the contextual relevance of selected instances by using the quantity information. The algorithms will use quantitative measurements to derive the needed information and provide a "weighted semantic context".

6.1.2.2 Weight-Analysis of Semantic Relations

The gathered quantitative information about the retrieved semantic relationships allows us to derive a contextual weighting of semantic relations. The context of semantic relations can be described by various variables that influence the relevance of the semantic neighbors. A semantic neighbor might be a hierarchical or an arbitrary relation between instances or concepts and provide information about the relevance of the neighbor regarding a particular data entity. To measure the concept relevance, from which a certain instance is inheriting, we introduce the *Inverse Instance Frequency* algorithm (iIf) that was adopted from the Term Frequency algorithm with a frequency probability based on the instances in correlations to their concepts. The Inverse Instance Frequency (published in [15]) proposes that the more instances inherit from a particular concept, the less relevant it is in context of a particular instance. Let us explain this scenario by a short example: we assume that the particular instance "Barack Obama" inherits (isa-relation) from three concepts: "Person", "President", and "Author". Let us further assume that the data base consists of a representative amount of data entities. In that case the concept "Person" has the most inherited instances, followed by "Author", and "President'. With the Inverse Instance Frequency algorithm, the concept "President" is weighted higher than "Author", and "Author" is weighted higher than "Person" (see Fig. 6.10). With other words, the inverse number of instances that inherit from a certain concept is an indicator for the relevance of that concept regarding the particular instance.

For a better comprehension of the above described example, we formalize the example by introducing a triple that describes the inheritance relations as $IR = \langle T, R_c, C \rangle$, where $T = \{t_1, t_2, \ldots, t_n\}$ is a set of instances, and $R_c = \{r_1, r_2, \ldots, r_m\}$ is a set of inheritance-relations (e.g. "isa" or "instanceOf") and a subset of all relations $R_c \subseteq R$, and $C = \{c_1, c_2, \ldots, c_k\}$ is a set of concepts. Thereby R is a set of all possible relations. Let us assume that based on our example the database contains following semantic data:

- $T = \{$Obama, Kennedy, Shneiderman, Fraunhofer IGD$\}$, a set of instances.
- $C = \{$Person, Author, President$\}$, a set of concepts.
- $R_c = \{$isa, instanceOf$\}$ is a set of concept relations. These are the relations between instances and concepts. Thus, each element $r \in R_C$ is a particular relation between the instances and the concepts, namely $r \subseteq T \times C$.

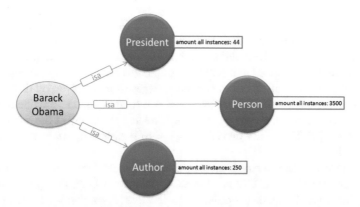

Fig. 6.10 Example for concept relevance in context of instances: The instance "Barack Obama" has inheritance relations to the concepts Person, Author, President. Each of these concepts has a total number of inherited instances as "isa" relations. Thus the concept "President" has the least number of instances. The fact that "Barack Obama" belongs to the concept "President" is most important inheritance relation

Based on the introduced exemplary instances, concepts, and inheritance relations, the inheritance relation of our example can be expressed by:

- isa = {(Obama, Person), (Obama, Author), (Shneiderman, Author)}

To illustrate the *Inverse Instance Frequency* algorithm and later the *Direct Relation Frequency inverse Relations Frequency* algorithm, we introduce some general definitions for a better comprehension of the introduced sets, terms, and definitions. Given two sets $A = \{a_1, a_2, \ldots a_n\}$ and $B = \{b_1, b_2, \ldots b_k\}$ and a set of relations $R = \{r_1, r_2, \ldots r_m\}$, where each element $r \in R$ is a relation between A and B. Let $\mathcal{P}(M) = \{N | N \subseteq M\}$ denote the *power set* (the set of all subsets) of a set M. We define an *edge function* $E : A \times \mathcal{P}(R) \rightarrow R$. It provides the set of relations from a given set $R_0 \subseteq R$ that relate an object $a \in A$ to any object $b \in B$. Therewith, we define $E(a, R_0) = \{r \in R_0 | \exists b \in B.(a, b) \in r\}$ for $a \in A$ and $R_0 \subseteq R$. Applied to a semantic graph, this function can be imagined as a function that provides all relations that imply an edge that starts at a.

Further we define a *source function* $S : B \times \mathcal{P}(R) \rightarrow B$. It provides us the set of objects from A that are related to a given object $b \in B$ via any of the relations of a given set $R_0 \subseteq R$. The definition is $S(b, R_0) = \{a \in A | \exists r \in R_0.(a, b) \in r\}$ for $b \in B$ and $R_0 \subseteq R$. Applied to a semantic graph, this function can be imagined as a function that gives us all source nodes of the edges from the given set that end at b. A *destination function* $D : A \times \mathcal{P}(R) \rightarrow B$ is further defined to provide us the set of objects from B that a given object $a \in A$ is related to via any of the relations of a given set $R_0 \subseteq R$. The definition is $D(a, R_0) = \{b \in B | \exists r \in R_0.(a, b) \in r\}$ for $a \in A$ and $R_0 \subseteq R$.

Further we define a *source function* $S : B \times \mathscr{P}(R) \to B$. It provides us the set of objects from A that are related to a given object $b \in B$ via any of the relations of a given set $R_0 \subseteq R$. The definition is $S(b, R_0) = \{a \in A | \exists r \in R_0.(a, b) \in r\}$ for $b \in B$ and $R_0 \subseteq R$. Applied to a semantic graph, this function can be imagined as a function that gives us all source nodes of the edges from the given set that end at b. A destination function $D : A \times \mathscr{P}(R) \to B$ is further defined to provide us the set of objects from B that a given object $a \in A$ is related to via any of the relations of a given set $R_0 \subseteq R$. The definition is $D(a, R_0) = \{b \in B | \exists r \in R_0.(a, b) \in r\}$ for $a \in A$ and $R_0 \subseteq R$.

For measuring the instance frequency distribution, the "number of all instances" of a certain concept $c \in C$ is of great interest. This number can simply be defined as the cardinality of the set that is provided by the *source function* for a given concept and a set of relations, thus, $|S(c, R_C)|$ is the number of instances that have a relation to concept c, for example:

- $S(\text{Author}, R_C) = \{\text{Obama, Shneiderman}\}$.
- $S(\text{President}, R_C) = \{\text{Obama, Kennedy}\}$.
- $S(\text{Person}, R_C) = \{\text{Obama, Kennedy, Shneiderman}\}$

The inverse frequency distribution of a concept c for a given set of concept relations R_C is defined as a function $\omega_0 : C \to \mathbb{R}$ (see Eq. 6.1).

$$\omega_0(c) = \frac{1}{|S(c, R_C)|} \tag{6.1}$$

The measured weight provides a non-normalized inverse frequency distribution of the instances in inheritance correlation to their classes. Thus the weighted ω_0 will be used for further weight-measurements, in particular for the interaction behavior of users. We further calculate the empirical probability ω_{ilf} of all the concepts, from which the instance t inherits. This weighted Inverse Instance Frequency (iIf) is a function $\omega_{ilf} : T \times C \to [0, 1]$, defined in Eq. 6.2, where $D(t, R_C)$ is the destination function, which in this case provides the set of concepts to which the given instance t has a concept relation.

$$\omega_{ilf}(t, c) = \frac{\omega_0(c)}{\sum\limits_{c_i \in E(t, R_C)} \omega_0(c_i)} \tag{6.2}$$

With the normalized measurement of the empirical probability ω_{ilf}, the related concepts, from which the particular instance is inheriting can be notated in a normalized way and used for further measurements on empirical probabilities or probability distributions. Figure 6.11 illustrates our examples with the weighted values.

The Inverse Instance Frequency provides a relevance weighting of instances in correlation to their own concepts. This weighting helps in particular to derive information about the relevance of the related contextual information on an abstract level of concepts, e.g. by visually illustrating the relevance of classes. Semantic data provide

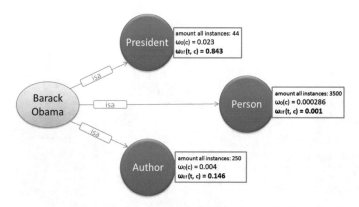

Fig. 6.11 Computed example for the Inverse Instance Frequency algorithm, including $\omega_0(c)$ and the normalized $\omega_{ilf}(t, c)$

another aspect that is particularly interesting for weighting the contextual information by measuring the relevance of instances that are correlated to an instance of interest. The inheritance relation to a concept is just one way of describing semantic relationships between semantic entities and provides a weighting of related concepts through the introduced *Inverse Instance Frequency* algorithm. Another way of semantic relationship is the relation between instances and may lead with a proper weighting to measuring the relevance of the directly related instances.

We propose the *Direct Relation Frequency inverse Relations Frequency algorithm* (dRf-iRf) to measure how "important" a certain direct relation and its corresponding related instance are for a focused instance (published in [15]). The main idea behind this algorithm is as follows: We consider all instance relations that a focused instance has to any other instance. For each of these other instances, we measure how "unique" the particular relation is for this instance. This "uniqueness" is the basis for the computation of how important the relation from the focused instance to the other instance actually is.

Let us explain the introduced idea with based on the Fig. 6.12, where the focused instance *Ben Shneiderman* has three relations to other instances:

- the relation *hasAffiliation* to the instance *University of Maryland*
- the relation *worksAt* to the instance *UMD HCI*
- the relation *isAuthorOf* to the instance *Designing the User Interface*

Each of the instances that *Ben Shneiderman* is related to has other relations as well. For example, other instances may be related to the book *Designing the User Interface* via relations like *hasCited* or *hasPublished*. In our example, there are 1200 other instances related to the book. However, out of these 1200 instances, only 3 have the relation *isAuthorOf* to the book. Therefore, the fact he is one of the few authors of this book is very important and relevant for the instance *Ben Shneiderman*. In contrast to that, there are 45,000 instances that are related to *University of Maryland*

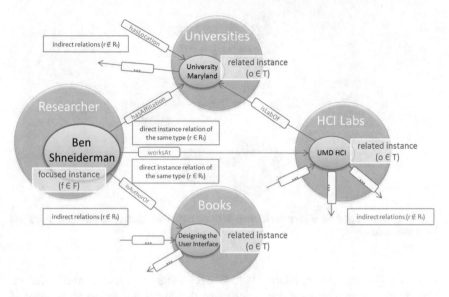

Fig. 6.12 Example for the direct and indirect relations of a focused instance

via any relation, and 9000 of them via the same relation as *Ben Shneiderman* (namely via the relation *hasAffiliation*). Since the ratio in this case is much higher than for the relation *isAuthorOf* and the book, the fact that *Ben Shneiderman* has the relation *hasAffiliation* to *University of Maryland* is less important.

The Direct Relation Frequency inverse Relations Frequency does not work on concept relations, which relate instances to concepts as a form of inheritance, but on instance relations, which relate instances to another instances. For a given set of instances $T = \{t_1, t_2, \ldots, t_n\}$, the set of relations $R_I = \{r_1, r_2, \ldots, r_m\}$ denotes the set of all available instance relations between the instances. That is, each $r \in R_I$ with $R_I \in R$ is a set of tuples defining the related instances: $r \subseteq T \times T$. The set R contains all types of relations as introduced before, therewith is $R_C \subseteq R$ and $R_I \subseteq R$. Let $F = \{f_1, f_2, \ldots, f_n\} \subseteq T$ denote the set of focused or "selected" instances.

We define the *Direct Relation Frequency inverse Relations Frequency* (dRf-iRf) as a function $\omega_1 : R_I \times T \to \mathbb{R}$ as illustrated in Eq. 6.3. The function S here is the *source function*, that returns the set of all instances that are related to a particular instance via one of the given relations.

$$\omega_1(r, o) = 1 - \frac{|S(o, \{r\})|}{|S(o, R_I)|} \quad | \; r \in R_I \text{ and } o \in T \qquad (6.3)$$

The measured weight provides a non-normalized inverse frequency distribution of the instance-relations in correlation to their type. Thus the weighted $\omega_{dRf-iRf}$ will be used for further weight-measurements, in particular for the interaction behavior of users, we calculate further the empirical probability (*weighted dRf-iRf*) as a function

$\omega_{dRf-iRf} : T \times R_I \times T \rightarrow [0, 1]$ as illustrated in Eq. 6.4.

$$\omega_{dRf-iRf}(t, r, o) = \frac{\omega_1(r, o)}{\displaystyle\sum_{o_i \in D(t,R_I)} \omega_1(r, o_i)} \quad | \ t \in T \ \text{and} \ r \in R_I \ \text{and} \ o \in T \quad (6.4)$$

The computation of the normalized dRf-iRf algorithm is performed with the focused instance "Ben Shneiderman". We can calculate based on the example graph for ω_1:

- ω_1(hasAffiliation, University of Maryland) $= 1 - \frac{9000}{45,000} = 0.8$
- ω_1(worksAt, UMD HCI) $= 1 - \frac{23}{15,000} = 0.99847$
- ω_1(isAuthorOf, Designing the User Interface) $= 1 - \frac{3}{1200} = 0.9975$

and the normalized dRf-iRf including the focused instance for $\omega_{dRf-iRf}$ as:

- $\omega_{dRf-iRf}$(Ben Shneiderman, hasAffiliation, University of Maryland) $= \frac{0.8}{2.79597} = 0.2861$
- $\omega_{dRf-iRf}$(Ben Shneiderman, worksAt, UMD HCI) $= \frac{0.99847}{2.79597} = 0.3571$
- $\omega_{dRf-iRf}$(Ben Shneiderman, isAuthorOf, Designing the User Interface) $= \frac{0.9975}{2.79597} = 0.3567$

The normalized *Direct Relation Frequency inverse Relations Frequency* algorithm enables us to measure the relevance of instances in correlation to a focused or chosen instance. Let us take again our example with the graph from Fig. 6.12: we have in this very simple graph three instances that are directly related to the focused instance "Ben Shneiderman": "University Maryland" (instance of the concept Universities) has the relation "hasAffiliation" with our focused instance, "UMD HCI" (instance of the concept HCI Labs) has the relation "worksAt" with our focused instance, and "Designing the User Interface" has the relation "isAuthorOf" with our focused instance (instance of the concept Books). Let us imagine that the book *Designing the User Interface* has some kind of further types of relations, e.g. "citedBy", "publishedBy", or "hasEdition". Let us further assume that our focused instance, *Ben Shneiderman*, has a direct relation of the type "authorOf" and there exist two further instances that have the same type of relation to the same related instance. But a high number of instances have other types of relations, e.g. "citedBy" to that particular instance. The dRf-iRf would weight this particular instance higher in context of the focused instance and would therewith enable a contextual semantic view on the focused instance. On the other hand, if we assume that the focused instance lives in a city named "Baltimore" and has the relation-type of "livesIn" to the instance "Baltimore" and the data are more or less representative, there would be huge number (estimated 600,000) of instances, which will have the same type of relation to that particular instance. Our dRf-iRf algorithm would rank the relevance of the place lower than the book that was written by the focused instance. Figure 6.13 illustrates the already introduced graph with some numbers of relation types and the measured dRf-iRf weight for the focused instance.

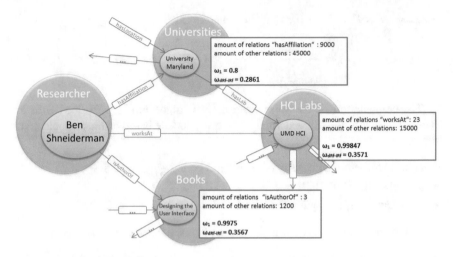

Fig. 6.13 Calculated example for Direct Relation Frequency inverse Relations Frequency algorithm (dRf-iRf)

6.1.3 User Model

An essential part of user-centered adaptive systems is the formal representation of users' behavior with a particular system. Users' behavior may contain various information about users, their interaction or usage behavior, or even demographic or skill aspects that may lead to adaptation effects. As described in the high-level design, the visual adaptation refers to two main aspects: data and user. The previous sections described the way how data and their features are gathered and analyzed to provide a sufficient data model and data feature model for adaptation. This section will describe how user information are gathered, analyzed, and structured to enable an adaptive behavior of visualizations. We will investigate in this thesis only that user information that can be gathered as natural consequence of users' interaction with the system. Explicit information gathering, e.g. by asking users for certain information, will not be considered, thus they may be conceived as obtrusive. This section will first introduce the formal representation of users' interaction to enable on the one hand a unique identification of the interactions and on the other hand further measurements, weightings, and modeling of users' behavior. Based on the formal representation the analysis of user interaction behavior will be introduced that provides already a useful model of users. Thereafter the prediction of users' next possible actions will be described. The section concludes with the formal description of the user model and the way how the probabilities are computed. In this context we will introduce two main components that are used in the following sections, the canonical user model and the individual user model. These models will be used in Sect. 6.2 to compute similarities and deviation between users, the average users of groups, and the canonical user as the average user of the entire models.

6.1.3.1 Formal Representation of Users' Interactions

The adaptation of information visualization requires the acquisition of users' informational context. The context information of interaction events constitutes a useful knowledge source for understanding and analyzing the behavior of the user [16, 17]. Thus in information and semantics visualization, the underlying data with which the user interacts may change due to the character of visualization systems that are designed to visualize different and heterogeneous content. Further the set of visual layouts may be enhanced, changed or reduced, due to the character of an enhanceable visualization approach. For these reasons we will introduce in the next parts of this thesis an interaction analysis algorithm that allows the analysis of users' interaction to model a behavioral pattern and provide predictions on next actions for changeable visual environments and data. The interaction analysis algorithm described in the next parts was elaborated and enhanced in [12, 17–22].

In order to allow the analysis of users interactions in changeable environments with changeable data, the context information is defined with adjustable domains. A domain is according to the *Relational Markov Models* a tree, representing an abstraction hierarchy [23] and indicates a certain field of knowledge about interaction events. An interaction event I is described as a relation instantiated with leaf values of the domains equivalent to *RMM* (see Sect. 4.3.2) as illustrated in the Eq. 6.5 [17, 18].

$$I = r(k_1, \ldots, k_n), \qquad k_i \in leaves(D_i) \quad with \quad 1 \le i \le n \qquad (6.5)$$

Thereby $leaves(D_i)$ are the leaf nodes of the domain D_i and r is a relation over the domains $D_1, \ldots, D_n$ [17].

In context of our visualization adaptation approach, we define three main domains, the *type* of users' interaction, the interaction with the *visual layout*, and the interaction with the underlying *data* [20]. The type of users' interaction is based on the device and the target to be achieved. This domain is not changing continuously, thus the devices and the intended interaction types remains commonly in the process of information visualization. Therewith this domain can be predefined as a taxonomy of interactions. An example of such an interaction taxonomy on the domain *type* is `device.mouse.select`, `device.mouse.removeVis`, or `device.keyboard.search`. The intention to include the domain in the interaction analysis process is twofold: First the interaction with data and visual layout can be identified quite clearly and disambiguated by including this domain and second, in future scenarios our visualization will probably be enhanced to be used on mobile devices or ambient environments. So it is useful to gather the behavioral information, if an interaction is performed with a standard device or is a result of alternating interaction devices. The structure of the domain type is gathered from the system as a result of the events. Each event performed by the user is modeled and named with the intention type, e.g. *select*, whereas the information about the device is a natural consequence of the interaction events.

The second domain of *visual layout* is representing the users' interaction with different visual layouts. The visualization environment contains an enhanceable set of visual layouts that can be used to visualize (semantic) data. The domain starts always with the tag *SemaVis* and indicates that the interaction is performed within the visualization environment. The domain's nodes are followed by a given name for the visual layout, e.g. *SeMap*, *SemaSpace*, or *SemaContent*. The next leave indicates if the interaction was performed on valid area of the visualization. This is to identify users' interaction with the different valid or invalid areas within a visual layout. In case of interacting with a valid and intractable area of a visual layout the next node is the area of interaction that is depending on the *Data* domain, e.g. *concept*, *instance*, *relation*. If the interaction was not performed with a data entity representation, the next leaf provides information about the area of interaction, e.g. *whitespace* or *config*. Not every interaction of users is within a visual layout. Users are able to *add* a new visualization into the screen. In these cases the second leaf indicates via the tag *noVis* that the interaction was not performed within a visual layout. Examples for the domain of visual layout interactions are SemaVis.SemaGraph.instance or SemaVis.SeMap.concept and for interactions outside a visual layout *SemaVis.noVis.semaGraph*. Each new visualization that is included in our visualization environment needs a taxonomical structure of all its functions. This step is performed once and leads to ensure that the interactions with the *visual layout* are gathered in an appropriate way. Further this domain contains all interactions with the main visual interface, which may consist of interacting with the recommendation interface or searching a term in the search-bar.

The third domain, the domain of *Data* contains the semantic hierarchy of the data entities with which the user interacted. The semantic hierarchy of the data is gathered as described in Sect. 6.1.1 and used as taxonomy for this particular domain. With the automatic inclusion of the semantic hierarchy and the generated taxonomy on inheritance-level, any changes of the data-base can be performed without restrictions, thus the domain data always provides appropriate structure for the formal representation of the user interactions. The taxonomic structure is generated as follow: Given are the set $T = \{t_1, t_2, \ldots, t_n\}$ of instances, the set $C = \{c_1, c_2, \ldots, c_k\}$ of concepts and the set $R = \{r_1, r_2, \ldots, r_m\}$ of relations, where each $r \in R$ is a relation between concepts and concepts, or between instances and concepts. Based on this, we define the global set of entities $E = T \cup C \cup R$. The set of relations R contains a subset of concept inheritance relations between concepts and instances $R_I \subseteq R$. It also contains a subset of "sub-concept relations" $R_S \subseteq R$ $\subseteq C \times C$. These sets of relations are combined into a set of *hierarchy relations* $R_H = R_I \cup R_S$. These relations define a hierarchy on the entities: The set of *child nodes* of a certain entity $e_i \in E$ that are contained in a set $E_0 \subseteq E$ is defined as $U_{E_0}(e_i) = \{e \in E_0 | \exists r \in R_H . (e, e_i) \in r\}$. Each node of this domain is built with the hierarchical relations of entities. An example for this domain, containing the hierarchical structure, could be Data.thing.persons.politicians.obama, where *obama* is an instance of *politicians* and this is a subconcept of *persons* and this is a sub-concept of *thing*. The root node *Data* remains to indicate that the followed terms are part of the data domain.

The formal representation of users' interaction enables to model each interaction in a unique way and analyze them to model the behavior or measure predictions and probabilities. The following example (6.6) illustrates an interaction of user with our system.

$$I = r(\texttt{Device.mouse.select, SemaVis.SeMap.concept, Data.thing.persons}) \tag{6.6}$$

The user's interaction in the example (6.6) indicates that the *concept person* was *selecting* in the visual layout *SeMap* by using the mouse [20].

6.1.3.2 Deriving Users' Interaction Behavior

Users' interaction behavior can be defined as the way how users are interacting with a particular system. Different users may have different way to interact and solve tasks with systems. Experts of some system for instance make use of keyboard shortcuts to achieve their goal faster. The interaction behavior can give us information about preferences in system use or even indicates the expertise level of the user. We described in Sect. 4.3.2 different approaches and algorithms that derive information about preferences or users' knowledge through an implicit analysis of users' interactions. The users' interaction behavior can be described as the probability distribution of users' interactions in contrast to the entire possible interactions of the system.

To compute the probability distribution of users' interactions, we first determine the Steady State Vector (SSV) as a relative measurement for the occurrence of interaction events. The Steady State Vector is defined as $\vec{s} = (p_1, p_2, \ldots, p_n)$, where n is the number of all possible interaction events and p_i is the probability that an interaction i occurs. Further the SSV is a normalized probability distribution with $\sum_{i=1}^{n} p_i = 1$ and therewith a probability distribution over the entire possible interactions [17, 18]. Commonly the SSV is computed based on the transition matrix [24]. Thus we are computing a new probability distribution after each interaction sequence; we use instead of the transition matrix the frequency distribution of the interactions. The frequency distribution is computed based on the quantitative occurrence of an interaction i in contrast to the entire interactions performed by a user. The probability for the occurrence of an interaction i is defined as:

$$p_i = \frac{v_i}{|A|} \tag{6.7}$$

where v_i is the amount of all occurrences of the interaction i and $|A|$ is the number of interactions the user performed previously. Commonly a set of training-data, e.g. interaction data from an expert is used to compute the SSV. Thus in case of visualization the domain data is changing continuously, the use of training data make not really sense, thus this data would not represent the users' behavior. We use for the set of all previous interactions A either the set of interactions of the individual user or the set of interactions of all users with a certain data domain A_c. Thereby the interaction behavior is modeled in a canonical user model that can be used to determine the common behavior of all users in context of a data domain [12].

The formal representation of the interactions (see Eq. 6.5) provides context information of the interaction events. Analogue to Relational Markov Models [23], abstractions of interaction events are defined as sets of interaction events by instantiating the relation r with the inner nodes of the domains [17]. A frequency distribution and thereby a weighting or probability of all domains can be computed on each degree of the domain abstraction [23]. Based on the defined quantitative occurrence measurement, we define the function $quant(depth_{D_1}, \ldots, depth_{D_k})$, where $depth_{Di}$ is the level of abstraction for every domain D_i as the hierarchical level of the domain, starting with 0 for the highest level of abstraction. With each occurrence of the function $quant(depth_{D_1}, \ldots, depth_{D_k}))$ a set L of the abstraction levels is generated illustrated according to the abstraction levels of Anderson et al. [23, p. 3] in Eq. 6.8.

$$L = \left\{ r(\delta_1, \ldots, \delta_k) \mid \delta_i \in nodes_i(depth_{D_i}), \ 0 \le depth_{D_i} \le maxDepth(D_i), \ 1 \le i \le k \right\}$$
(6.8)

Thereby $nodes_i(depth_{Di})$ are the nodes of the domain D_i with the depth $depth_{Di}$, and the maximum abstraction level of the domain D_i is defined as $maxDepth(D_i)$. With the defined set of abstractions all interaction events are partitioned into subsets of interaction events which account some similarity in respect of their context information [17, 23]. In our case we instantiate this function with $k = 3$ thus we have the three domains of *Device*, *Visual Layout*, and *Data*. Therewith the function is used as $quant(depth_{D_1}, depth_{D_2}, depth_{D_3})$

The probability p_α for each abstraction $\alpha \in L$ is calculated with the probabilities from the Steady State Vector $\vec{s}$ [17, 18] as illustrated in Eq. 6.9.

$$p_\alpha = \sum_{q_i \in \alpha} ssv(q_i)$$
(6.9)

Thereby $ssv(q_i)$ is the probability of the interaction q_i from the Steady State Vector. Hence the result is a probability distribution over sets of interaction events. The calculated probabilities permit statements concerning preferences of users. This can be in particular used to determine the use of visual layout concerning the hierarchical level of the semantics but also give indications of general preferred visualizations. With the underlying abstraction level, assumptions like preferred visualizations, preferred knowledge domains, or preferred interaction devices can be performed. Further the user's knowledge about a particular domain can be determined due to the enclosed semantic hierarchy and the different abstraction level of the introduced algorithm.

Let us explain the algorithm based on the following example: Let us assume that the visualization provides the interaction with the devices *mouse* and *keyboard* and there are two different visual layouts that can be used. Let us further assume that the data that can be visualized is already gathered with the introduced iterative querying and contains two concepts, *Persons* and *City* and the concept *Persons* inherits two instances *Obama* and *Kennedy* as illustrated in Fig. 6.14. We further assume that the

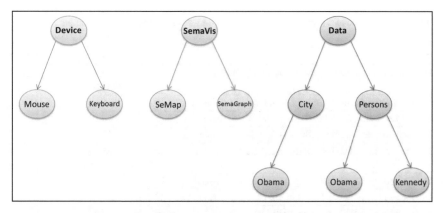

Fig. 6.14 Example of domains for the measuring the probability distribution based our algorithm

user has already interacted with the domains and values for the Steady State Vector exist, whereas the user had always only the interaction possibility with instances. Therewith no values for the concepts are given and these are not included in our SSV. For modeling the preferences, we want to know the probability distribution over the given concepts, thus they are not gathered explicitly. The device and the visualization the user interacted with are in this example not of interest for us. Therefore the function *quant* is formulated as *quant*(0, 0, 1)

The function *quant*(0, 0, 1) gives us the probability distribution of the interactions with each *Device*, each *visual layout* and the abstraction level one of the domain *data*. In case of our example there are the two concepts, *City* and *Persons* for the first abstraction level of the third domain. Table 6.1 illustrates the values for our example and the measured probability for the concepts.

The probabilistic distribution of users' interactions over the different levels of abstraction enables us to measure various values for preferences and knowledge of the users. Modeling users' can be performed on a detailed level by investigating all abstraction levels of three identified domains. It should be outlined that the sum of all probabilities of each abstraction level can differ from the total probability of 1 and should be normalized according to the number of existing relations. Thus our approach aims at providing a visual environment for heterogeneous data-bases and model users' previous interaction behavior based on the different data bases. The use of such abstraction levels and degrees are necessary. Further the interaction behavior with particular visual layouts can be gathered through the same abstraction level, whereas a differentiation between interacting in invalid area and valid areas are made for determining the interaction behavior. The main aspects of this section were published in [17–20]. Real case instantiation were performed in [8, 12].

Table 6.1 Example of a probabilistic distribution for an abstraction level of a domain

Steady state vector		quant(0,0,1)	
Interactions	p(%)	Abstractions	p(%)
Device.Mouse, Semavis.SeMap, Data.City.Obama	7	Device, SemaVis, Data.City	25
Device.Mouse, Semavis.SemaGraph, Data.City.Obama	10		
Device.Keyboard, Semavis.SeMap, Data.City.Obama	3		
Device.Keyboard, Semavis.SemaGraph, Data.City.Obama	5		
Device.Mouse, Semavis.SeMap, Data.Persons.Obama	20	Device, SemaVis, Data.Persons	75
Device.Mouse, Semavis.SeMap, Data.Persons.Kennedy	10		
Device.Mouse, Semavis.SemaGraph, Data.Persons.Obama	20		
Device.Mouse, Semavis.SemaGraph, Data.Persons.Kennedy	10		
Device.Keyboard, Semavis.SeMap, Data.Persons.Obama	3		
Device.Keyboard, Semavis.SeMap, Data.Persons.Kennedy	4		
Device.Keyboard, Semavis.SemaGraph, Data.Persons.Obama	3		
Device.Keyboard, Semavis.SemaGraph, Data.Persons.Kennedy	5		

6.1.3.3   Predicting Users' Actions

The probability distribution of users' interaction with the possibility to compute different abstraction levels based on Relational Markov Models enables a weighting of preferences or prior-knowledge based on the interaction behavior. The probability distribution does not consider the sequence of interactions to determine next possible actions of the user. We introduced in Sect. 4.5.3.3 the approach of Gotz and Zhou [25–27], an adaptive visual analytics approach that determines based on recurring identical interaction patterns the next possible action and recommends an alternating visualization. We further outlined that the recommendation of visualization they perform is based on fixed and predefined rules that just compare the last interactions with a library (knowledge model) and provide the recommendations. In contrast to their approach, our visual environment has the main goal to be extendable in terms of visual layouts and provide far more a reliable and contagious approach without the need of experts to design rules for each visual layout implemented into our visual environment. To reach the intended goal, we applied a derivation of the KO-Algorithm, originally proposed by Künzer et al. [28] that was introduced in Sect. 4.3.2 [17, 18].

The original version of the KO-Algorithm from Künzer et al. [28] calculates predictions of interaction events by the search of recurring identical sub-sequences with maximal length of three interaction events. By the recognition of recurring sub-sequences it is possible to identify usage patterns, which can be used to identify changes in user behavior and to recognize user activities [17, 18]. The limitation of three interaction events for predicting the next interaction reduces the information content of possible activities in particular in visual environments. The behavioral patterns should therefore not be limited to a fixed and predefined length. This would not only enable a more information content for the behavioral pattern of the users, but also to predict and recommend alternating visual environments that was used by certain or all users in a particular stage of activity. Therefore we extended the proposed algorithm so that with every prediction calculation the longest sub-sequence of interaction events is identified. We call this extension *KO*/19-Algorithm* [17–19]. It should be mentioned that the first attempt of the changed algorithm was proposed by Stab (diploma thesis supervised by the author) in context of an encryption tool. The algorithm was enhanced, evaluated, and applied to visualizations in further works [17, 19].

The KO*/19-Algorithm [17–19] is based on the KO-Algorithm that calculates a probability distribution for predicting the next possible interaction based on an observed sequence of interaction events $O = i_1, i_2, \ldots, i_n$. According to the action regulation theory proposed by Hacker [29], user interactions occur in hierarchical and sequential structured action sequences.

Based on this assumption the KO-Algorithm calculates a probability distribution by searching recurring sequences up to a predefined length in O and comparing them with the interactions lately executed by the user. Therefore the algorithm considers not only sequences of the predefined length but also shorter sequences. Thus the algorithm uses a variable Markov order for predicting events. Additionally the KO-Algorithm uses a weighting function $w(i)$ dependent on the length of the identified interaction sequence. Künzer et al. proposed the function $w(i) = i^{19}$, that leads to the best prediction quality in their studies. The maximum Markov order and hence the maximum length of the regarded sequence is limited to a fixed value. The original algorithm uses a maximal Markov order of three. For this reason the algorithm proposed by Künzer was called KO3/19-Algorithm.

For recognizing behavioral patterns in form of recurring action sequences the limitation of the considered sequences to a fixed length is not suitable. Short sequences which contain two or three actions are less meaningful and don not reveal much information about users and their current behavior. Therefore we modified the KO-Algorithm so that in every calculation the recurring behavioral pattern with maximal length is considered.

Given the interaction sequence $O = i_1, i_2, \ldots, i_n$ and the set of all possible interaction events A the prediction problem can be formulated as a probability distribution $p(i_{n+1}|i_1, i_2, \ldots, i_n)$, where $1 \leq n \leq \infty$ and $i_m \in A$ for $1 \leq m \leq n+1$ [30]. For every occurrence of a possible forthcoming event $a = i_{n+1} \in A$ at the position v in O, the KO*/19-Algorithm compares the last observed events in O with the events previous to v. In the first step i_n is compared with i_{v-1}. If they are identical, i_{n-1} is compared with i_{v-2} and so on (see Fig. 6.15). Therewith for every possible forthcoming event $a \in A$, sequences of different lengths are identified in O, which correspond with the actual user behavior. The lengths of these sub-sequences are weighted with the weighting function $w(i) = i^{19}$ proposed by Künzer [28] and summed. The resulting values for every possible forthcoming interaction event constitute the probability

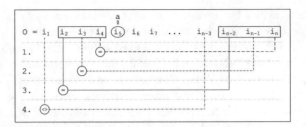

Fig. 6.15 Procedure of the KO*/19-Algorithm [18], [19, p. 609]

```
01:   function calcPrediction(o:Sequence):Vector;
02:   begin
03:     result := ZeroVector;
04:     for each a in A do
05:     begin
06:       for each occurrence of a in o do
07:       begin
08:         length = calcSeqLength(o, occurrence);
09:         result[a] := result[a] + length * w(length);
10:       end;
11:     end;
12:     Normalize(result);
13:     return result;
14:   end;
15:
16:   function calcSeqLength(o:Sequence, occurrence:Integer):Integer
17:   begin
18:     length := 0;
19:     while occurrence-length-1 > 0 and
20:           |o|-length > occurrence and
21:           O_{occurrence-length-1} == O_{|o|-length}
22:     begin
23:       length++;
24:     end;
25:     return length;
26:   end;
```

Fig. 6.16 The KO*/19-Algorithm [18], [19, p. 609]

distribution for predicting the next possible interaction. The algorithm of the described procedure is illustrated in Fig. 6.16.

The described KO*/19-Algorithm computes a probability for the next interaction. This weighting is amongst others in particular used to recommend visual layouts that were chosen by users after certain interaction steps. Further it leads predict the focused data elements to be selected and if this element is not loaded in the background of our data model, it leads to loading and computing the various quantitative and probabilistic measures on data level. This procedure leads to a more performant system behavior.

Another mentionable aspect of the algorithms is the ability to recognize users'
activities. Thus the behavior of user may change in the course of time [31], it is recommended that systems which collect user information are able to adapt themselves
to these changes. Beside the adaptation of the user information to the changes of
user behavior, our method is able to identify these changes by recognizing transitions between behavioral patterns. For this task we define an user activity as recurring
similar sequence of interaction events. With every new interaction event that is sent to
the interaction analysis system the KO*/19-Algorithm identifies a behavioral pattern
as the longest recurring sub-sequence in the interaction sequence of the user [17].

These behavioral patterns are compared and grouped to similar sequences and
constitute a user activity. To identify an activity, the first recognized sub-sequence
is added to a new activity. For every new sub-sequence a similarity function is used
to measure if the new sequence matches to the activity created in the first step. If
the similarity value exceeds a threshold, the sequence is added to the activity. If
the similarity value does not exceed the threshold a new activity is created and the
sequence is added to it. In this process the current activity of the user is detected as
the activity that matches to the sub-sequence or a new created activity, if no matching
activity could be identified. The choice of the threshold has direct influence on the
number of detected activities. If the threshold is too high, it is possible that for
every detected sequence a new activity is created and so no similarities between the
sequences are recognized. If the threshold is too low, all sequences are added to the
same activity and the differentiation of user activities is not possible [17, 18].

To face the problem of defining the threshold in the visualization environment,
the common process is to define a special training-file (Fig. 6.17). This file contains
predefined behavioral patterns, as proposed by Gotz et al. [25, 27, 32] that correspond
with possible user activities. Since the order and number of interactions may vary
for the same activity, the trainings file contains not only one predefined pattern
for an activity but also different variations. In order to avoid the recognition of
an activity-overlapping pattern we divide every activity variation sequence with a
special interaction event that is not observable. Therewith the behavioral pattern that
is recognized by the KO*/19-Algorithm in the trainings file corresponds with one
predefined activity and can be used to infer the actual user behavior. The trainings-file
is generated through the users' most frequent interactions. Therefore the canonical
user model that will be described in next sections is used. The canonical user model
provides the average user behavior of all users with our visual environment. Beside the
trainings file the observed user interactions are included for calculating predictions
and behavioral patterns. Thus the described interaction sequence O is a concatenation

Fig. 6.17 Schema of the initial trainings file for activity recognition [18], [19, p. 609]

of the training-file and the observed interaction events from the user and constitutes the knowledge base for inferring user information [17, 19].

For the cognition of the semantic meaning of the recognized activities, we propose the same method as for user preferences. Therefore the Steady State Vector of every activity is calculated from the interaction events of every sequence belonging to the activity. The context information of the activity interaction events is used to calculate probabilities of abstraction sets which are defined by the parameters of the query as described above. So for the derivation of the semantic meaning every activity can be also queried on different grades of abstraction [17].

Evaluation of KO/19-Algorithm*

One main question that may arise in this context could be the choice of the interaction analysis and prediction algorithm. We have illustrated in Sect. 4.3.2 various methods, algorithms, and applications that analysis users' interactions to provide a kind of adaptation. So it is more than obvious to illustrate why the KO-Algorithm was used as foundation of our interaction analysis and prediction algorithm. It should be outlined that the choice was performed after a conducted evaluation of different algorithms.

This part illustrates the evaluation that was performed and lead to the choice of the KO-Algorithm. The entire part was published without any major changes in [19]. The main goal was to evaluate the prediction quality based on two established criteria and compare the results of the KO*/19-Algorithm with those of other prediction methods. The first criterion we use for evaluating the prediction quality was the *Mean Prediction Probability* (MPP), which is defined as the average probability that is calculated with the prediction algorithm for really occurring interactions. Formally the MPP is defined as illustrated in Eq. 6.10.

$$MPP = \frac{1}{|o|} \sum_{i=1}^{|o|} P(o_i|o_{1:i-1}) \qquad (6.10)$$

Thereby o is the interaction sequence and $P(o_i|o_{1:i-1})$ is the calculated probability that the interaction o_i occurs after the sequence $o_{1:i-1}$.

The second criterion we used for the estimation of the prediction quality was the *Mean Prediction Rank* (MPR). Formally the MPR is defined as illustrated in Eq. 6.11

$$MPR = \frac{1}{|o|} \sum_{i=1}^{|o|} rank(o_i|o_{1:i-1}) \qquad (6.11)$$

Thereby $rank(o_i|o_{1:i-1})$ denotes the position of the occurred interaction o_i in the sorted probability distribution. So the most possible interaction has a rank of 1, the second a rank of 2 and so on. Hence, the lower the MPR of a prediction algorithm is, the better is the quality of the prediction calculation.

In our evaluation scenario we compared the prediction quality of the KO*/19-Algorithm with the reference files from Greenberg [33] (used with author's permission). These reference files contain the interaction traces of 168 users, which are divided into four different user groups with different grades of knowledge about handling computers (non-programmers, novice-programmers, experienced-programmers and scientists-programmers). Averaged every file contains a sequence of 1730 containing 78 different user actions. For our evaluation we removed the commando parameters in every file and replaced every user action with a uniform action id.

For every user and accordingly every file we accomplished a sixfold cross-validation. For that purpose we divided every file into six parts and use every part of the file as initial trainings data and the other five parts for determining the MPP and MPR. Hence, for every user and accordingly every file in the reference set we accomplished 30 test cases and 5040 test cases for every prediction method.

The results of the evaluation with the Greenberg reference files confirm the results of Künzer et al. [28]. In respect of the MPP, the KO3/19-Algorithm. The KO3/19-Algorithm achieves a MPP of 38.75 % and thus performs 3.56 % better than Markov Chains. Our KO*/19-Algorithm achieves a better MPP compared to the original KO3/19-Algorithm. The KO*/19-Algorithm achieves for all user groups the best MPP (Table 6.2). Averaged our algorithm achieves a MPP of 39.98 %. That denotes in spite of the extension an enhancement of 1.23 % compared with the KO3/19-Algorithm.

Remarkable are the results concerning the MPR. In contrast to Markov Chains, IPAM, and KO3/19, which achieve an average MPR of about 25, the KO*/19-Algorithm outperforms the other methods with a MPR of 10,5 and thus performs 15 ranks better than IPAM (Table 6.3). Thus the KO*/19-Algorithm improves the MPR about 19.23 % relative to the average number of possible interactions.

The results of the evaluation revealed that the KO*/19-Algorithm calculates predictions with high quality in real application scenarios. In particular the KO*/19-Algorithm outperforms the other prediction algorithms concerning the MPP and improves the MPR of the original algorithm significantly. So the algorithm is well suited for the prediction of interactions and well applicable for adaptive visualization environments. This evaluation was published in [19].

Table 6.2 Comparison of the MPP for different prediction methods

Data				MPP				
	Files	Avg size	Avg actions	MC (%)	IPAM0.8 (%)	LEV3 (%)	KO3/19 (%)	KO*/19 (%)
Non-prog.	25	1026	46	37.99	38.15	39.80	40.28	**41.52**
Novice	55	1409	57	43.78	47.13	47.49	48.22	49.44
Experienced	36	2107	104	29.84	31.48	32.70	34.20	35.55
Scientists	52	2379	105	29.13	30.40	30.79	32.29	33.42
Average		1730	78	35.19	36.82	37.70	38.75	39.98

Table 6.3 Comparison of the MPR for different prediction methods

Data				MPR				
	Files	Avg size	Avg actions	MC	IPAM0.8	LEV3	KO3/19	KO*/19
Non-prog.	25	1026	46	12.66	12.54	20.22	12.97	6.20
Novice	55	1409	57	14.62	14.46	22.47	14.92	6.40
Experienced	36	2107	104	36.79	35.41	54.48	36.24	12.32
Scientists	52	2379	105	38.18	37.81	60.41	38.71	14.27
Average		1730	78	25.31	25.05	39.39	25.71	10.05

6.1.3.4 Modeling Users

The main goal of analyzing users' interaction behavior with data and visualizations is to represent this for generating an abstract model of users' behavior and provide sufficient visual adaptations. As we outlined in Sect. 4.4.2 the main aspects in modeling users are the *nature* and the *structure* [34]. Thus *nature* refers to user' characteristics or features that are in our approach gathered just implicitly (R4) from users' interaction, we constrain these features to users' knowledge, users' interest and tasks [34–36]. In general, the nature can be further summarized in preferences for visualizations, interest in general knowledge topics, and knowledge in a particular topic. The structure of the user model should be transferable to other domains of knowledge (data-sets) according to the defined requirement (R3) and should therewith enable the use of the model in various data domains.

With the introduced Steady State Vector and the various levels of abstraction, we already defined an abstract model of the user. The SSV represents the probability distribution of users' interaction in different levels of abstraction. Further it refers to three dimensions: the used device and type of interaction, the visual layout, and the data [20]. In case of adaptive visualizations it is necessary to model users' interest and users' knowledge for both, the available visual layouts and the data domain. With our approach the user should be enabled to change the data sources, that are commonly data bases on Web, without any knowledge about their structure and every time he wants. These changes may occur during one single session with the visualization environment. This means that in particular the data dimension of the user model has to be highly dynamic and support at various abstraction levels the new and unknown data domains. Although, the visual layouts may change during the time, e.g. by adding new visual layouts or removing existing one, these changes are less frequent.

Individual Users' Interest
In common it is important to generate from the taxonomic structure in the SSV of the *Data* domain, generalized terms that represent users' interest data-base independent. As the semantic structure of data may differ in different data-bases, the taxonomy in the SSV would not be sufficient enough to determine users' interest data-base

independent. Users' interests can be determined on a general level by using the concepts of the semantic data. Let us explain this process with an example. Let us assume that a user is interested in "visual perception". Let us further assume that he or she searches and interacts with different data bases for different terms related to visual perception. The users' model based only on the SSV would provide us the probability distribution of the interactions in the particular data base containing the taxonomical structure. Users' search and interaction with the term *preattentive* would lead to differing taxonomies that are domain dependent and would not enable to measure the interest in a particular domain. Such taxonomies for users' interest *preattentive* may result as follow in our SSV:

- *Data.Science.Cognition.Cognitive_Science.Perception.preattentive*
- *Data.Perception.VisualPerception.pre_attentive*

This example illustrates that if only the taxonomy is investigated, the two different data-bases would not have any common terms that lead to determine users' interest in visual perception, even not for the concepts *cognitive science* or *perception*. Although these two terms (Cognitive_Science and Perception) are used in both data-bases, these terms would not be recognized, due to their differing hierarchical structure. To face this problem and ensure that at least the users' interest in concept-level (semantic concepts) can be modeled, we introduce the ssv_I for users' interest. This contains all semantic concepts with the measured probability distribution without the taxonomical structure. Therefore we first introduce the function $path(D_i, r)$ that returns the path *pa* of the domain D_i from the interaction relation r. Thus we want have the leaf-node of a particular path without the taxonomic structure, we further define the function $leaf(pa)$ that returns just the leaf-node of the path *pa* without any further hierarchical information. To gather the intermediate nodes of a path *pa* the function $intermediate(pa)$ is defined. Based on these functions we are able to determine terms that are explicitly *semantic concepts* for modeling users' interest on a general level of abstraction. To identify concept in a SSV ssv_u the second domain is used. The domain D_2 provides at the abstraction level 2 information about the item of the visualization, with which the interaction was performed. The leaf at this abstraction level provides information-types such as *instance, concept,* or *whiteSpace* and enables therewith the identification of semantic concepts. Thereby, we determine a SSV of users interests ssv_I that contains just terms with their probability distribution as illustrated in Eq. 6.12.

$$\forall r \in ssv_u : leaf(path(D_2, r)) = concept \ \wedge \ s = leaf(path(D_3, r)) \ \rightarrow \ s \in ssv_I$$
(6.12)

If the leaf-node on a particular abstraction level in the *Data* domain D_3 is in the visual layout domain *SemaVis* D_2 a concept, the particular leaf-item s is an element of the ssv_I. It may be, that the interaction is not made directly with a particular concept. Instead of this, users may interact with an instance without navigating through the concepts. In these cases the domain visual layout *SemaVis* just provides the information that the interaction was performed with a certain semantic instance. To retrieve users' interest on an abstracted level, even if the interaction was performed with an instance, we use the higher abstraction levels of the instance. Let us explain the procedure with an example: Let us assume that the users' search resulted in a number of entities and the user interacts directly with the instance *preattentive*. Our SSV contains this instance with its entire taxonomy, e.g. *Data.Science.Cognition.preattentive*. The visual layout domain D_2 would indicate in this case that the interaction was performed with the semantic instance *preattentive*. Based on the taxonomy, we can determine that all higher abstraction levels are semantic concepts, due to our description in Sect. 6.1.3.1. In our example, *Science* and *Cognition* are concepts. Thereby ssv_I can be enhanced with the intermediate nodes of instances as illustrated in Eq. 6.13.

$$\forall r \in ssv_u : leaf(path(D_2, r)) = instance \wedge s \in intermediate(path(D_3, r)) \rightarrow s \in ssv_I$$
$$(6.13)$$

The Steady State Vector ssv_I contains all the terms that are explicitly concepts. Further the ssv_I contains the related probabilities for each of the terms from the original Steady State Vector ssv_u.

$$p_s = \sum_r ssv_u(r) \mid s = leaf(path(D_3, r)) \vee s \in intermediate(path(D_3, r)), r \in ssv_u$$
$$(6.14)$$

With this procedure a new SSV ssv_I is generated that contains just concepts and their probability distribution. In the next step all identical concepts are identified and merged to one concept, whereas their probability is summed. To get again a normalized SSV, the probability distribution of all concepts is divided by the reduced number of concepts in the SSV. This vector is further added to the users model and builds the foundation for determining his or her interest in a general manner that can be used in different data-bases. Table 6.4 illustrates the new SSV with the abstract concepts for determining the users' interest in contrast to the SSV that was originally built based on users' interactions.

The introduced procedure enables to determine users' interest in general manner without a dependency to the data-base and meets the defined requirement (R3). It

Table 6.4 Example of a SSV on concept level for modeling users' interests

Steady state vector ssv_u (example)		SSV interest (ssv_I)	
Interactions	p (%)	Concepts	p (%)
Device. ..., SemaVis. ..., Data.Science.Cognition	15	Science	10
Device. ..., SemaVis. ..., Data.Science	20	Cognition	15
...	20	Cognitive_Science	25
Device. ..., SemaVis. ..., Data.Perception. VisualPerception	10	Perception	45
Device. ..., SemaVis. ..., Data.Perception	30	VisualPerception	5

enables a data-base independent measurement of users' interest and is an essential part of our proposed approach. But this procedure has some disadvantages too. It may occur that some concepts are labeled identical but have different meanings and contexts. But the probability that such labels are defined in the semantic data-base is rare. Commonly semantic concept labels are sophisticated enough for measuring the general interests of users. Further the new SSV does not provide any information or correlations between users and their interaction behavior with different visual layouts. Although, the above described procedure can be used to generate a "Preference SSV" for visual layouts too by extracting the underlying information for second domain D_2 it does not really make sense, thus the number of users' SSV will increase without any real benefits. Further the precise information are lost about the exact information object, the users interacted with. Therewith the applied approach is an add-on to model the users' interest. Users' knowledge is still modeled based on the gathered SSV that is data-base dependent but provides exact measures for both knowledge entities in form of instances and concepts. The abstraction levels introduced in Sect. 6.1.3.2 provides already sufficient information about the users' behavior with information objects and visualizations to determine the knowledge.

Individual Users' Previous Knowledge and Task Modeling
For a more comprehensible illustration for modeling users, in particular on the precise level of knowledge and correlation with visual layouts, and to provide for the next steps of measuring users' similarities and deviations with the canonical user model, we introduce some general definitions, that are used throughout the section and in the Sects. 6.2.1 and 6.2.2. We define the set $U = \{u_1, u_2, \ldots, u_n\}$, where each u is a user and U the set of all users. Additionally, we define the set $V = \{v_1, v_2, \ldots, v_k\}$ with each v being a visual layout of all visual layouts V from the first abstraction level of the visual layout domain D_2 and $D = \{d_1, d_2, \ldots, d_l\}$ with each d being a data element from the set of all data elements D of all abstraction levels from the data domain D_3. For measuring the interest, we can replace this with $D \in ssv_I$. For a more comprehensible illustration the users interest SSV ssv_I is not used anymore for the definitions. It should just be outlined that for determining interest always the ssv_I is used in combination with the values from the original SSV $\vec{s}$ with three domains.

For considering the users' behavior on individual user level, we extend the Eq. 6.9 from Sect. 6.1.3.2 by allowing the extraction from the Steady State Vector (SSV) ssv_u of each individual user $u \in U$ as follows:

$$p_{u,\alpha} = \sum_{q_i \in \alpha} ssv_u(q_i) \tag{6.15}$$

Furthermore we introduce $p_{u,v,d}$ as a short form to extract the probability of an individual user for the correlation of a visual layout v and a data element d.

$$p_{u,v,d} = p_{u,r(device,v,d)} \tag{6.16}$$

Although, the interaction type *Device* D_1 is gathered in each users' interaction, we dismiss this information and use the abstraction level 0 of the device domain D_1. This lets us extract the relevance value of the data element d in combination with the visual layout v for a specific user u.

In the next step, we introduce two relevance vectors for each user in the user model. The *visual layout usage-vector* contains the relevance values of visual layouts according to their usage of each user and provides us information about users' "visual layout preferences" that is again a probability distribution of the interaction behavior with the visual layouts. The *data interests-vector* contains the relevance values of the data elements according to the interest and previous knowledge of the individual users. The previous knowledge is determined by the interaction behavior with semantic entities. The more a user interacted with topics from the same knowledge domain, the more previous knowledge can be assumed for this knowledge domain [34, 37, 38]. Further these vectors are used later to form user groups and therewith identify users' similarities. Each entry $p_V(u, v)$ in the *visual layout usage-vector* of an individual user $u \in U$ contains the normalized relevance values of each visual layout $v \in V$ and is calculated as follows:

$$p_V(u, v) = \frac{\sum\limits_{d \in D} p_{u,v,d}}{\sum\limits_{d \in D} \sum\limits_{v_i \in V} p_{u,v_i,d}} \tag{6.17}$$

The creation of the *data interests-vector* for each user $u \in U$ uses the semantic information graph in addition to the relevance values between visual layouts and data elements. Using the advantages of content based recommendation systems by treating semantic relations as features helps with the problem of data sparseness.

Let, as previously stated, $p_{u,d,v}$ be the relevance value of an individual user $u \in U$ for a data element d in combination with a visual layout v and let $S_d \subseteq D$ be a set of all data elements, which have a semantic relation with data element d. The relevance value $p_D(u, d)$ of an individual user $u \in U$ for a data element d is calculated as follows:

$$p_D(u, d) = \max_{v \in V} p_{u,v,d} + \sum_{d_i \in S_d} \frac{\max\limits_{v \in V} p_{u,v,d_i}}{|S_{d_i}|} \tag{6.18}$$

These relevance values of the individual data elements form the *data interests-vector*. In contrast to the *visual layout usage-vector*, the individual relevance values between data elements and visual layout are not summed while creating the vector. Instead, the visual layout relevance values are used, which has the highest value for the corresponding data element.

With the introduced definition so far, the users' interest, previous knowledge and preferences for visual layouts and data are modeled. For determining the tasks, we use as described in Sect. 6.1.3.3 the occurrences of similar interaction sequences O as behavioral patterns. We already illustrated that for recognizing these activities a threshold has to be defined by a special training-file that contains pre-defined activities of users as similar interaction sequences [17]. Thus the processes in interacting with visualizations has more an exploratory character, the pre-defined training file would not lead to sufficient task or activity recognitions. Another way to define such a training file that is continuously updated by users' behavior and would lead to a more efficient way of identifying the occurrence of similar and frequent sequences is the use of a *canonical user model* [12] that contains the interaction behavior of all users, even if they are interacting with the system anonymously. With the introduced activity recognition process in Sect. 6.1.3.3 the most frequent similar occurrences of interaction sequences are automatically determined and used as the training file.

The interaction with visualizations is data elements leads already to sufficient modeling of users' behavior. The measured probability distribution leads to model users' interest, preferences, and in a limited way users' previous knowledge on certain knowledge domains. Our definitions and algorithms make use of the interactions as relations between data and visual layout, whereas the interaction type *Device* is investigated too as the first domain of the SSV D_1. This domain can be used to determine different dependencies on transition level. As described in the previous sections commonly the starting point of the exploration process in our visual environment is the search task followed by the selection of a *focused object*, commonly a focused instance (see Sect. 6.1.2.2). The users' interactions with data and visualization may have different relevance. This differing relevance appears in particular if a user selects a new visual layout or removes a visual layout from the screen of the visual environment. Further the fact the selection of a data entity in a particular visualization were performed and led to a successful new data entity or to a step, where the user has to go back and select another item. To gather this information the first domain $D1$ of the SSV is used that provides information about the type of users' interaction, e.g. as *Device.Mouse.select, Device.Mouse.removeVis, Device.Mouse.selectVis*. The procedure allows to weight successful interactions that leads to achieving the goal or explicit selecting visual layout higher than those interactions that lead to removing visual layouts or interacting backwards, thus the information were not found.

The Canonical User

The canonical user model represents the average users' behavior with the visualization system. This user model is the baseline for adapting the visual layout and data for all users and improves the general usage of the visualization system. Thus it is used on the one hand for the general adaptation and on the other hand for measuring certain deviations and anomalies in behavior for individual user, it is one of the core components of our conceptual model. Every user that interacts with the visualization environment pulls one's weight to the canonical user model. The interaction of each user, even if the adaptivity of the system is disabled, contributes to this model. The canonical user model is in particular used to determine the overall usage behavior with the system and provides with the integrated adaptation features of our approach an improvement on choosing visual layouts in correlation to data and data characteristics, recommending visual layouts, and choosing and recommending data to be visualized from the data model. Further it enables to provide the most occurred similar sequences of interactions that lead to identify tasks, activities, or even roles.

One other main aspect of the canonical user model is that it serves as the user model for new users. As the first steps of a new user is accompanied by the canonical user model, the individual user model for the particular user is generated and the "weight" of the canonical user model for that particular user is decreased, while his individual user model serves for the adaptive features more and more [12, 21].

Our canonical user model, models the behavior of all users by analyzing the interactions with the system. Therefore users' interactions are transformed in a numerical, internal representation and the Steady State Vector is determined as a relative measurement for the occurrence of interactions [12]. The model involves the interaction quantity with each data element, visualization element and the choice of visualizations to enable a learning system that considers the behavior of the majority of users. Further it provides general usage information of the visualizations to enable the recommendation and automatic selection of visualization-algorithms.

The canonical user model does not require personal information about the user because the model itself provides a general "initialization". To overcome an overgeneralization of visualization choice, the canonical user model is counterbalanced with an additional user grouping, based on individual interactivity preferences and behavior. Thus, the system provides the capability to respond to individual users. For this, we applied an algorithm that computes the deviation of the user interaction behavior. Therefore the user-interaction behavior of the current user is compared with the canonical user model with respect to the number of users. This enables to estimate, if the same or a similar user is interacting with the system and can be enhanced to group the users with diverting intentions. The approach provides two different modi for user-oriented adaptation. First, the canonical user model that investigates the behavior of all users, and second an individualized user model. The individual user model is an instantiation of the canonical user model with certain preferences and interaction history of a certain user as described above. If a user is interested in getting behavior-based visual adaptation, he is able to log-in as individual user. The default user model in our approach is the canonical user model. It is activated, if an individual user is not logged-in.

The canonical user model is used as already mentioned to measure similarities in particular deviations of users' behavior from the average behavior of users. This enables the identification of certain behavioral anomalies. To compute similarities, deviations, and in particular the average usage behavior, we describe the canonical user model in a more formal way. To describe the interaction behavior and therewith the probability distribution for the canonical user in context of visual layouts, we use the probability values $p_{u,v,d}$, which represents the probability of the interaction of the user $u \in U$ with the visual layout $v \in V$ in combination with the data element $d \in D$. Based on these probabilities the interaction behavior of a canonical user can be computed as illustrated in Eq. 6.19. Where the sum of interaction probabilities of each user $u \in U$ with a certain visual layout v is divided by the amount of all users $|U|$.

$$can_V(v) = \frac{1}{|U|} \sum_{u \in U} p_V(u, v) \qquad (6.19)$$

Fig. 6.18 Abstract
illustration of the canonical
user model

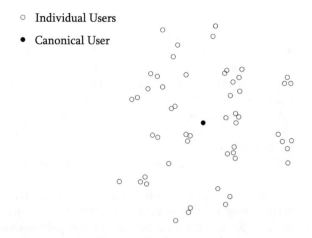

A similar correlation can be built between users and data. The main difference is
that the leaf nodes are investigated in different levels of abstraction. Thereby either
a data or information entity can be a leaf node or an intermediate node of the entire
taxonomic structure. We previously defined $d \in D$ as a data element from a set of
all data elements in all abstraction levels from the data domain D_3. Additionally,
for the canonical user, only users who actually interacted with the specific data or
information entity are considered. Therefor we do not divide by the number of all
users, but instead only by the amount of users, who interacted with this data entity.

$$can_D(d) = \frac{1}{|\{u|u \in U, p_D(u, d) > 0\}|} \sum_{u \in U} p_D(u, d) \qquad (6.20)$$

Figure 6.18 illustrates the canonical user amongst all individual users. The angu-
lar dimension from the canonical user between two individual users represents the
similarity between those. The distance of each individual user from the center (and
therefor from the canonical user) represents the similarity and deviation between
the individual user and the canonical user. This distance measure can be used to
determine anomalies in users' behavior.

User Group and Role Definition
Similar interaction behavior of users can be used to define certain *User Groups*. The
identification of these groups leads to define certain user roles based on the interaction
behavior of the users. Rule-based approaches for identifying the intended task of a
user is error-prone. The execution of a specific task is performed by interactions that
are associated with other tasks. A user model, which automatically assigns users
to groups based on their interactions and refines them with every new performed
interaction leads to identify user roles. Even without the explicit labeling of these
roles, it is able to support users with similar interaction behavior by considering the
previously analyzed interactions of other users from the same group.

To define user groups and roles, we use two methods: In the first method, users are clustered based on their usage of the visual layouts V, in the second method, their interest in certain data or knowledge dimensions is the basis for the clustering.

A specific user is assigned to a cluster based on the following definition. Let $sim(c, u)$ be a function, which provides the similarity of a user u to a cluster c of all clusters $C = \{c_1, c_2, \ldots, c_n\}, c_i \subseteq U, \forall c_i, c_j : i \neq j, c_i \cap c_j = \emptyset$. Here, a higher value means stronger similarity. A user is assigned to a cluster c, if there is no other cluster $c_i \in C, c \neq c_i$, that has a stronger similarity with the users individual previous interactions:

$$\forall c_i \in C, c_i \neq c : sim(c_i, u) < sim(c, u) \rightarrow u \in c \qquad (6.21)$$

The average value of each cluster c of all visual layout cluster C_V is calculated in the same way as for the canonical user. With the main difference that of all users, only the users in their respective clusters are taken into account for the measurement. The normalized value $p_V(c, v)$ of a visual layout v of a cluster c is calculated as follows:

$$p_V(c, v) = \frac{1}{|c|} \sum_{u \in c} p_V(u, v) \qquad (6.22)$$

The average value of each individual cluster c of all data domain clusters C_D is also calculated similar to the calculation of the canonical user. Again, only users in the cluster are taken into account. Additionally, the normalization only considers users, who actually contributed to the calculated value. The measurement of this normalized average value $p_D(c, d)$ of a data entity d of a cluster c is calculated as follows:

$$p_D(c, d) = \frac{1}{|\{u|u \in c, p_D(u, d) > 0\}|} \sum_{u \in c} p_D(u, d) \qquad (6.23)$$

Figure 6.19 illustrates the previously presented abstract illustration of the user model with additional clustering. The $+$-signs represent the center of each cluster and therewith the average user behavior in the particular cluster. The individual clusters are aligned in form of radial rays around the canonical user in the center of the figure, because the similarity between users is measured by the angle (see Sect. 6.2.1). With this procedure different user clusters can be determined automatically, even if the clusters are not labeled. Our visualization technology is in some use case embedded in a Web application, where users log-in and provide some demographic information and further information about themselves. In these cases the cluster can be labeled according their role or any other common information. Figure 6.20 illustrates the user grouping in a more hierarchical way to outline the relationship to the canonical user model. Every user, regardless, if he or she belongs to a group, provides interaction information to the canonical user model and their models inherit from the canonical user model. Grouped users inherit further from the average group user model and provide interaction information to the average group user models.

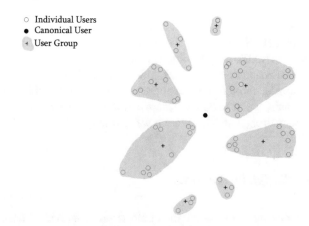

Fig. 6.19   Abstract illustration of the user groups and their center (average)

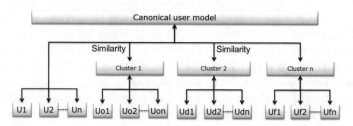

Fig. 6.20   Abstract illustration of the user groups and their relation to the canonical user model

6.2   Process of Adaptation

The analysis and modeling of influencing factors provides structured and formalized models for the adaptation process. We introduced so far various and heterogeneous factors that can be used to adapt the visual environment. The introduced models and algorithms can be subdivided on an abstract level in two main categories: Persistent models that contain various information about a certain influencing factor and are enhanced and over the time and volatile models that provide adaptation information about momentary effects that can be applied for a particular point of time or interaction. A user model, for instance provides information about the entire user interaction history. It is a persistent model that can be applied to different stages and in different contexts. Examples for volatile models are the results of data query that effect the momentary visual environment but is dispersed after a new user search. A more volatile model is the weighting of semantically neighbored entities that may change even by a single user interaction. To bring the persistent models and volatile models together, it is necessary to define an adaptation process that investigates all models and applies the adaptation effect. This section will introduce in particular the adaptation process. Therefore, we start with the formal description of the user similarity and deviation analysis that enhances the user models with further volatile and persistent models. Thereafter the formal annotation of visual layout with our defined *Semantics Visualization Markup* language will be described as a formal persistent model. The adaptation process as "rules" builds the core of this section. We will guide the reader through the adaptation process and describe the main subprocesses. The main goal is to outline where and when, which models are applied to adapt the visual environment. This section was partially published in [21].

6.2.1   User Similarity Analysis

User similarity measurements allow the comparison of individual users through a numerical quantity value [22, 39, 40]. This value can be used to measure how similar different users are according to their behavior or interests. These measurements are being used for the calculation of the similarity between a user and a user group in addition to the similarity of two different users. The here introduced user similarity measurements were initially proposed by Retz [22] (diploma thesis supervised by the author) and further enhanced and applied to visualizations [21]. Figure 6.21 illustrates an abstractly the similarity between users and user groups. The angle between two objects represents their similarity to each other. A smaller angle means more similar objects. The green gradient illustrates regions of high to low similarity of the selected individual user (blue icon) to other users in these regions.

The basis for the calculation of the similarity is the two previously described vectors, *visual layout usage-vector* and *data interests-vector* (Sect. 6.1.3.4). They are also used to model the canonical user and to create the user groups.

The composition of these two vectors differs greatly. The *visual layout usage-vector* is normalized and not very sparse after a short usage of the visual system by the specific user, because most of the time the range of available visual layouts is very limited and new visual layouts are only added in large intervals. On the other side,

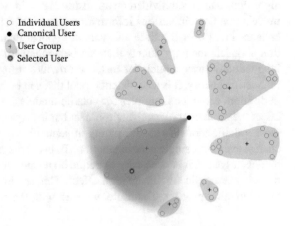

Fig. 6.21 Abstract illustration of the user similarity with the region of high similarity between one selected individual user and other users (adapted from [21, 22])

○ Individual Users
● Canonical User
+ User Group
◉ Selected User

there is high probability that the *data interests-vector* is very sparse, even after a very thorough usage of the visual system by the specific user. There are too many data elements, new data elements can be added continuously, and users may interact in various data-bases. This is the main reasons, why different similarity measurements are used to determine the similarity between users and other users or user groups.

For the calculation of the similarity between users on the basis of their data interests relevance values, the *Pearson Correlation Similarity* [22, 39–41] metric is used. Let $u_a \in U$ and $u_b \in U$ be two users and $p_D(u_a, d)$, $p_D(u_b, d)$ their respective relevance values for the data element $d \in D$. The similarity between these two users $sim_D(u_a, u_b)$ is calculated as follows:

$$sim_D(u_a, u_b) = \frac{\sum\limits_{d \in D_{ab}} \left(p_D(u_a, d) - \overline{p_D(u_a)} \right) \left(p_D(u_b, d) - \overline{p_D(u_b)} \right)}{\sqrt{\sum\limits_{d \in D_{ab}} \left(p_D(u_a, d) - \overline{p_D(u_a)} \right)^2} \sqrt{\sum\limits_{d \in D_{ab}} \left(p_D(u_b, d) - \overline{p_D(u_b)} \right)^2}} \quad | D_{ab} = D_{u_a} \cap D_{u_b}$$

$$(6.24)$$

Here, $\overline{p_D(u)} = \frac{1}{|D|} \sum_{d \in D} p_D(u, d)$ is the mean value of all values in the *data interests-vector* for an individual user $u \in U$.

The calculation of the similarity between users on the basis of their visual layout relevance values also uses the *Pearson Correlation Similarity* metric. But here, no normalization with the mean value of the respective vector occurs, because these vectors are already normalized. The value for the similarity of two users $u_a \in U$ and $u_b \in U$ and their respective relevance values $p_V(u_a, v)$ and $p_V(u_b, v)$ for the visual layout v is calculated as illustrated in Eq. 6.25.

$$sim_V(u_a, u_b) = \frac{\sum\limits_{v \in V_{ab}} p_V(u_a, v) p_V(u_b, v)}{\sqrt{\sum\limits_{v \in V_{ab}} p_V(u_a, v)^2} \sqrt{\sum\limits_{v \in V_{ab}} p_V(u_b, v)^2}} \quad | V_{ab} = V_{u_a} \cap V_{u_b} \quad (6.25)$$

6.2.2 User Deviation Analysis

The user deviation represents the difference in user behavior of each individual user to the average behavior of the canonical user. It is assumed, that users can also be similar to each other, if they differ similarly in their interaction behavior from the average behavior, whereas their direct similarity to each other is not measurable. This can happen, if e.g. the adaptive system could not yet determine the overlapping interests for the particular user (new user), or if the user interacts with a completely new data-base (new situation). The here introduced user deviation measurements were initially proposed by Retz [22] (diploma thesis supervised by the author) and further enhanced and applied to visualizations [21].

Fig. 6.22 Abstract
illustration of the regions
with similar deviations of a
user to the canonical user
(adapted from [21, 22])

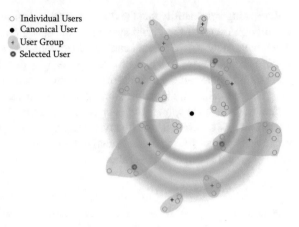

○   Individual Users
●   Canonical User
⊕   User Group
◉   Selected User

Figure 6.22 illustrates abstractly the deviation of users' behavior from the average behavior of the canonical user in addition to the previously calculated user groups from the similarity analysis [22]. The distance of the user to the canonical user and accordingly the radius represent the aforementioned behavioral deviation. For a selected user, the gradient of the ring symbolizes the region with similar deviation in behavior.

The calculation of this behavioral deviation is also based on the two previously described vectors, *visual layout usage-vector* and *data interests-vector*, which were also used for the modeling of the canonical user and for the creation of the user groups.

Unlike the calculation of the similarity between the users, we used the *Cosine Similarity* metric [40–43] for calculating the behavioral deviation. This is because the consideration of the relevance values, which are not common for both users are relevant for the measurement of the deviation. Since the *Cosine Similarity* metric does not perform a normalization, the calculation of the *visual layout usage-vector* and the *data interests-vector* are identical.

Let $p_D(u, d)$ be the relevance value of a data element d of the data element set D for a user $u \in U$ and $can_D(d)$ the relevance value of the canonical user for the data element d. The information deviation (interest-deviation) $dev_D(u)$ of a user u can be calculated as follows:

$$dev_D(u) = \frac{\sum_{d \in D} p_D(u, d) can_D(d)}{\sqrt{\sum_{d \in D} p_D(u, d)^2} \sqrt{\sum_{d \in D} can_D(d)^2}} \qquad (6.26)$$

This leads to the definition of a similarity between two users based on their interest-deviation from the canonical user $sim\_dev_D(u_a, u_b)$, which can be expressed as follows:

$$sim\_dev_D(u_a, u_b) = 1 - |dev_D(u_a) - dev_D(u_b)| \qquad (6.27)$$

Equivalently, the similarity between two users based on the deviation in the usage of visual layouts to the canonical user can be calculated as follows:

$$dev_V(u) = \frac{\sum_{v \in V} p_V(u, v) can_V(v)}{\sqrt{\sum_{v \in V} p_V(u, v)^2} \sqrt{\sum_{v \in V} can_V(v)^2}} \qquad (6.28)$$

$$sim\_dev_V(u_a, u_b) = 1 - |dev_V(u_a) - dev_V(u_b)| \qquad (6.29)$$

The returned value is in the range between one (identical distance) and zero (completely different distance) [22]. The analysis of deviations is performed between the canonical user and either one individual user or the average user of a certain identified group. With the measurement of the distance between the canonical user and the individual user, anomalies in user's behavior can be detected that leads to reduce the adaptation effects. If the distance between the user's behavior is similar to that of the canonical user, the canonical user model can be taken as user model for new situations, visual layouts, users, or data-bases. Further the distance between the canonical user and the average user of a user group can be used to determine the application of the canonical user to new situations, visual layouts, or data bases for the entire group. In this case the average user model of the group inherits the canonical behavior for the new aspects, if the distance is small. Vice versa, if the distance between the canonical user and the average user of a group is large, the application of the canonical user model does not make sense and the adaptation effects are reduced until enough information of the users' in that particular group are gathered. Further the measured similarities between same distanced users (Fig. 6.22) are used to detect similarities as described in the Sect. 6.2.1. If the distance of two users or average user groups is similar according to the canonical user model, the similarity algorithm is applied to measure certain similarities between the groups or users. If similar behavioral patterns are detected and one of the user groups contains certain information that is missing in the other group, this information is applied to extend the average user model of that group (Fig. 6.23).

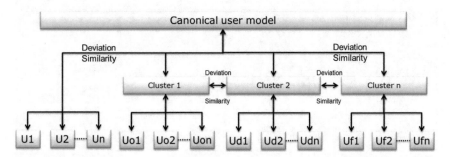

Fig. 6.23 Abstract illustration of the user groups and their deviation and similarity relations to the canonical user model (adapted from [21, 22])

6.2.3 Adaptation Process

The general adaptation process (see Fig. 6.24) considers various influencing factors
to adapt certain visual features in an appropriate way as described in the previous sections. Beside the "learning" capabilities of the system and the continuously
changing factors, some rules have to be defined to achieve the adaptation effects
and support users. The rules make use of the models and measurements described in
Sect. 6.1 Knowledge Model and describe the priorities and appliance of the models.
The described models of users' behavior, data, and data features illustrate the way
how influencing factors are gathered, measured, and structured. Their appliance to
visual attributes requires a notation of the visual attributes. This notation is in particular important for the data and their features, thus the user models with the measured
similarities and deviations already build a correlation to at least the visual layouts.

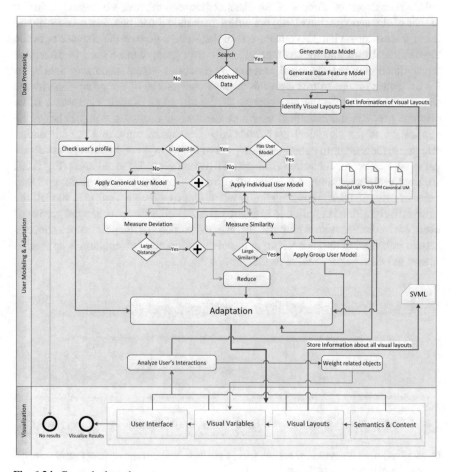

Fig. 6.24 General adaptation process

Table 6.5 Presentation layer of SVML

Visual variables on presentation layer	
Hue	Color-value of the presented entities
Saturation	Saturation of the presented entities
Brightness	Brightness of the presented entities
Size	Size of the presented entities
Order	Order of the presented entities
Icon	Can icon or glyphs be used for the brightness of the presented entities?
Orientation	Orientation of the presented entities
Shape	Shape of the presented entities
Transparency	Transparency of the presented entities
Distortion	Distortion of the presented entities
Background	Can the background color be changed?

The core components on the visualization side are the visual layouts that should be annotated with further information about their capabilities, variables and so on. The notation of the visual layout has to be readable by the core adaptation engine that processes the required information and models to apply the adaptation effect. Thereby the precise information about visual layouts is appropriate enough to determine information about the visual variables, visual interface, or semantics and content. To provide such a notation for all included visual layouts, we integrated our *Semantics Visualization Markup Language* (**SVML**).

SVML-Semantics Visualization Markup Language for Visual Layouts

The SVML includes beside general information about the user interface or data-bases, detailed annotation of the capabilities of each visual layout. Therefore three main categories are annotated: *presentation* that provides information about the visual variables, *layout* that provides information about the visual layout's characteristics, and *semantics* that provide information about the visualization capabilities regarding data quantity. Thereby SVML is an important part of the visual adaptation, thus it provides information about the visual capabilities of each visual layout. We investigated for the *presentation* layer all the visual variables that were used in the entire set of visual layouts. Therefore, we identified based on the work of Oelke et al. [44] and the perception studies [45–47] a number of visual variables that are investigates for each visual layout and refers to Bertin's model of *Implantation* [48]. Table 6.5 illustrates the visual variables in the presentation layer of our approach.

For each variable the SVML indicates, if the particular visual layout makes use of the variable and if the variable is changeable in terms of adaptation. Commonly all used variables are changeable and can be used for the adaptation. The second layer of *layout* indicates the general capabilities of the visual layout. It is considered as skeleton and refers to Bertin's model of *Imposition* [48]. The layout tag of the SVML contains two general attributes that indicate the main characteristics of the visual layout: *type* and *structure*. *Type* provides information about how the underlying

Table 6.6 Type and structure in layout of SVML

Type and structure in layout of SVML

Type		Structure	
Spatial	Area-based geometric transformation	Hierarchical	Illustrating in particular hierarchies
DensePixel	Point-based geometric transformation	Arbitrary	Illustrating arbitrary relationships
Lines	Line-based geometric transformation	Temporal	Illustrating temporal aspects of data
Graph	Graph-based transformation	Geographic	Illustrating geographical aspects of data
Combination	Combined use of transformations	Entity	Illustrating semantic entities
		Document	Illustrating textual documents
		Source	Illustrating the source of entities commonly as HTML

data are transformed (use of the screen) and *structure* provides information of the way they are illustrating data. These two attributes are essential parts of the layout and use the different taxonomies introduced in Chap. 2 in particular the outlined classification in Fig. 2.7. Table 6.6 summarizes the different included *types* and *structures*.

The annotation of *type* and *structure* provide general information about the layout layer of the visual layout and are commonly further annotated with more detailed information about the capabilities of the layout. Therefore a number of data and visual properties were investigated and elaborated in [49] to provide a more detailed view on the visual capabilities and the applicability to certain data-sets or search-results. Our SVML includes these information as properties for each the layout layer of each included visual layout. The main goal is to select or recommend in case of different visualizations that provide similar *structure* and *type* information the best suited ones by investigating the detailed layout information. Table 6.7 illustrates a possible set of such information that were partially elaborated in [49].

The third and last layer *semantics* annotates the visual layout with information about data capacity and constraints. Commonly the integrated visual layouts are designed to visual a particular perspective on data. The combined view should enable a multi-perspective view in a comprehensible way. The main attribute of *semantics* tag in the SVML is *datatype* that indicates which types of data can be visualized with the particular visual layout. Although, commonly semantic data or metadata with a light-weight semantics are visualized with our approach, the visualization environment is not constrained to this kind of data. It is further possible to visualize data-tables with particular statistic or numeric data or results of text-mining approaches to visual-ize just topics and their occurrences over time. Further the semantics tag includes

Table 6.7 SVML properties on layout layer

Properties on layout layer	
Hierarchical_depth	Ability to visualize the hierarchical path on deep levels
Multiple_hierarchy	Ability to visualize the multiple hierarchies
Hierarchical_level	Ability to visualize all entities on one hierarchical level including the path
Multiple_inheritance	Ability to visualize multiple inheritances in one hierarchy
Overview	Ability to visualize the overview of semantic relationships
Hierarchy_navigation	Ability to navigate through the hierarchical structure and get detailed information
Show_concepts	Ability to visualize semantic concepts on a abstract level
Show_instances	Ability to visualize the semantic instances
Show_relations	Ability to visualize the semantic relations
Show_entities	Ability to visualize entities
Show_source	Ability to visualize source of entities
Concept_relations	Ability to visualize relations between concepts
Instance_relations	Ability to visualize relations between instances
Geo_relations	Ability to visualize geographical relations
Temporal_relations	Ability to visualize temporal relations
Crossing_relations	Ability to avoid crossing relations in arbitrary graphs
Instance_clusters	Ability to visualize clustered instances
Concept_clusters	Ability to visualize clustered concepts
Concept_instance_clusters	Ability to visualize clustered instances in concepts

parameters for the particular visual layout. These parameters are commonly used to determine, if a visual layout is able to illustrate the amount and type of data. Table 6.8 illustrates the parameters of the SVML semantics tag.

With this introduced information about visual layouts, the main attributes and characteristics of the visual layouts can be annotated and used to determine the appropriateness of the visual layout for particular data and data sets. Besides the introduced three layers that describe the visual layout, the SVML includes some general information about the visual layout that is used for certain aspects. Table 6.9 introduces the general attributes of the visual layouts in SVML.

With the general information and the three layers of visual layout all needed information are provided to determine the best suited visual layouts. Figure 6.25 illustrates exemplary a part of the SVML annotation of a particular visual layout.

Each SVML contains a set of visual layouts that are annotated as described above. Beside the visual layouts some general information about the user interface and the data providers are annotated as configuration for the visualization environment. These aspects of the SVML will be introduced later in context of the visual interface adaptation. For the adaptation process it is just necessary that a visual layout is

Table 6.8 Parameters of the semantics layer in SVML

Parameters of the semantics layer	
Preferred_amount	Numerical value of preferred amount of entities that can be visualized
Max_amount	Numerical value of maximum amount of entities that can be visualized
Preferred_pathlength	Numerical value of preferred path length that can be visualized, with "null" no path visualization and "infinite" for infinite path length
Max_pathlength	Numerical value of maximum path length that can be visualized (only if not "null" or "infinite")
Constraints	Predefined tags for constraints, e.g. *only_concepts*, *only_instances*, or *only_time*. Visual layouts with no constraints are tagged with "false"

Table 6.9 General parameters for the visual layout in SVML

General parameters for the visual layout	
Id	Identifies and labels a visual layout
Source	Provides the taxonomic structure as described in the previous section
Preferred_orient	Indicates how to place the visualization on the screen in a preferred way
Orientations	Provides possibilities to place the visual layout on the screen
Similar_to	Labels of visual layouts that are similar in their capability and appearance

once annotated with its features by the developer to enable it as part of our adaptive semantics visualization environment.

Data Processing and Visual Layouts
Commonly the visualization process in our visual environment starts with user's search. Although, it is possible to start the visualization process by processing an entire set of data and navigate through this to achieve a certain goal, we explain the adopted rule and processes with users' search as starting point. This will enable to illustrate the complex procedure more comprehensible and adapt it to the visualization of initial given data.

By starting a new search process, the user enters a term that is queried on the given data-bases. If the search process returns no results, the visual environment alerts that no results were found and the process ends (see Fig. 6.26). If the search results in a set of data, the iterative processes described in the Sect. 6.1.1 Data Model and Sect. 6.1.2 Data Feature Model starts and the data attributes and characteristics are analyzed. Based on this analysis process the characteristics of data and those of the available visual layouts annotated in the SVML are compared and a set of appropriate

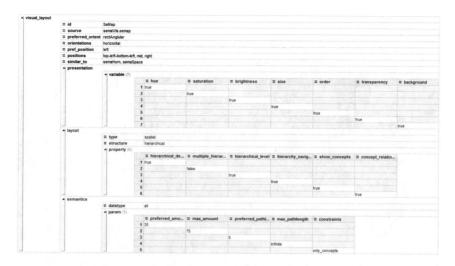

Fig. 6.25   Example of the Semantics Visualization Markup Language (SVML)

visual layouts are identified. The identification of appropriate visual layouts for a certain data-set as the result of search is determined by applying the Euclidean norm between the data characteristics and the visual layout properties [49]. In this step visual layouts can be identified that are able to visualize the data and further those that are more appropriate, thus the data characteristics matches more to the visual layout characteristics. An important issue in this step is to identify those visual layouts that have not the capability to visualize the data. The set of the appropriate visual layouts is in this step ranked based only on the data-matching and data-properties. The identified set is then compared with the user's profile. Figure 6.26 illustrates this process so far.

Choice of User Model
In the following the user's profile is investigated to gather information about previous behavior and interaction. Therefore the adaptation process first controls if the user is logged-in as an individual user. This is to find out, if the user wants an individual adaptation. If the user is not logged-in, the canonical user model is applied that commonly contains enough information about the usage behavior of all users. If the user is logged-in, the process checks first, if the user has already an individual user model. In case of an existing individual user model, this is applied. Else, if the user is logged-in but has not yet an individual model, the canonical user model is applied. In this case the user's interactions train both, the canonical and the particular individual user model. Figure 6.27 illustrates the described procedure.

Choice of User Model based on Similarity and Deviation Analysis
A given and adequate individual user model enables a user-specific adaptation on all levels of adaptation. One main challenge occurs, if the user is logged-in to get such

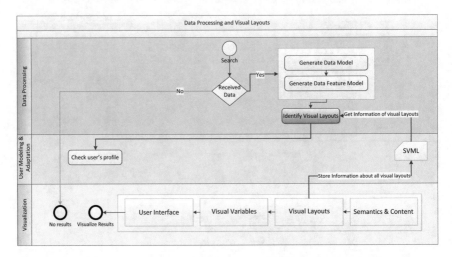

Fig. 6.26 Determining visual layouts by data processing in the adaptation process

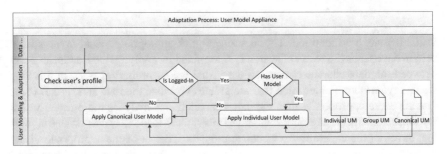

Fig. 6.27 Application of individual or canonical user model

individual adaptation effects, but the system has not yet enough information about the particular user. As described above, the canonical user model is the first instance of our approach that is applied for the adaptation effects. It enables to use the general and average user's behavior to provide adaptations on all visual levels and layers. Commonly the average usage behavior is appropriate enough to provide adaptations until enough information are gathered about the particular user. But the average behavior does not always fit to the interaction behavior and perception levels of the users. It is important to check continuously the user's behavior and measure similar and in particular deviated interaction patterns to avoid unsuitable adaptation effects for the particular user. This is performed in the adaptation process as follow. If the user is logged-in and indicates that he or she wants actively the adaptation options, but an adequate user model does not exist for that particular user, the canonical user model is used as initial model. With each interaction pattern the deviation between him and the canonical user and the similarity between him and possibly given groups are measured. If the system determines that the distance between the particular user

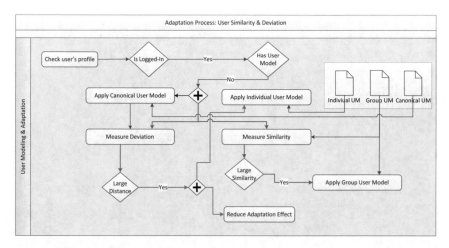

Fig. 6.28   Application of similarity and deviation for the adaptation effect

and the canonical user is large (and gets larger), the adaptation effects based on the canonical user model are reduced linear to the distance and his individual user model is applied, even if there exist not enough information yet. This process is further performed for all users, if the individual user model contains enough information about the users. Parallel to measuring the deviation between individual user and canonical user model, the similarity between user and certain user groups are measured. If a similar behavior is detected, the user is getting part of that group. This means that the group user model is used for him and his interaction behavior trains the group user model. Figure 6.28 illustrates the described procedure of user similarity and deviation measurement to apply different user models.

Application of User Models to Visual Adaptation
With the introduced processes and rules, the user models are chosen to be applied for the adaptation process. In the next step the iterative process of human information gathering and user model appliance is performed. Based on one or more of the determined user models, the adaptation on the entire visual environment is applied. The adaptation on each level layer has different effects that will be described in the next section. In this part it is important that the interaction analysis is providing information for the used user model and trains the particular user model. If only the canonical user model was used to adapt the visualization, the user's interaction just trains the canonical user model. Otherwise, if the group or individual user model were used for the adaptation process, the user's interactions train these models too. The user models build therewith an essential part of the entire adaptation process. Based on the preferences, knowledge and interest in visualization and content, which are modeled in each user model, the different layers are adapted. Therefore the visual capabilities of each visual layout and the entire visual interface are investigated. Beside the user model that provides a persistent knowledge model of the user, group, or average

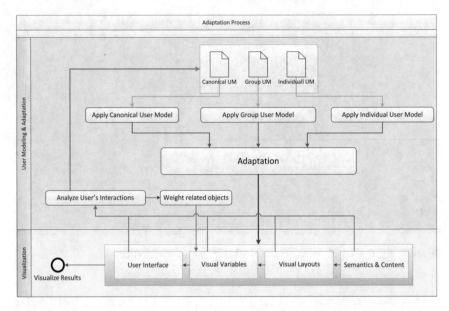

Fig. 6.29 Adaptation of visual layers and the interaction analysis process including the weighting of related-objects

user and the already investigated data model and data feature model, there is further information that affects the adaptation. The introduced approaches for weighting related objects as calculated with the *Inverse Instance Frequency* algorithm (iIf) and the *Direct Relation Frequency inverse Relations Frequency algorithm* (dRf-iRf) from Sect. 6.1.2.2 Weight-Analysis of Semantic Relations are not persistent knowledge models. The interaction of users with particular semantic entities enables the weighting of related object, while interacting with the system. These measurements and algorithms are highly dynamic, thus both, user's interactions and data relations may change. In this process the weighting of the related objects is directly applied to the *visual variable* layer of the visualization environment. If a visual layout has the capability to visualize the related objects, visual variables, e.g. hue, brightness, or size is used to indicate the relevance of the related objects by calculating the *iIf* and *dRf-iRf*. The direct weighting is added to the already measured weight by the user model and is thereby not perceived as obtrusive. However, the indication that the related objects are more relevant can be visually perceived through the changed visual variables. Figure 6.29 illustrates the last step of the adaptation process.

6.3 Visual Adaptation

The process of adaptation described how the influencing factors that are measured and formally modeled are applied in our visualization environment. Beside the measurements of the different models, the application of the gathered knowledge in adaptive environments is of great importance, due to the heterogeneous character of the models. This heterogeneous knowledge models may have various and differing effects on the visualization. The relevance measure may affect the content that is visualized or the visualization layout. Another measurement on users' behavior may affect the color of some visual entities and enable a faster guiding to those entities. So it is more than important to provide a model for visual adaptation that describes the values and parameters that can be changed or should be changed for supporting the user in her work. This section introduces our layer based reference model of adaptation. The reference model uses outcomes from studies of human visual perception and the related models to define appropriate layers that can be adapted. To illustrate the model, we will give first an overview of the applied models and approaches in our reference model followed by the overview of the core contribution, our reference model of adaptation. Thereafter each layer will be described separately. To describe the layers in an appropriate way, we subdivide our measured models from the previous sections into volatile and persistent models and describe how they influence each layer. In this context, we will introduce a set of integrated visual layouts for our adaptive semantics visualization. This section was partially published in [12, 20, 50–54].

6.3.1 Layer-Based Reference Model of Adaptation

Interaction with visualizations enables the dialog between user and the visual representation of the underlying data. The interactive manipulation of data, the visual structure or the visual representation provides the ability to solve various tasks and discover insights. One goal of interactive visual representations still remains the acquisition of knowledge and insights [51, 55]. As illustrated in Sect. 2.4, the term "task" in context of information visualization is often used ambiguously. A dissociation of interactions and tasks in visualizations is rarely performed, whereas the knowledge about the task to be solved with the visualization is of great importance for its design and for the adaptation. We have investigated in Sect. 2.4 various task and interaction classifications to outline an abstract view on visual tasks for a mapping to the human information processing. Therefore all the tasks and interactions were categorized into three abstract levels: "search", "explore", and "analyze". Figure 6.30 illustrates the identified high-level tasks and their assigned interactions and subtasks derived from the existing visual task classifications [20, 51].

We further investigated in context of semantics visualization and human interaction with semantics, different models that distinguish and describe the process of

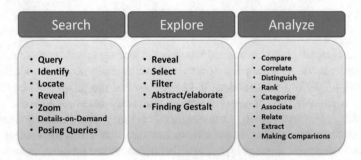

Fig. 6.30 Abstracted tasks in information visualizations (published in [51, p. 33])

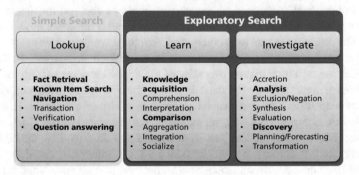

Fig. 6.31 Exploratory versus simple search (adapted from [6])

exploratory search in Sect. 3.4.2. The main aspect here was to outline and differentiate the main characteristics of the search process in contrast to exploration. Based on
the model of Marchionini [6], we could illustrate that the three main kinds of search
activities can be identified as: *Lookup*, *Learn*, and *Investigate*. *Lookup* is the lowest
level of a search activity and results with discrete and well-structured information
based on precise and well-formulated queries. This step can be investigated as *simple
search* thus the main objectives are fact retrieval (declarative knowledge [56]) based
on known items or question answering, whereas the questions are well-formulated
with certain expected results. The exploratory search steps of *Learn* and *Investigate*.
The exploratory search activities are more to acquire new knowledge in particular to
refine and precise the search, to compare certain results and discover new, unknown
or unexpected issues. Figure 6.31 illustrates the exploratory search model of Marchionini, whereas related sub-activities are highlighted to enable a correlated view
on the high-level task classification.

The different classifications of tasks show clearly a differentiated view on
processes of search and exploration. However, there are similarities that should be
considered throughout a search or exploration process. The "simple search" (Lookup
or search) shows the most common attributes in both classifications and processes.
The main aspect here is to find or locate a known entity based on appropriate prior

knowledge in the particular domain. This process involves mainly two steps: the formulation or posing a query that is commonly related to an expected result and revealing, identifying or locating the result in the visual set of given items. The process is related more to *Lookup* which can be seen from the visual perception point of view as the visual "pop-out" of certain results with the main goal to perceive them faster. Different and independent studies illustrated a rapid and parallel processing of the retinal or visual variables, e.g. color, size, shape by the low-level human vision as outlined in Sect. 2.2.1. In particular the outcomes of studies on pre-attentive visual information processing [45, 47, 57] play an essential role in this context. The so called "pop-out effect" makes use of the human's parallel vision processing and guides the attention to the related location on the screen [47, 51].

Those search activities and tasks that rely on more complex cognitive processes, e.g. comparing certain informational issues, analyzing certain topics, learning to refine or precise the query and in particular finding certain unexpected information that leads to amplifying cognition, has more an exploratory character. In this context the two illustrated classification and processes of tasks in the Figs. 6.30 and 6.31 shows some essential differentiations. The process of search explicitly involves a step of *learning*, due to the human intervention in the search process. With this intermediate step, where the acquisition of knowledge for refining and precising the query plays an essential role, the exploratory search gets more a repeated interaction with the search system. With each intervention of user by refining the search, the results get more precise, unexpected issues turn into expected, and the user acquires implicitly knowledge about a certain domain [8]. The process of learning is categorized in our high-level task classification (see Fig. 6.30) as *explore*. This aspect may be a little confusing due to the fact that exploration is classified by Marchionini [6] as the general process of *learning* and *investigating*, which includes all higher level cognitive tasks in the search process. This can be argued with the different characteristics of the search process in visual and non-visual systems. In a visualization system, in particular with semantics, where data are related in a meaningful way with each other, the exploration process is predominantly a process of implicit learning [8]. Exploring information means revealing the meaning, their relationship to other data objects and associate them with the own knowledge in terms of human knowledge. It can be conferred to the assumptions of *Constructivism* with the main attitudes of self-directed and somehow experiential learning [58, p. 107]. The user of visual environments *explores* the underlying information by navigating through the data. The navigation process itself consists of implicit and self-direct learning and constructs knowledge that leads to problem solving. In new situations, which may be in visual environments new tasks, new data, or even new visual layouts, users are able to transfer the previous knowledge and construct knowledge in new situations. Therefore we use in context of visual environments the term exploration as a process of self-directed knowledge acquisition and construction. The last step of the two models shows in contrast to *Explore* and *Learn* some kinds of similarities. The *Investigate* process is assigned to analytical tasks that require complex cognitive activities [6]. One main step of this process is analysis, which build in the task classification the last most complex step of processing information sequentially by comprising attention [47,

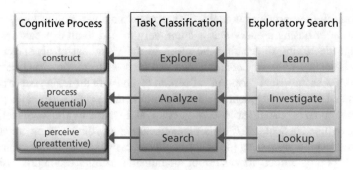

Fig. 6.32 Assignment of tasks to cognitive activities

51]. Figure 6.32 illustrates on an abstract level the described assignments of tasks, search activities, and cognitive complexity.

The different levels of tasks require the adaptation of visualization on different levels as we proposed in [51]. The differentiated investigation of *visual layers* allows a more goal-directed adaptation to users' needs and tasks [51]. We introduced in Sect. 4.4.3 the work of Bertin [48], who differentiated between two elementary aspects of visual mappings: *Implantation* and *Imposition*, visualization attributes that use the two dimensions of a plane (screen) to encode information through graphical marks and those, which encode information through their relationship to each other. This differentiation is of great importance for adapting visualizations, which is also supported by results in cognitive science (e.g., Feature Integration Theory [57] or Guided Search Model [47] (introduced in Sect. 2.2)) [51]. Different and independent studies illustrated a rapid and parallel processing of the retinal variables by the low-level human vision (see Sect. 2.2). The so called "pop-out effect" makes use of the human's parallel vision processing and guides the attention to the related location on the screen [47, 51]. Ware proposed a three-tiered model by considering both the preattentive parallel processing and attentive stages of human vision [59]. He subdivides the attentive processing of visual information into a serial stage of pattern recognition and a further stage of sequential goal-directed processing. While the preattentive stage refers to the retinal (or visual) variables, the attentive stages (or post-attentive stages) require a serial (or sequential) processing of information, which can be provided by visual information of object relationships [51]. This aspect of attentive serial processing, in particular by separating the visual retinal variables and layout information was also investigated by Rensink [51]. In his *coherence theory* and the *triadic architecture* the strict differentiation of *layout* and the low-level retinal variables was proposed in terms of the dynamic generation of a visual representation. Rensink's triadic architecture starts with the low-level vision (pre-attentive) and is generally similar to Ware's model. The most important aspect in this context is the unification of *layout*. Rensink proposed that one important aspect of the scene structure is *layout*, "without regards to visual properties or semantic identity" [51], [60, p. 36]. Further the representation is limited to the *amount* of displayed information.

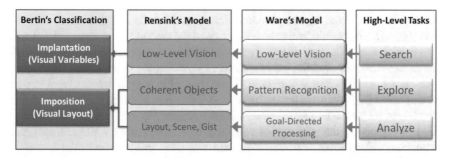

Fig. 6.33 Visual Layout and Visual Variables in correlation to visual perception model (published in [51])

The described processes of visual perception in relation to human attention are the foundation of our differentiated visual layers for adaptation. Based on the introduced models and the results on research of parallel and serial (or sequential) processing we introduce first a model for visual adaptation based on two major visual layers: *Visual Layout* and *Visual Variables* as illustrated in Fig. 6.33.

This model includes in particular the differentiation of Bertin's *Implantation* and *Imposition* and maps these to the visual perception models described in Sect. 2.2 and the abstracted task model. The main value of the differentiated adaptation of these layers is that simple search tasks, such as locating or identifying can be supported by the low-level vision in terms of visual variables. The visualized items can be highlighted using the pre-attentive vision models to guide the attention of users to certain or one data entity that is queried. In contrast to that processes, which require complex cognitive activities make use of the serial vision processing and can therewith be supported best in adapting *Imposition* and therewith the *Visual Layout*. Therefore the arrangement of the entities on screen can be changed by various visual layouts and placement algorithms. The attentive or post-attentive [45, 47, 57] processes require a more serial processing of visual information. The placement in terms of their arrangement and relationship to each other enables solving more complex cognitive tasks, such as comparing or analyzing information, acquiring knowledge through the exploration and navigation process, or getting unexpected insights.

The differentiated adaptation of these two layers already provides an essential enhancement in contrast to the existing works on adaptive visualizations (see Sect. 4.5). It leads to a finer adaptivity based on the perceived visual information and enables the measurement of the effects, such as *interpretation or task completion time, distinction of graphical entities,* or *correctness of tasks* [1, p. 23] of adaptive visual variables in contrast to adaptive visual layouts. However, the main scope of this thesis is to investigate the entire process of information or semantics visualization to provide the most effective way of adaptation. In this context, the separated adaptation of visual variable and visual layout would lead to a more adequate adaptation but would investigate the entire process. Rensink proposed in his works on *Coherence*

Theory that the (visual) representation is limited to the *amount* of displayed information [60, 61]. This would lead to an adaptation of the amount of the displayed entities and would refer to the question *how many entities are displayed*? The aspect of content adaptation was investigated in context of visualization in some previous works (see Sect. 4.5.3.2), whereas commonly the content is tailored in terms of quantity to be displayed [20]. We go one step further and claim that the content adaptation relies on various factors to be investigated in the adaptation process. In particular for any kind of semantics, the structure of data (e.g. hierarchy, arbitrary, dimensions), the domain of data in terms of what are the data about, the ability to order the data entities (e.g. nominal, quantitative, ordinal), and the amount of the displayed data, may play a key-role for the adaptation [20]. With semantics and the related concepts, the underlying data can be aggregated to an abstract concept-level, which includes categorical information about the data. This abstraction is another kind of data reduction that leads to another kind of content adaptation. Thus our content adaptation includes not only the adaptation of the content itself and provides further adaptation possibilities, such as structuring, abstracting, or categorizing, named this layer of adaptation *Semantics Layer*. The Semantic Layer makes use of the introduced measurements in Sect. 6.1 and the introduced processes in Sect. 6.2 and adapts all data-related aspects and their mapping to the visual environment.

With the inclusion of the three layer *Semantics*, *Visual Layout*, and *Visual Variable*, we separated all main aspects of the visualization transformation process and enable a further step of granularity for the adaptation process. The visual transformation process starts commonly with the Semantics layer that adapts the content, structure, order, and amount of entities to be displayed [20]. Thereafter, a visual layout is selected based on the two introduced influencing factors of data and human behavior on the *Visual Layout Layer*. For guiding the attention to certain data-entities, the layer of Visual Variables adapts certain retinal or visual features. The entire process involves at all steps the persistent and volatile models that was generated and computed for the adaptation process (see Sect. 6.2) [20, 51]. Figure 6.34 illustrates schematically the described and so far introduced layers.

The above described procedure involves and separates all visual layers and enables their adaptation on each level to provide a guidance of attention through the visual variables, the information processing by different and adapted visual layouts, and the appropriate mapping of the visual layouts by adapting the semantics in terms of content, structure, amount etc. The main limitation of the described process is that the visual layout provides just one view on the data. Commonly already the Semantics layer provides not only one main characteristic, e.g. only hierarchical structures, to provide the best fitting single interactive view on the data. Further the processes of that require more complex cognitive tasks would not be supported by single view on data in an appropriate way [53, 62]. To face this limitation, we introduce a further adaptation layer that manipulates based on the introduced persistent and volatile models and computation (see Sect. 6.2) of the adaptation process, the entire user interface. The user interface adaptation is the last layer of the visual adaptation layers. Due to the fact that we only investigate the visual appearance and orchestration of visual layouts on screen [8, 12, 20, 53, 62], the user interface refers more to the layer

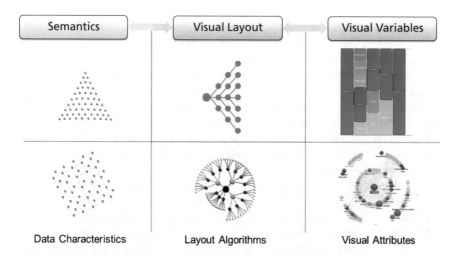

Fig. 6.34 Schematic illustration of the introduced visual layers with examples

of *Visual Interface*. The main value of adapting the *Visual Interface* is that complex tasks can be supported with the differentiation and reduction of data characteristics. This differentiation is commonly performed by orchestrating the visual layouts in a juxtaposed manner and provides separated visual perspectives without losing the context or certain information [53]. The visual perspective are either interlinked with each other by using the brushing and linking interaction (see Sect. 2.3) or decoupled from each other to support comparative tasks on different views and levels. The process illustrated in Fig. 6.34 can therewith be enhanced as illustrated in Fig. 6.35.

The differentiated adaptation of the identified visual layer enables a fine-granular tailoring on users' demands. Therefore each visual layer can be adapted based on the underlying volatile or persistent knowledge models as described in previous sections. The differentiation of visual layers can be adopted to the reference model of Card et al. [1] (see Sect. 2.1) to provide a reusable and transferable reference model for visual adaptation. The transformation process of the reference model is mapping the data tables to *Visual Structures*. Here the work of Bertin [48] builds the foundation

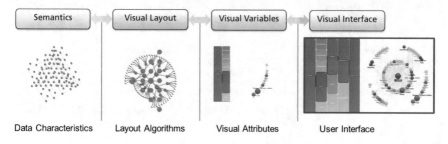

Fig. 6.35 Schematic illustration of visual layers with tentative examples

of visual variables and structures to provide an effective mapping. The reference model proposes that two main factors are important to provide an effective mapping to visual structures. The mapping should preserve the data with their type of variables and be perceived well by human. Human should be enabled to interpret faster, distinct graphical entities, or leads to fewer errors [1, p. 23]. Although, the reference model uses and enhances the differentiation of Bertin [48], a separation of layers for supporting different tasks is not proposed. Further the reference model outlines that the visual encoding of data for *uncontrolled processing* (pre-attentive visual information processing (see Sect. 2.2.1)) and for *controlled processing* (sequential or attentive visual information processing (see Sect. 2.2.2)) [1, p. 25] are different. The reference model does not propagate or investigate this separation. It mainly focuses on a general transformation of data and their sequential characteristics to visual structures and closes the loop to data with direct human interaction and intervention.

In contrast to the established reference model, we propose a reference model for visual adaptation investigating the main outcomes of perception studies. Our model investigates in this particular case mainly semantic data as starting point of the transformation process see Fig. 6.36. However, the initial transformation step can be raw data too, we use semantic data, thus this thesis investigates the adaptation of semantics visualization. The first transformation step is similar to the reference model of Card et al. and transforms the underlying data to a structured data model as described in Sect. 6.1.1 including common semantic information [20]. The main difference here is that our model does not start with a static pool of data. The data are commonly results of user or system queries. Further this step includes already the volatile and persistent model in the transformation process. Querying data by users is already a volatile model, thus the resulted data are a subset of the entire data-set. Further the weighting of semantic relationships and the amount of each semantic entity-type as illustrated in Sect. 6.1.2 is considered. The persistent models influence the transformation step too. The data feature model provides structural information

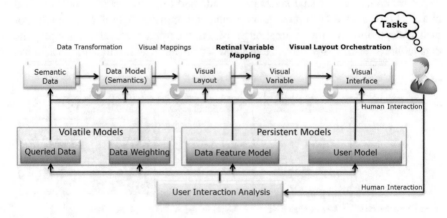

Fig. 6.36 Layer-based reference model for adaptive visualizations

that is used to generate one or more data models as internal representations of data. The user model that influences the amount of data and the "data of interest" can be either a canonical, group, or individual user model. This aspect of user model can be applied to all transformation steps of the proposed reference model as illustrated abstractly in Fig. 6.36.

The generated data model provides our Semantics (Content) layer. The visual mapping to the *Visual Layout* layer is performed based on the capabilities of the visual layouts (see Sect. 6.2.3) and the volatile and persistent models. This step chooses based on the underlying information the most appropriate visual layout. The retinal variable mapping sets the visual variables in terms of color, shape, size etc. based on the volatile and persistent models. The strict bisection of visual layout and visual variable allows guiding the attention of users to certain data entities, anomalies and so on, regardless which visual layout is chosen in the previous transformation step. The visual layout orchestration builds based on the volatile and persistent models a juxtaposed arrangement of different visual layouts. Further the interlinking of the visual layouts with each other is set. With this further layer for visual adaptation the transformation of data to a visual user interface covering one or more visual layouts is completed. The loop to user is performed in two various ways: each interaction of the user is analyzed to provide a more efficient adaptations and the interaction of users' are manipulating the transformation process too as proposed by Card et al. [1]. Figure 6.36 illustrate our reference model of visual adaptation and illustrates the described process [12, 36, 51].

6.3.2 Semantics and Content Adaptation

Semantic or structured data are transformed in the first step of our reference model to the *Semantics Layer* of adaptation. The semantics layer defines in particular *what* is visualized [20]. Commonly the use of the visualization environment starts with a user query. As already stated in the description of our data model (see Sect. 6.1.1) the resulted queries consists either data that are instantiated as objects to be visualized or not instantiated objects that are present in the data model but not visible for the user. The differentiation of which data-part is visualized and which part remains in the model, is computed based on the previous individual user or canonical user behavior and the overall structure of data that is in any case given. If a user model does not exist even on a canonical level, only the structure of the data is considered in this transformation process.

The process of differentiation for this transformation step is quite easy, due to occurring delays of higher computations: First the overall structure and amount of data is determined as described in Sects. 6.1.2 and 6.1.1. Here the amount of data plays an essential role. If the amount of data is higher than the given visual layouts are designed to visualize, the structural information of the data are used as indicators for abstraction or even reduction. The structural information contains information about data hierarchy, arbitrary relations with information about incoming or outgoing nodes

[20], amount of entities on each hierarchy-level, and further semantic information such as properties, which may contain geographical or temporal information. In this context the type of data is important for the abstraction and therewith a reduced and categorized view on data. In the next step the structural information, in particular the hierarchical information, are used to generate "knowledge categories" based on the semantic concepts. This enables to visualize all queried data on an abstract level. The named categories are used to determine the interests of users in certain field of the domain knowledge. Based on the weighted probability, the data entities with higher relevance are instantiated in the data model. The geographic, temporal, and other properties and aspects of data provide a secondary relevance for the instantiation of the data. This aspect is determined by two main factors: do the queried data have any geographic or temporal information and are amount of these information high enough for visualization, and does the user have any preferences on visualization of these properties?

Based on our reference model the Semantics layer is a structured and categorized subset of the queried data with relevance values for the abstracted categories and data entities. Further the structure itself, e.g. hierarchical or arbitrary, and the properties are assigned with relevance values. The relevance measurements are performed based on the introduced volatile and persistent models. The preferred domain of data is determined through the introduced measurements based on the user model. The preferred structure and data properties are determined through the overall data structure and user's preference in the user model. The users' interactions lead to changing all criteria of the Semantics layer. Based on the *Inverse Instance Frequency* (iIf) and the *Direct Relation Frequency inverse Relations Frequency* (dRf-iRf) algorithms (see Sect. 6.1.2.2) the relevance values of neighboring entities are manipulated. Further the interaction with domains leads to new measurements and loads the not instantiated data on demand [20]. Figure 6.37 maps the described procedure to our reference model and provides an overview of the main steps.

6.3.3   Visual Layout Adaptation

The *Visual Layout* layer defines the visual appearance of data in terms of their relationship to each other and in relation to the screen according to Bertin's Imposition [20, 48]. Therewith the placement of visual representations of data is defined in particular with their relations to other entities and refers to the questions *where* to visualize data entities. Based on the Imposition model, the placement of entities in relations to the screen (or plane) plays an important role as information carrier. In particular, if the axes of the screen provide information, e.g. in a temporal or hierarchical manner. In our reference model the Data Model of the *Semantics* layer provide after a reduction and categorization of the data, a subset of the queried data to Visual Layout layer and initiate the visual mapping transformation. The visual mapping is performed in the first based on the information about data. This includes the amount of instantiated entities with their type. For each semantic entity, e.g. concepts, instances, incoming

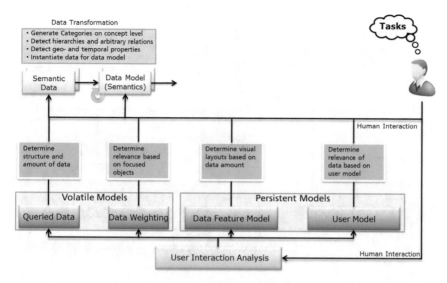

Fig. 6.37 Semantics layer in the reference model

and outgoing relations, or properties, quantitative information are provided. Further the main structural information of the data or provided in terms of their dominance, e.g. dominant hierarchies, arbitrary relations, dominant existence of knowledge entities, or no relational information. Further this step provides information about data properties in terms of their characteristics, e.g. temporal or geographical data. All these information are provided with their quantity, which is investigated as relevance value for the visual mapping transformation. The main aspect in this step is to identify visual layouts that are not capable to visualize the underlying data (see Sect. 6.2.3). The mapping is performed with the specifications of the visual layouts in the *Semantics Visualization Markup Language* (see Sect. 6.2.3) a persistent model that is describing the capabilities of each visual layout. In our reference model this model is not assigned as an exclusively, but is part of the data feature model. The volatile models plays just indirectly a role, thus they manipulate the *Semantics* layer and therewith provide implicitly information about the data to be visualized. The main model here, beside the capability of each visual layout, is the user model. The user model at individual, group, or canonical model, contains information about the preferred layout of the user. The combination of the two persistent models enables to determine the identification of the most appropriate visual layout based on the underlying data and user preferences [20].

The user model contains commonly just relevance values for each visual layout. Therefore the restriction of visual layouts that are not capable to visualize certain data is an important issue. Even if a visual layout has the highest relevance value based on the user model, but is not able visualize the underlying structures or data, the particular visual layout is excluded from the adaption process. A simple and comprehensible example is the visualization of temporal data. If the user model

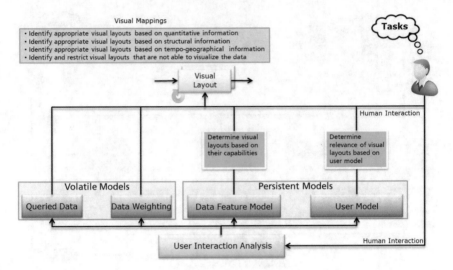

Fig. 6.38 Visual layout layer in the reference model

contains high relevance values for temporal visual layouts, but the data do not provide any temporal information, the next visual layout with a high relevance value from the user model is investigated for the adaptation process [12, 20, 51]. Figure 6.38 maps the described procedure to our reference model and provides an overview of the main steps.

6.3.3.1   Classification of Visual Layouts for Semantics

Semantics data consist of a set of different and varying data and structural components. It contains hierarchical structure, which we used in the Sect. 6.1.1.1 to generate a categorization of search results. The instances have relations to each other as arbitrary relations and contain additional information in form of properties. These properties consist of varying data types, which includes number, date, time, geographic, text and so on. Semantics provides a variety of heterogeneous data and according to the introduced models, a variety of ways to visualize the semantic information. We illustrated in our review on related works (Sect. 3.5) that a great number of semantics visualizations focus on visualizing one main characteristic of the semantics. The separated view on semantics makes sense, thus the variety of included information leads to very complex and incomprehensible views, whereas the main limitation is in the information lost. In context of our survey, we identified three main classes of semantics visualizations: *hierarchical semantics visualizations, relational semantics visualizations*, and *entity-based semantics visualizations* (Sect. 3.5.2). For introducing the integrated visual layouts of our reference model, we use our classification of semantics visualization and enhance the class of *Property-based Visual Layouts*, thus our approach investigates aspects of geographical and temporal visualizations too. Further the definition of *Entity-based Visual Layouts* is adapted to our approach, where all the visualization process is focused on information search and retrieval. Therefore, we rename this class of visual representation to *Resource-based Visual Layout*, due to the character that the targets URIs (Resources) are visualized.

Hierarchical Visual Layouts
One main class in for the visual layouts of our reference model still remains the hierarchical view on semantics. In context of our visual layouts it is important to outline that hierarchical layouts enables to categorize domain-specific resources in inherited concepts and allow a topic-related access to the domain knowledge. With the categorized views they provide a substantial element for visualizations, namely overview on abstracted levels of categories. The overview on hierarchical structures provides the ability to find the starting point and locate the knowledge-resource of interest. Further these layouts provide commonly an overview of the entire hierarchical structure of the data and data-entities, which enable an implicit knowledge acquisition by retrieving the ancestors of semantic entities [20, 53].

Relational Visual Layouts
As illustrated in the definition of relational visualizations (Sect. 3.5.2) aim to visualize the semantic context of information and provide therewith navigation and browsing abilities within an information space. Common approaches for visualizing semantic relationships are usually based on graph-based visualization techniques and provide navigation through the nodes and semantic neighborhood. Relational Visual Layouts visualize the in particular the arbitrary and non-hierarchical structure of the semantics data. The main goal of these visual layouts is to provide a view on the semantic neighborhoods and enable the navigation through the arbitrary structure.

Property-Based Visual Layouts
Property-based Visual Layouts aims at visualizing certain explicit or implicit properties of the semantics: explicit in terms of explicitly annotated properties and implicit by means of those properties that are retrieved during the data transformation process. Examples of explicit properties are temporal and geographical properties that are part of the semantic annotation, whereas the number of entities in a time period or in certain categories is gathered implicitly. This type of visual layout commonly makes use of the screen axes as information carrier, e.g. to visualize the spread of entities over time.

Resource-Based Visual Layouts
Resource-based Visual Layouts are responsible for the presentation of the content referenced by URIs in the semantics. Their main goal is to visualize the content behind a semantic entity. The content view enables a detailed view on the information and builds therewith the most focused view on data. These visual layouts are commonly not using any algorithms for placement. They are just to visualize the content in form of text, pictures, or any other multimedia object [20].

6.3.3.2   Integrated Visual Layouts

This part of the thesis introduces some of the integrated visual layouts that are part of our visual environment and provide the described adaptive features. The description of the visual layouts will follow the introduced classification, whereas some visual layouts may combine the outlined visualization characteristics. In these cases, we will categorize the visual layout based on the most dominant visual capability and describe further features.

Hierarchical Visual Layouts
Hierarchical visual layouts can be used in various ways to convey information. One main aspect in this context is to provide an overview of the subset of the queried and instantiated data. To support such a visual overview, we used the force-directed algorithm [63] to visualize explicitly the result set of concepts and therewith categories in our *SemaSpace-Concept* visual layout. SemaSpace-Concept visualizes only the concepts that were retrieved from the Semantics layer. It illustrates the concepts on an abstracted level with the main intention to provide an overview of the entire domain knowledge. With an integrated *mini-map* as help for orientation and the possibility of Panning and Zooming (see Sect. 2.3.2), the visual layout provides the acquisition of knowledge about a certain queried item for gathering more information and refine the information search. The nodes are not connected with each other. They just provide information about subclasses in terms of annotated numbers to each node. The navigation through the concepts allows viewing and interacting with other concepts or sub-concepts. Figure 6.39 illustrates the visual layout *SemaSpace-Concept* providing an overview of the entire concepts and zoomed-in to provide further information about certain concepts.

The introduced visual layout may provide an overview on an abstract level for the searched term, but the hierarchical characteristics of the data are not explicitly outlined. It has more the character of gathering domain knowledge on an abstract level rather than navigating through the hierarchy to the concept or instance of interest.

(a)   **(b)**

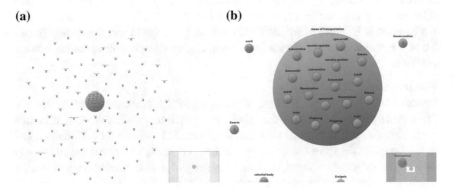

Fig. 6.39 *SemaSpace-Concept*: **a** overview and **b** zoomed hierarchical visual layout

(a)   **(b)**

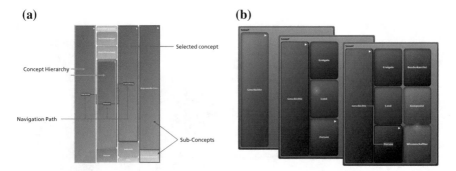

Fig. 6.40 *SeMap*: **a** categories, and **b** incremental navigation

To provide a real hierarchical view on data and enable the navigation through the sub-concept hierarchy, we developed the *SeMap* visual layout [7].

SeMap combines two different visual layouts to provide on the one a more goal-directed navigation through the conceptual hierarchies and on the other hand an overview on the contextual hierarchical path that leads to comprehend the *knowl-edge path* [7]. Therefore, we used Shneiderman's Treemap [64, 65] and Treeview that leads to navigate through certain levels of hierarchies. Thereby the layout starts at a root node of a tree and displays the next and more detailed level by users' inter-action. The main advantage of SeMap is that the users are not overcharged with the entire hierarchy at the initial point of search. They build their hierarchy through the interactions, whereas the neighboring concepts on the same hierarchical level are visualized too. With the space-filling approach of Shneiderman the navigation and therewith the inclusion of implicit knowledge gets easier as evaluated in preliminary study [7, p. 90]. This is due the rigorous support of different visual variables in the layout algorithm. Figure 6.40 illustrates SeMap with the different levels of hierar-chical abstractions. Further the successive buildup of the hierarchies is illustrated that lead to less complex view on the hierarchical data. The visual layout helps to build up a map of semantic concepts based on the interactions of users. With users' navigation through the semantics, SeMap only visualizes the relevant aspects and information.

SeMap enables a simple navigation and goal directed navigation through the hier-archical structure of semantics but has one limitation: it does not support multiple-inheritance and a view on different concept at the same time, due to the integrated navigation path. To provide a visual layout that supports these two features, we developed and integrated the *SemaSun* visual layout [66]. SemaSun uses the Sun-burst metaphor [66] to provide similar to *SeMap* an incremental "building" of the navigation path. At the startup of the visual layout, the root concept of the data is illustrated (Fig. 6.41: 1a). The user is able to incrementally explore the hierarchy by expanding entities of interest (Fig. 6.41: 1b). The radial layout of the sunburst visual-ization offers thereby the expansion of multiple paths (Fig. 6.41: 1c) so users are able to gather an overview of the whole inheritance structure and are not limited to the

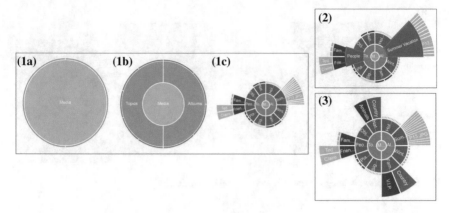

Fig. 6.41 *SemaSun*: *1* Incremental exploration, *2* multi-focus feature, and *3* multi-inheritance support of the visual layout

exploration of a single path [66, p. 914]. To maintain the informational context and to avoid the need for scrolling, entities that users visited earlier move closer to the center and are reduce in their size (Fig. 6.41: 1c). A distortion technique combined with mouse-over reveals the entire label in reduced entities. The number of child nodes is denoted as the number of arcs around the parent node (Fig. 6.41). During the exploration process, users may find entities they are interested in. In order to not loose these relevant entities during the further exploration of the knowledge space, SemaSun integrates multiple focus support. With this feature users are able to mark entities of interest, so that the labels of these focused entities are always visible and are not reduced in their size by the distortion effect (Fig. 6.41: 2) [66]. Further the multiple inheritance of semantic entities are supported with the multi-focus feature, if an entity is selected that inherits from more than one concepts all duplicates of this entity are focused by the multiple focus feature (Fig. 6.41: 3) [66].

The introduced visual layouts make use of spatial or space-filling approaches to convey the semantic hierarchy. The *SemaZoom* visual layout [67] combines space-filling approaches with node-link a layout to illustrate in particular the semantic hierarchy. SemaZoom makes use of distortion and focus plus context techniques to provide in particular an overview of the entire queried subset of the semantic data. In contrast to the introduced *SemaSpace-Concept*, the layout is not limited to a view on concept level. It makes use of two main components: a schema layer that provides in a space-filling manner the concept hierarchy and an instance and relation layer that orders instances according to their concepts. The layout focuses on visualizing the hierarchy of instances and heir relations. An assignment of instances to their concept provides an orientation for users. The interaction with the layout can be commonly performed without mouse-clicks. Here, a fish-eye layout is integrated that responds to the mouse movement and allows gathering the entire information of the semantic hierarchies. The selection of an instance leads to a stop of the distortion technique.

(a) **(b)**

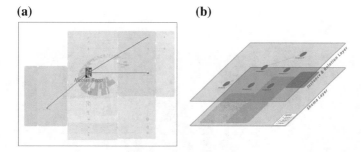

Fig. 6.42 *SemaSun*: *1* Incremental exploration, *2* multi-focus feature, and *3* multi-inheritance support of the visual layout

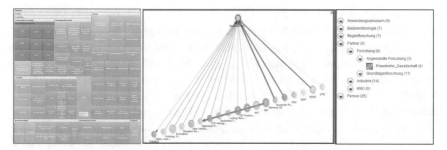

Fig. 6.43 Treemap and hierarchical node-link layout

Figure 6.42 illustrates the introduced visual layout and the two layers of schema and instance and relations.

We further integrated for the hierarchical view on the semantics a Treemap version according to Shneiderman [65], a node-link graph, and an indented list with hierarchical view as illustrated in Fig. 6.43. The main goal here was to evaluate these visual layouts against the introduced ones.

Relational Visual Layouts

Relational Visual Layouts commonly aims at providing navigation ability through the semantic relation and context of the underlying data. The integrated *SemaGraph* [53] visual layout enables such navigation through the semantic relations with a concentric-radial node-link layout. The main value of this simple visual layout is that the degree of complexity can be adjusted through the level of details and the level of semantic relations. We chose the concentric radial algorithm for conveying such relational information due to the fact that the focused object is always placed in the center of the graph-layout. This is to convey implicitly the information that the graph does not provide any hierarchical relations or the hierarchical relations are not focuses in this layout and to provide a more comprehensible way to perceive the focused objects. Although, the aspect of identifying the main and focused object is more a task of the visual variables layer, the positioning in the center of a graph supports this visual feature. SemaGraph's main visual layout just provides infor-

(a)   **(b)**   **(c)**

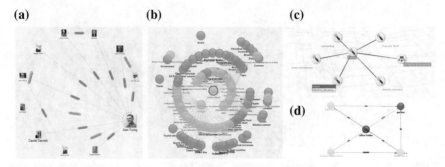

(d)

Fig. 6.44 The derivation of the *SemaGraph* visual layout

mation on the instance level. It illustrates the instances of the semantic data with an arbitrary graph that leads to comprehensible views of the arbitrary semantic relations for small entity-amounts. In case of high amounts of entities the graph-based visual layout overcharges users due to the complex arrangement of entities and relations. Therefore this visual layout is not design to be used for huge graphs or amounts of semantic entities. Further, SemaGraph provides a varying set of different features and integrated layouts. It supports, e.g. loading-on demand from not yet instantiated entities in the data model or supports different derivations of the concentric-radial layout. In particular it supports the navigation through semantic instances and provides a view on the semantic relations of a particular semantic entity. Figure 6.44 illustrates some derivations of SemaGraph and illustrates the complex view (Fig. 6.44b), if the number of entities are high.

The introduced version of *SemaGraph* does not provide the underlying concept information of the semantic data. This kind of information reduction leads to a more efficient navigation of the semantic relations that are commonly arbitrary and provide relational information. In some cases it might be necessary to add the concept information directly to the graph. This is in those cases more efficient, if the relational information is limited in their number and the screen space can be used to provide another schema layer as provided by *SemaZoom*. For providing this information, we enhanced *SemaGraph* with an additional layer of concepts. The concentric-radial layout places all instances according to their concepts in a radial way. The limitation of the enhanced *SemaGraph* is that not further levels of detail can be added. The advantages are that the conceptual information is provided in the same layout and a kind of interaction history is integrated. The interaction history illustrates in a vertical manner all previously illustrated instances and in particular the previously selected instance. The enhanced version of *SemaGraph* includes further a paging functionality, due to the limited visual capabilities. With the paging, hidden instances are revealed. Figure 6.45 illustrates the enhanced version of *SemaGraph* with the integrated and described features.

A similar approach for visualizing relational information according to their concepts was proposed by Bhatti in *SemaSpace* (formerly SemaVis) [68]. The first version of *SemaSpace* was designed to visualize formal ontologies as a standalone

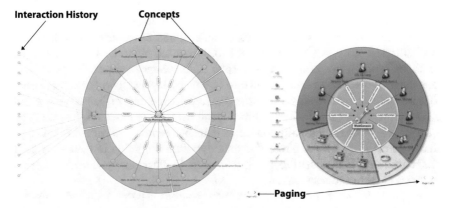

Fig. 6.45 The enhanced *SemaGraph* visual layout with its features

visualization using the force-direct algorithm. The main value of the first *Sema-Graph* version was that the force-directed algorithm was used to pull the instances to their concepts (called knowledge spaces [68]) and provide similar information as the advanced *SemaGraph* layout. We enhanced *SemaSpace* according to our approach by separating the layout from the entire system and the other described layers [53]. Further we integrated a more simple a comprehensible view of sub-class of relations in the visual layout and class relations. The enhancements on *SemaSpace* led to the ability to open or close the concepts and provide a more comprehensible view on semantics data. *SemaSpace* is one of the visual layouts that is theoretically able to visualize all aspects of semantics, namely hierarchies, concept-relations, instance-relations, inheritance-relations on concept and instance level, and sub-class relations. Figure 6.46 illustrates the initial version of *SemaSpace* and the enhanced version with the enhanced features.

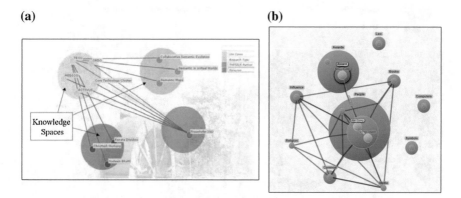

Fig. 6.46 *SemaSpace*: **a** The initial version (according to [68, p. 314]) and **b** the enhanced and integrated version

To support in particular the search and exploration process in semantic data, we investigated the relations between search queries and semantic entities as results [69, 70]. Therefore, we combined the two introduced algorithms to visualize the relations between the terms and the results. *Term Nodes* represent the recognized search terms. These nodes are placed by a concentric layout algorithm at the startup. Users are able to move and order them on the screen according to their wishes. *Result Nodes* represent the hits in the combined result list that are found for the given user query. These nodes are visually connected to related term nodes with directed edges and are treated by a force-directed layout algorithm according to their weights to the surrounding term nodes [70, p. 139]. Figure 6.47 illustrates the introduced visual layout with term and result nodes.

A similar layout that enables the view on relations between searched terms and results of the query is *SemaPrism* [50]. *SemaPrism* uses a placement algorithm with three different levels of abstraction. The results of one or more search terms are placed on an outer circle. A circle in the middle illustrates all concepts of the underlying semantic data as areas. The semantic related entities are placed in the center circle according to their concepts. The queried terms or term-nodes are placed in the inner circle of the visual layout. This way of layouting search results enables to gather the relation between the searched term and the resulted entities. In particular, the semantic relation in the center circle provides a comprehensible way to view how the resulted entities are related to the search terms. Further the relations to other entities are illustrated according to their concepts. These entities enable an exploration and navigation to other semantically related entities and gather new and unexpected information. *SemaPrism* provides further similar to the previous layout, a paging feature that enables to visualize a huge amount of entities, whereas each page is limited to a result set of twenty entities. Figure 6.48 illustrates the visual layout and the described circles.

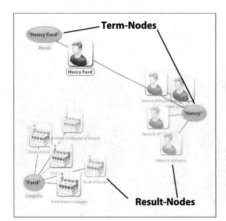 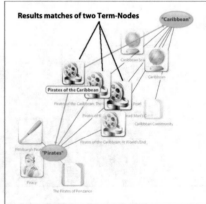

Fig. 6.47 Combined visual layout for visualizing relations between search term and results (adapted from [70, p. 142])

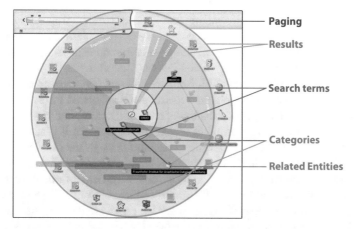

Fig. 6.48 *SemaPrism* with three main *circles* for queried terms, semantically related entities, and result entities [50]

Property-Based Visual Layouts

The relational visual layouts enable a navigation and exploration in particular through the semantic relations of knowledge entities. The knowledge or semantic entities are commonly annotated with further properties. We differentiate for the *Property-based Visual Layouts* two kinds of properties, explicit properties that are annotated in the data and are part of the semantic data entities and implicit properties that are gathered through different computations, e.g. amount of entities in relation to time. Figure 6.49 illustrates a simple way to reveal explicit properties of entities in textual manner. It includes all the information that a single semantic entity contains.

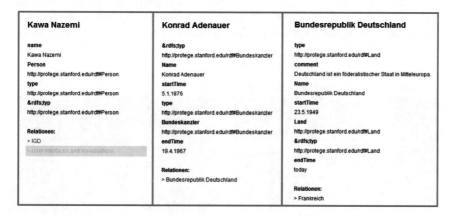

Fig. 6.49 Textual view on annotated (explicit) semantic properties

For visualizing semantic information two kinds of explicit properties are important: Temporal and geographical, thus commonly other properties can be visualized with other visual layouts. For investigating the temporal aspects in semantics, we integrated the *SemaTime* [71, 72] visual layout. Therefore we used the horizontal screen axis for providing temporal information. *SemaTime* consists of main view that contains semantic entities placed according the annotated the annotated time. Further it provides relational information of semantic entities and if given temporal relations of the entities. The temporal navigation component is placed at the bottom and enables with a time-slider the navigation through the time-period and an overview of the instances in the main view as thumbnails. *SemaTime* is provided as visual layout with two different versions. One of these provides further a hierarchical view on concept level and enables filtering on concept level [71]. The hierarchical view is similar to the already introduced *SeMap* and was adapted from this visual layout. Further this version provides a detailed view that enables the illustration of the underlying resource. The other version of *SemaTime* reduces the complexity of the visual layout by dismissing the hierarchical view on the semantic concepts. Figure 6.50 illustrates both versions of *SemaTime* [71, 72].

To support the geographical properties of semantic entities, we use the geographical visualization *Google Maps*. To support the placement of semantic entities and relations, we build up a further layer of semantic entities and relations, similar to the introduced visual layout *SemaZoom*. Each semantic entity is placed on the map according to the geographical coordinates or geographical attributes. Further the countries of the entities are highlighted. The semantic relations are visualized through an arbitrary node-link graph. Figure 6.51 illustrates the *SemaGeo* visual layout.

The introduced visual layouts make use of explicit annotated properties. In some cases, the properties are not explicitly part of the semantic entities but can be gathered through the introduced approaches of iterative querying (Sect. 6.1.1.1) and quantitative analysis (Sect. 6.1.2.1). The most common and useful way is to retrieve the amount of certain semantic entities. Commonly these are assigned with further data properties in the data set, such as date of creation or place of creation. In contrast to the

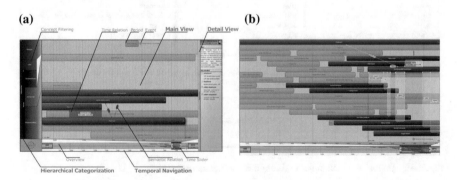

Fig. 6.50 The two versions of *SemaTime*: **a** advanced version with hierarchical and detailed view (from [71, p. 523]) and **b** reduced version for a better comprehension [71, 72]

Fig. 6.51 *SemaGeo* visual layout for geographical properties

introduced visual layouts not only one dimension of screen is needed to provide this information. The provision of temporal information in correlation with the amount of data entities requires at least two dimensions of the screen. To provide such information that are built up by explicit (temporal annotations) and implicit (amount of entities), we introduced the *SemaRiver* [73] visual layout. *SemaRiver* represents the temporal occurrence of semantic entities in an extended stacked-graph layout. The advanced overlay techniques integrated in the stacked graph enable the exploration of topic specific and temporal co-occurrences [73]. Similar to *SemaTime*, *SemaRiver* provides a times-slider with an overview of the entire temporal set. The slider enables navigation through the temporal spread. Figure 6.52 illustrates the introduced visual layout.

Resource-Based Visual Layouts
The so far introduced visual layouts aimed at providing structural, hierarchical, temporal, or geographical information through the visual placement of semantic entities. According to the tasks to be solved with semantics visualizations and to the *Visual Information Seeking Mantra* of Shneiderman [13], it is necessary to provide a detailed view on the data. In case of semantics data it is commonly the resource of each entity that is referenced as an URI. In case of metadata, similar references consists that enable the retrievement of the data entity. Thus commonly our visualization approach accesses data from the Web. It is obvious to provide the data using the most common markup-language on Web. We therefore integrated the content layout *SemaContent* [20, 53, 74] to illustrate the resource as content in the visual environment. *SemaContent* supports the most common Web-based markup languages as well as the illustration of pictures or other multimedia objects. Figure 6.53 illustrates some derivations of *SemaContent*.

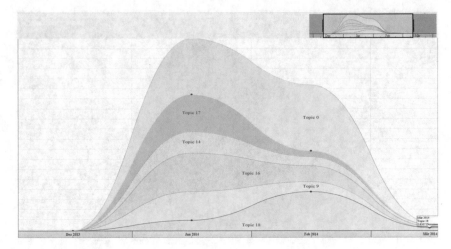

Fig. 6.52 *SemaRiver* visual layout for temporal spread of entity amounts

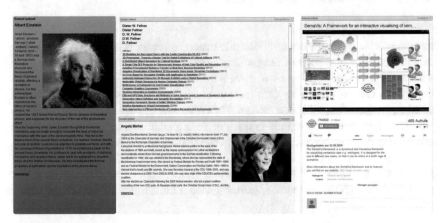

Fig. 6.53 Various derivations of the resource visualizer *SemaContent*

The integrated visual layouts and content illustrations are an important part of our conceptual design. In particular the fact that heterogeneous classes of layouts are investigated to provide different perspectives on semantics is essential for the adaptation process. However, not all the introduced layouts provide the full adaptive behavior in terms of content adaptation and visual variable adaptation, this aspect is not limiting our general approach. Our application scenarios in the following chapter and the evaluation of the entire approach will consider only the full adaptive visual layouts. Furthermore, we propose to enhance the set of visual layouts and even the integrated classes for new application scenarios and a broader range of users.

6.3.4 Recommending Visual Layouts

The process of adaptation of the visualization according to our reference model does not differentiate in the first attempt a differentiated level of adaptation according to the identified layers. Commonly the adaptation can be performed initially before the user starts with the work or interaction on a system or during the work and inter-action. This differentiation is particular for visualization adaptation essential. The focused attention according to the illustrated models requires a sequential processing of information. Human are involved and work focused on various aspects or tasks by interacting with visualization and perceiving visual information. The initial change on visualizations would not affect this interaction and perception process. But if the user is in a focused information or acquiring process, the unexpected changes on the visualizations may be destructive and miss their main target: to support the user. To face a destructive adaptation behavior, the adaptation process follows an automatism level that investigates process of adaptation during the work with a visualization system. Changes during the work on *Semantics* layer can be compared to recom-mendation systems. The changes occur in a more hidden and not perceivable manner [75]. Loading data on demand, after an interaction of user, leads to a more compre-hensible way of understanding the relations and correlations of data. This "hidden" recommendation or adaptation is thereby not disturbing the user in his work process.

The adaptation of the *Visual Variable* layer is affecting the layer that is next to users' attention. An automatic adaptation of visual variables, e.g. color or size, leads to guiding the attention to certain elements or data representations. The adaptation process in this case has to be performed in a smooth and comprehensible way and coupled with user's interactions. The direct coupling of users' interaction with certain changes on the *Visual Variable* layer leads to the user assumption that the result of the interaction and the reaction of the system was the appliance of certain visual variables. Thus the visual variables just changes a small amount of the visual capacity compared to the entire adaptation possibilities of visualizations the changes during the work with the system as reaction of users' interaction are not disrupting the working process of the users.

A similar constellation is the adaptation process of the *Visual Interface* layer. Commonly the visual or user interface consists of dynamic and therewith adaptable components and static components that remain during the entire process of work or either during all sessions of users. The visual interface adaptation refers commonly to the adaptation of different visual layouts, their order, and their amount. Changes as system reaction of users' interaction by adding new visual layouts can be performed in a full automatic way. From the users' perspective it is more or less just another perspective on data. The user may interact with it or work with other visual layouts. The similarity of the constellation can be argued by the users' attention. If a new layout is added on the screen as a result of users' interaction, the attention of the user is directed to that layout. In some cases it might make sense due to the new or alternating perspective and in other cases the user may ignore the change or even put

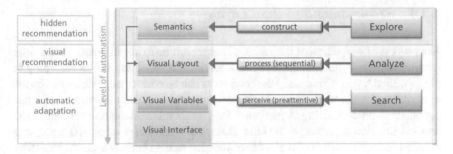

Fig. 6.54 Levels of automatic adaptation in the visual environment

the layout away from the screen. Here, a real disruption of the working process is not really given, thus the focused visual layout is not changing.

In contrast to the described adaptations, a change of the focused visual layout may lead to context lost, disruption of focused attention, or even overcharging effects. In particular, if the system changes the visual layout, the user is currently working with. To avoid the mentioned effects, we integrated three level of visual adaptation as illustrated in Fig. 6.54. The visual layout is commonly recommended to the user. The full adaptation is just performed if data are loaded that cannot be anymore visualized by the particular visual layout or the user is expecting a change in visual layout, e.g. for a detailed view. This adaptation effects are further constrained to "similar" visual layouts, e.g. replacing a node-link graph layout by another node-link graph layout. In this cases the user does not lose the context, thus the selected node still remains selected and the visual changes are performed slightly. Further each visual change is a system reaction of users' interaction.

The recommendation of visual layout is performed by arranging icons of alternating visual layouts on the screen. The slight movement of the visual layout recommendation guides the attention of the user to the alternating and better fitting visual layouts in a scant way of attention guidance. The visual recommendation through the representing icons makes use of two main visual variables. The order indicates based on the user model (canonical, group, or individual model) the best fitting visual layout, while the size of the icons indicates the relationship of visualizable content in contrast to the instantiated content. This is due that the preferred visual layouts may only have the capability to visualize just a sub-set of the instantiated data. The user should be aware that the chosen visual layout has limitation for the given subset of data. Further each icon illustrates the number of visual layouts that are placed on the screen. Figure 6.55 illustrates the described procedure as a paper-prototype of the recommendation functionality.

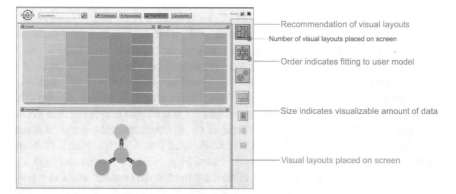

Fig. 6.55 Recommendation of visual layouts

6.3.5   Visual Variables Adaptation

The *Visual Variable* layer defines the visual appearance of visual layouts and the underlying data in terms of visual parameterization by setting e.g. the texture, color, or size according to Bertin's Implantation [48, 51]. The visual variables enrich the geometry of the visual layout layer to indicate information about relevance, quantity, and data value related order ability [20]. According to the various studies on visual perception as illustrated in Sect. 2.2 the visual variables guide the attention of users [12, 47, 51]. Certain visual variables throughout various distractors can be perceived preattentively and reduce therewith the visual search on low level task such as locating an item on the screen. The use of visual variables to enable a preattentive information processing requires a visual separation of the item or items that should be focused by the users and the distractors on screen. In this context the visual separation refers to the works of Treisman et al. [45, 46] and Wolfe et al. [47, 76], in which the visual difference of the items leads to the *pop-out* effect and guides the attention of users to certain visual representations. Beside the pop-out effects of visual variables, these are used in our model to indicate information on abstract and detailed levels. To avoid the bottom-up activation in a preattentive manner, this visual information use commonly similar variable, e.g. size or saturation, with differing parameterization and without a pop-out effect. This leads to indicate for example the amount of entities in a concept by using the size of the visual representation of the concept. In such cases a pop-out effect would occur automatically if significant differences between the numbers of entities in the concepts are given. The *Visual Variable* layer defines the visual appearance of visual layouts and the underlying data in terms of visual parameterization by setting e.g. the texture, color, or size according to Bertin's Implantation [48, 51]. The visual variables enrich the geometry of the visual layout layer to indicate information about relevance, quantity, and data value related order ability [20]. According to the various studies on visual perception as illustrated in Sect. 2.2 the visual variables guide the attention of users [12, 47, 51]. Certain visual variables throughout various distractors can be perceived preattentively and reduce therewith the visual search on low level task such as locating an item on the screen. The use of visual variables to enable a preattentive information processing requires a visual separation of the item or items that should be focused by the users and the distractors on screen. In this context the visual separation refers to the works of Treisman et al. [45, 46] and Wolfe et al. [47, 76], in which the visual difference of the items leads to the *pop-out* effect and guides the attention of users to certain visual representations. Beside the pop-out effects of visual variables, these are used in our model to indicate information on abstract and detailed levels. To avoid the bottom-up activation in a preattentive manner, these visual information use commonly similar variable, e.g. size or saturation, with differing parameterization and without a pop-out effect. This leads to indicate for example the amount of entities in a concept by using the size of the visual representation of the concept. In such cases a pop-out effect would occur automatically if significant differences between the number of entities in the concepts are given.

Our reference model investigates all volatile and persistent models to adapt the visual variables. Each model is handled in a different way and provides different information for the *Visual Variable* layer (see Fig. 6.56). The queried search term results as illustrated in a sub-set of instantiated data that are visualized on the screen. Here the recognition of the queried search term is in particular of interest for the user, thus this term is the starting point of the navigation and exploration through the visualization. The matching of search term and the resulted semantic entities are highlighted by using a visual variable to ensure that the attention of users is guided to the term first. In cases, where the search term does not provide a unique matching to the results, the resulted entities are all highlighted to differentiate them at least from their semantic neighbors and provide therewith a faster information processing. The weighting of semantic neighbors by our introduced *ilf* and *dRf-iRf* (see Sect. 6.1.2.2) is used adapt visual variables based on the calculated weights of focused objects. Therefore, visual variables are commonly used that indicates the quantitative calculations, e.g. saturation. In a similar way, the quantity of instantiated data is visualized. Therefore the number of instances of the concepts or number of relations of instances is visualized through a visual variable that is capable to visualize such quantitative measures, e.g. size (see Sect. 2.5.1: classification by data value). An important role for the *Visual Variable* layer plays the user model. The reference model refers to just one user model as an abstract model that stands for the canonical, group, or individual user model. As illustrated in Sect. 6.1.3 the user model contains information about users' interest and previous knowledge on data entity level. Visual variables are here used to guide users' attention to their domain of interest on an abstract concept level and to certain entities of interest. This adaptation makes use of distinguishable visual variable to ensure that the user is supported in the exploration process. Therefore the measured relevance values are transformed in visual variable values, e.g. the computed relevance value is mapped directly to the saturation of instances in a visual layout. Figure 6.56 illustrates the described adaptations.

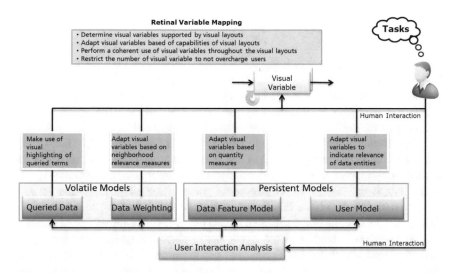

Fig. 6.56 Visual variable layer in the reference model

The visual variables are commonly dependent on the capabilities of the visual layout in the reference model. As already mentioned the visual layouts are annotated with their capabilities in our SVML. One aspect of the SVML is the information about supported visual variables for each visual layout. The transformation step of *retinal variable mapping* first determines the supported visual variables of the visual layouts. There are a number of visual variables that are supported by all visual layouts: saturation, hue, brightness, size, and transparency. Beside this set of visual variables, visual layouts may support further and enhanced visual variables, e.g. distortion, order, icon and glyphs, orientation, and background that are annotated in the SVML of each visual layout. Based on the supported visual variables the above described adaptation based on various factors is performed. The main aspect in the adaptation process is to provide a coherent picture of the visualized data. Therefore only those visual variables are used for adaptation that are supported by all visual layouts on the screen. Dismissing, selecting, or replacing single visual layouts has no affect to visual variables thus the least common number of visual variables are used for the adaptation. In case of creating a new visual interface with other visual layouts, the visual variables are again determining the least common numbers of the supported visual variables and use them for the adaptation purposes.

Another aspect is the number of used visual variables. The effect of visual variables is reduced with each new visual variable. If a visual layout adapts the entire range of visual variables to support the adaptation effect, the real effect of guidance and information carrier of visual variables may be lost or the user may be overcharged with the range of visual variables. To avoid this effect, the number of adaptive visual variables is commonly limited to three.

6.3.6   *Visual Interface Adaptation*

The layer of *Visual Interface* is the last transformation step in our reference model. It is responsible for creating a coherent visual interface by placing one or more visual layouts including the already defined visual variables on the screen. The visual interface or user interface is created by the dynamic visualization components of the reference model and some static components that may provide a dynamic behavior, e.g. the introduced recommendation model for visual layouts (see Sect. 6.3.4). The adaptation by orchestrating more than one visual layout leads to view various or the same data by various or the same visual layouts. Therefore first the overall diversity of the data is determined of the data model instantiated as query results. In case of more than one data-base, this step is performed for each data-base and resulted data-sets. The amount of the different data types and structures enable to measure how necessary or efficient a visual layout can be placed on screen investigating the already placed visual layouts. Further the investigation of the user model plays in important role for adapting the visual interface. Here not only the preferred visual layouts for the particular user model are investigated. Far more the placement of the visual layouts on the screen and the combination of those are determined from the user model. The user model indicates that in case of interacting with a certain visual layout another visual layout as extension may be useful, thus the user or user group combined this two visual layouts often with each other [20, 53].

The transformation step of visual layout orchestration is primarily responsible for the selection, placement and initialization of visual layouts. The changes in visual interface are commonly linked to user interactions. Therewith the occurred changes are perceived as reaction of the system to users' interactions. The transformation step is further responsible to provide a coherent use of visual layouts and visual variables. The coherency is achieved by providing a set of visual layouts that fits to the data structure and user model. The number of the same visual layouts on the screen is limited. Visual layouts that visualize the same data-base or sub-set of it are linked with each other. The linking is not only performed on the interactions that supports a brushing and linking (see Sect. 2.3) of the different visual layouts. The coherency affects further the visual variables of linked visual layouts that visualize the same data-set. The transformation step includes further a general restriction of the number of visual layouts that are place or can be placed on the screen. The number is determined by the user model as a result of previous interactions that led to performant task-solving. Here the canonical user model is the main baseline. Figure 6.57 illustrates the last step of our reference model, which ends with the iterative loop of users' interaction with the system.

The visual interface is the user interface of the entire model. The dynamic components of the visual interface are instantiated visual layouts with already adapted visual variables. The static components are any other components that enable to interact with the entire interface [20]. The configuration of the visual components is performed in our SVML. The SVML contains a set of all dynamic and static visual and interactive components. Further the placement of the static components is defined in

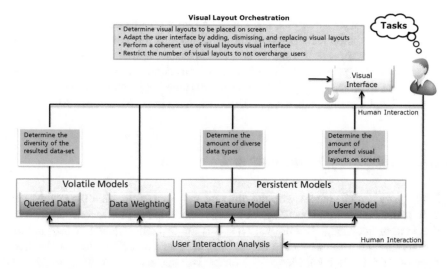

Fig. 6.57 Visual interface layer in the reference model

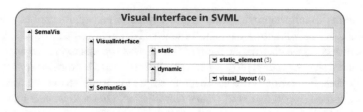

Fig. 6.58 SVML annotation of the visual interface layer

the SVML. As the visual interface may handle more than one data-base, the sources of the data bases can be configured in the SVML too. Figure 6.58 illustrates a sample SVML with the described annotation of the visual interface.

6.4 Support of Exploratory Search

One main purpose of our approach is to support the process of exploratory search as illustrated in Sect. 3.4.2 through the adaptive behavior of semantics visualization. As already described the process of exploratory search consists of different stages [6, 8, 77, 78] and requires different views and level of details on the data. This part of the thesis introduces our *visualization cockpit* model that supports different views on the same data or on different data with the visual interface adaptation. Therefore the placement of the visual layouts on the screen and their linking to other visual layouts and data-sets plays an important role. The section starts with an abstract view on the search process in a general manner. In this context, the search process is investigated from two different viewpoints: bottom-up search and top-down search. The two views on the search process allow enlightening the search process from the traditional search perspective and the adoption in visual environments. Thereafter, the *visualization cockpit model* will be introduced that supports both search processes with visual interfaces and goes beyond the existing approaches on visual search models. The following parts were partially published in [5, 53, 54, 62, 79].

6.4.1   Top-Down Versus Bottom-Up Search

Exploratory search enables with the different stages of exploration the acquisition of in particular implicit knowledge or information. Implicit in this context refers to that kind of knowledge or information that is not explicitly known or may not be formulated by the user explicitly, due the lack of knowledge. From the visualization point of view, implicit knowledge or information refer to that knowledge that is not explicitly modeled in the data but can be enlighten through the visualization of the modeled data [79].

Different disciplines provide technologies, systems, and approaches to enable the acquisition of implicit knowledge or information. For simplifying the investigation of these approaches, we classify the methods into *bottom-up* and *top-down* approaches. The standard search process [80], e.g. is a simplification of a bottom-up approach (see Fig. 6.59: left illustration). The approach attempts to formalize the iterative search process a three-stepped model of *Query Formulation*, *Query Refinement* and *Result Processing*. This model assumes that the search begins with the formulation of query of known knowledge. During the search process the subject gets more knowledge about a certain topic to refine his query and gather more knowledge about the certain topic. The main aspect of this model is that the search process starts with the ability to formulate a query and to reformulate the query during the search. During the search process new knowledge is adopted, which leads to a reformulation of the query. A more complex example for a bottom-up information gathering and search process is the *information-seeking process Marchionini* [81]. This process includes eight phases and encloses the internalized problem solving of subjects too. *Marchionini*'s model consists of eight phases in information seeking: *Recognize and accept an information problem*, *Define and understand the problem*, *Choose a search system*, *Formulate a query*, *Execute search*, *Examine results*, *Extract information and Reflect/iterate/stop* [79], [81, pp. 49–58].

The exploratory search approaches introduced in Sect. 3.4.2 are per se bottom-up approaches that start commonly with a search term and enable in different stages the investigation, reformulation, learning, and refining. The process of information exploration in information visualization is contrary to the bottom-up approaches of search interface. Commonly in this context a top-down approach is proposed [13]. The most famous example for a top-down information exploration or gathering model is the already introduced *Visual Information Seeking Mantra* (see Sect. 2.4.2) [13].

Fig. 6.59 Top-down versus bottom-up search

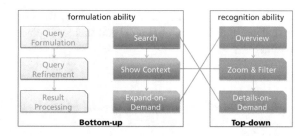

This model proposes the opposite of the bottom-up approach and is designed for the visual information seeking. The three-stepped model propagates to Overview the data first, then Zoom and Filter the relevant parts and finally gather Details on Demand. Beginning with the overview of data, this model premises not the verbalization ability, here the focus is on the recognition ability. If a subject detects in the overview step an area-of-interest, he can zoom into the area or filter this information out. After he gets enough information to recognize a seeking problem, details about the information can be fetched. The top-down model of search and information acquisition based on Shneiderman's work is applied to many visualization environments and is the main approach for gathering information in visual environments [79]. The investigation of the search process in a bottom-up manner plays an increasing role in visualizations. van Ham and Perer for instance proposed a bottom-up search approach in visual environments [82] that starts with *search*, by means of querying the data-set followed by *show context* that enables the contextual view on data and *expand on demand* that provides a detailed view on demand (see Fig. 6.59).

The described seeking approaches require different human abilities required for solving a seeking problem. In a bottom-up approach the formulation of the searched topic is important, whereas the recognition ability plays an important role in the top-down approaches. The mentioned top-down approaches are primary information visualization approaches, thus the overview of information and recognition of area-of-interest can be more supported with visualization systems. Figure 6.59 illustrates the two approaches. Thereby the left schema refers to the standard search process, the mid one illustrates the process as proposed by van Ham and Perer [82], and the right one is a simplified illustration of Shneiderman's model [13].

Semantically annotated data provides complex structures for seeking information in different ways. Both methods (top-down and bottom-up) of information seeking are supported by the formal structure of semantic data. A specific query on semantic data would provide results from a domain of interest, the way how the results are visualized appropriately depends on various factors, but in particular on the amount of entities and the previous knowledge of user'. The visual interface adaptation supports the exploratory search. But this process should investigate the amount of data-bases that are requested for results and the task of users. To support these aspects, we enhanced our visualization model with the *Visualization Cockpit* model [8, 12, 53, 62, 83] that makes use of adaptable and adaptive visual layouts to generate and enable the generation of visual user interface. The visual user interfaces relies on the visual interface layer of our model and enables to place visual layouts in a juxtaposed manner.

6.4.2   The Visualization Cockpit Model

Most of our visual layouts specialize upon one feature of semantics. This is because the visual layouts have advantages for a special feature, but disadvantages for others. We can easily show the relations between instances in an arbitrary graph-layout, which provides interaction methods for expanding or collapsing a node to gain a better overview, but we can hardly display a textual article, a picture or properties like geographical or temporal data in arbitrary graphs. On the other hand geographical visual layouts support the view and search for geo-related semantic properties, but their enhancement with relational or hierarchical layouts may lead to overcharging users and non-comprehensible visualizations. To face on the one hand the visual overflow and on the hand the reduction of visualizations, we introduce the *Visualization Cockpit* model [53, 62] that reduces the information overload by separating the visualized information in a visual interface of juxtaposed visual layouts.

Our visualization cockpit separates semantic information attributes and visualizes this information in separate visual layout without losing any information and without overcharging the user by complex visualizations. The advantage of the separation of complex information units is obvious; the user of is able to perceive the same information from several perspectives by the placed juxtaposed visual layouts. With this approach both, bottom-up and top-down approaches are supported. A bottom approach starts with the query formulation. If the formulated query is precise enough and the user model indicates high-ratings for previous knowledge and interested items, a semantic instance and the semantic neighborhood is presented. Otherwise, if the query is not specific or the user wants to have an overview, the abstracted schema of the semantics is presented. The different perspectives on data or semantics enable more comprehensible view. Thereby the visual layout are linked with each other and make use of a brushing and linking metaphor to support the comprehensible view and changes on users' interactions. Figure 6.60 illustrates a screenshot of the same data with different perspectives, where the visual layouts are linked with each other.

The introduced visual layouts (Sect. 6.3.3.2) can be integrated in the visual interface to provide different perspectives on the same information in abstracted and different ways. Users are able to rearrange or add visual layouts on the screen or dismiss the placed visual layouts. The introduced visual recommendation (see Sect. 6.3.4) supports users in this process based on user and data models. The view on different perspective or aspects on the same data and data-set with different visual layouts is one of the cockpit generation styles. The main purpose remains the support of exploratory search. In order to support this search we identify following *styles* for our Visualization Cockpit:

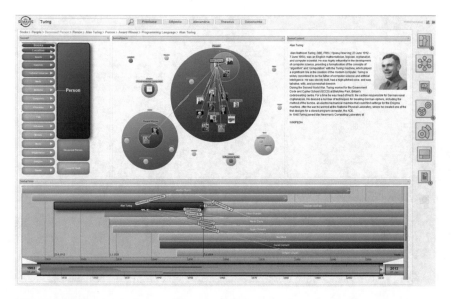

Fig. 6.60   Visualization Cockpit: different perspectives on the same data (perspective view)

- **Perspective view**: Visualization of the same data with different visual layouts.
- **Perspective-comparative view**: Visualization of different sub-set of data from the same data-base with different visual layouts.
- **Comparative view on level-of-details**: Visualization of the same data using the same visual layouts with different parameters.
- **Comparative view on data sub-sets**: Visualization of different data sub-sets from the same data-base with the same visual layouts.
- **Comparative view on data**: Visualization of different data-bases with the same visual layouts.
- **Non-linked view**: Visualization of different data-bases with different visual layouts.

With the different adjustments of the visualization cockpits, different goals can be achieved and different requirements fulfilled. As introduced the perspective view (Fig. 6.60) enables the exploration of a queried sub-set of data from different perspectives with different visual layouts. The layouts are linked with each other and the user is able to navigate through the different visual layouts and gather required information from other visual layouts. The perspective-comparative view allows solving comparative tasks by providing the free choice of visual layouts for different data-subset form the same data base. Here only one data-base is queried, e.g. by different search terms. The results for each sub-set of data are linked with each other, whereas the visual layouts are just linked through the data. If a user interacts within a visual layout, only those visual layouts react to the interaction that visualize the same data-subsets. Adding a visual layout leads to a coupling of this with the data sub-set of

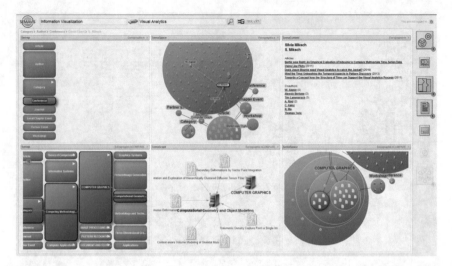

Fig. 6.61 Visualization Cockpit: perspective-comparative view on different data sub-sets

the last user interaction. The user is able to change the linking each visual layout. Figure 6.61 illustrates a screenshot of this cockpit style.

The perspective-comparative view enables to compare tasks with the freedom to choose the visual layout for each data-subset. This view is in particular efficient if the data sub-set has different characteristics. Thus this view is not providing at each level the same visual layout, it goes beyond comparison tasks and enables a more investigative view on various topics of the same data-set. A comparative view on a low-level is provided by the comparative view on level-of-detail. This view enables the visualization of the same data with the same visual layouts, but different parameterization for gathering on the one hand an overview and on the other hand a detailed view on the data. The parameterization of certain visual layouts allows controlling the level of detail as part of the zooming.

The zoom levels may vary from visual zoom, to semantic zoom with semantics based filtering. For example the level of detail can on the one hand be used to show a greater part of the semantics or information space for showing the structure of the information and on the other hand with small numbers of elements of interest to show detailed information. There are two main ways to combine the same visualization technique duplicated in a cockpit for providing more information. First the level of details can be provided as a zoom on a specific area of the semantics while the entire search results is displayed too (Fig. 6.62b) and second the semantic neighbors of a particular focused elements can be enhanced and reduced due to enabling an overview and detailed view (Fig. 6.62a). A reduction of the numbers of entities can be achieved by semantically filtering the information, e.g. based on the introduced relevance metrics.

With this kind of information visualization a similar effect can be achieved. Many information elements gives an overview about the whole structure of the semantics

(a) **(b)**

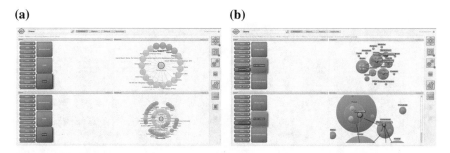

Fig. 6.62 Visualization Cockpit: comparative view on level-of-details. **a** Different levels-of-detail. **b** Different levels-of-zoom

and the information about the focused element can be revealed with a visual layout that visualizes a small number of elements. A similar approach with a more focus on comparative tasks is provided by the comparative view on data sub-sets. This view enables the visualization of different search or interaction results with the same visual layouts that are commonly placed upon each other. The usage of same visual layout supports the comparison and analysis process, thus a direct visual correlation is built. Visual layouts visualizing the same content or query result are linked with each other, while visual layouts that visualize other subset are not affected. The interaction coupling of visual layouts is depending on the data that are visualized. If a user interacts with the visual layout that visualizes a certain data-set, only those are changed by users' interactions that are visualizing the same content. With this procedure and the visualization through the same layouts, the users are able to navigate independently through the different sets of data and get insights, compare results, and investigate deeper search tasks on each data base. Figure 6.63 illustrates a screenshot of the visualization cockpit with the comparative view on data sub-sets.

The comparative view on data sub-sets enables solving comparative and analysis tasks in one domain of data. With the growing semantic data sources on Web, in particular as part of the Linked-Data bases, the combined search on different data sources gets more and more relevant. We mean with the combined search, a simultaneous search in different data bases on Web with the same search term. This enables a deeper search and investigation of certain entities or information of interest by considering not only one data base. One main side effect of this search is that the visualization of the results enables to validate and proof the quality and information value of a data-base. Our main goal remains the support of exploratory search by providing appropriate visualizations that enables an adequate and comprehensible result retrieval. Our comparative views on data enable the simultaneous search and visualization of search results from different data sources. Thereby the search results from each data base are visualized with the same visual layouts to enable a more comprehensible view on data. The visual layouts that are visualizing data from the same data base are linked with each other and enable the independent navigation in various data sources. Users are able to add, rearrange or dismiss certain visual

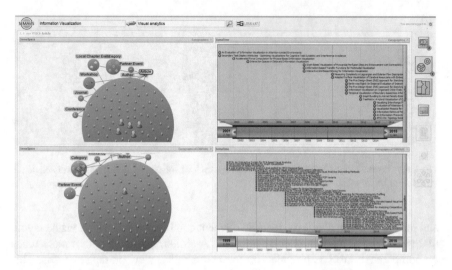

Fig. 6.63 Visualization Cockpit: comparative view on data sub-sets visualizing different data sets from the same data base with the same visual layouts

layouts. The affects the entire visual interface, e.g. if a user adds a new visual layout on the screen, the same visual layout appears twice for two data bases.

The visualization cockpit is not limited to certain number of data bases. Therewith the user is able to view retrieved results from various data bases simultaneously. Although the number of the data bases is not limited, the system limits the number of visual layouts based on the user model to not overcharge the user with visual information. The comparative view on data enables analysis tasks without querying different data-bases and changes the view. The results are presented in the same way, so that the process of investigation in exploratory search can be supported in one visual interface. Figure 6.64 illustrates a screenshot of the visualization cockpit with the comparative view on data. Thereby a searched term was found in three different data-bases. The same visualization enlightens different information on the same search term and enables a clear comparison of the search result.

The comparative view on data has the advantage that all results from all data-bases are visualized in the same way and enable therewith an easy comparison. The view is limited to the fact that only the same visualization can be used in this context for the various resulted data. These resulted data may have different attributes that cannot be visualized with the certain chosen layouts. In these cases information about the results are lost. To face this aspect, we introduce the non-linked view that has no limitations at all. It enables the visualization of data from different data-bases with various visual layouts. The main idea is to provide a non-limited view for the deeper exploratory search steps as described in Sect. 3.4.2 and proposed by Marchionini [6]. Thereby we use the visual layout linking for the data-bases too, as in other views, but the user is able to disable this linking even for the same data-base. This procedure enables the freedom of retrieving the search results from different perspectives and

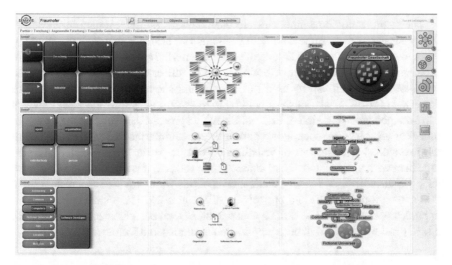

Fig. 6.64 Visualization Cockpit: comparative view on data visualizing different data-bases with the same visual layouts (comparative view on data)

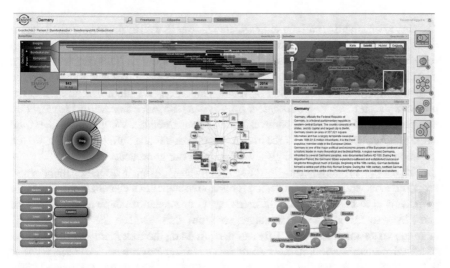

Fig. 6.65 Non-linked view: visualization of different data-bases with different visual layouts

different data-bases according to the assumption and theories of constructivism [58]. The user gets guidance and recommendations for the visual layout when he selects a data-base. The entire interaction behavior with all the visualization cockpit styles lead to train the user model. Figure 6.65 illustrates a screenshot of a non-linked visualization cockpit. Thereby different visualizations are selected for the different data-bases.

The visualization cockpit enables to view data from different data-bases or different sub-sets of the same data with the same or various visual layouts. With the juxtaposed arrangement and linking of visual layouts the approach supports the entire process of exploratory search. We introduced six different styles or views how the visualization cockpit can be used for the different stages of exploratory search or the given tasks. This six views should be seen as examples how the visual layouts can arranged and what kind of tasks and in which process they support the user. The model of visualization cockpit is a part of the visual interface adaptation. Although the juxtaposed arrangement of visualization can be performed manually and provide therewith a more 'adaptable' character, we focus in this thesis on the values for adaptive visualization. Fixed visualization cockpits were provided as a result of this thesis in various contexts and projects for different tasks. The visualization cockpit approach was published in various papers with several enhancements [5, 8, 12, 20, 53, 62] and was applied in other application scenarios and for other data [84].

6.5   Chapter Summary

The survey on adaptive and semantics visualizations as well as the review on literature on information visualization revealed both: potentials for improved adaptive and semantics visualizations and shortcoming in existing systems. This chapter faced the shortcomings of existing approaches by providing new approaches, models, and in particular a new reference model of visual adaptation. We identified already in the previous chapter *The Methodological Approach of Adaptive Semantics Visualization* the main aspects that should be investigated in our research. We therefore identified some requirements that are scarce in current approaches and solutions and illustrated based on the requirements a high-level design of our approach.

This chapter investigated the introduced high-level design in a detailed manner and proposed a holistic conceptual design for adaptive semantics visualization. The main purpose was to investigate the different layers or stages of adaptive visualizations and illustrate for each of these stages solutions that advance the general idea of adapting visual environments. This was performed on the four main identified layers of adaptive visualization, namely *Knowledge Model*, *Process of Adaptation*, *Visual Adaptation*, and *Support of Exploratory Search*, according to our referenced high-level design in the previous chapter. In each of the named layers the conceptual design was described in a detailed and replicable way. The structure of the chapter was according to the high-level design to enable a comprehensible view on the main contributions and determined requirements.

We described the entire conceptual approach in a bottom-up manner by starting with the knowledge model. The knowledge model consists of all the needed information for the adaptation process. We focused on the combined modeling of data and user, thus these influencing factors are the most influential ones for adapting visualization. In this context, we described how data is gathered from various Web-sources and introduced an iterative querying approach for generating semantics from source with unknown semantic schema. The iterative querying approach was further exemplary enhanced for metadata that does not provide any semantic formalism. The main part illustrated how the gathered data are stored in a data model that provides beside the data, structural information about the data. This structural information is one main adaptation factor of our approach. We identified that the quantitative measurement of each data-type that occurs commonly as types of semantic relationships plays an important role for characterizing the structure of data. We therefore applied the iterative querying to measure the quantity of each semantic type, e.g. number of concepts, number of incoming and outgoing relations and further structural information that leads to a more appropriate characterization of the queried semantic structure. The quantitative measurements were not only used to determine the structure of queried data. We introduced in this context two algorithms that measure the relevance of semantic neighbors and consequently the contextual information of a selected semantic instance. The relevance measurement of the introduced algorithms (iIf and dRf-iRf) provided a volatile model that enables the visual adaptation of the contextual information in relation to a selected semantic instances. The core element of our knowledge model is the user model. The user model combines the information about data and users' behavioral information. Based on the concepts of Relational Markov Models, we formalized the interaction of users and constrained the domain nodes to the three most influential: *type of interaction, visualization type and area of interaction*, and *data element*, whereas a hierarchical notation enables a unification of the interaction and further measurements on different levels. Based on the relational interaction formalization, the measurements of users' behavior were illustrated. Thereby two different measurements were described: a data-source independent and a data-source dependent fine granular measurement of users' behavior with the adaptive visualization. Further the interaction analysis and formalization were used to predict users' actions with an enhanced unsupervised predictive statistical algorithm. The enhanced algorithm was compared with commonly used prediction algorithms to illustrate the added value. In the following parts the user modeling was described in a formal and comprehensible way. In this context, we introduced the approach of the canonical user model that formalizes the average usage behavior of all users and enhances the entire adaptive behavior without the need of an expert to train the adaptive system. The canonical user model builds one of the main components of our approach. It is not only for modeling an average user and enabling the adaptation in a general manner, it is further the baseline for measuring similarities and deviations to create similar user-groups and determine anomalies in usage behavior.

In the following section, we described the process of adaptation and started with the similarity and deviation algorithms to bridge the gap between user model and the adaptation process. We introduced a similarity measurement of users' behavior that is permanently observing users to determine similar behavior among other users and assign the user to a more adequate group or enhances his model with information from another user model. The idea of the canonical user model that is applied for new users, in new situations, and for new data-sources, is enhanced with the similarity measurement. In case of finding a similar user or user group, the usage behavior of this user or user group is applied for a certain new situation or new data-source, instead of the canonical user. The deviation analysis that was introduced, attempts a contradictory approach. It measures the deviation between a certain user and the canonical user model. This approach leads to determine behavioral anomalies to the average usage behavior and consequently reduces the adaptation effect of the system, if such anomalies are detected. After the introduced algorithms for measuring similarities and deviations, we introduced the general adaptation process. The main goal was to provide a comprehensible view on the interplay of the different measured values and created models that affect the visual adaptation. We introduced in this context our *Semantic Visualization Markup Language* (SVML) that annotates each visual layout with its different capabilities and is thereby the model of the integrated visual layouts. Thereafter the general adaptation process was described step by step to provide a replicable way of the interplay of the different models and their effect to the visual adaptation. We introduced in the section of visual adaptation our layer-based reference model of visual adaptation. The reference model is one of the core elements of our entire conceptual design and enhances an established reference model for adaptation purposes. We used outcomes of studies in the field of visual perception and elementary works for defining the layers of adaptation. The layers separate the graphical characteristics of visualizations to enable a finer adaptation at each layer and support thereby the human visual perception. The model was created based on established models of visual perception and provides a twofold transformation process: the horizontal transformation starts with the data, which is in case of this thesis commonly semantic data. The semantic data are transformed to a data model that includes the structural information of the data. Based on this information a visual mapping is performed that determines one or more appropriate visual layouts for the data. The visual layouts in this stage can be seen as object placements without any visual representation (skeleton). With the retinal mapping, visual variables are assigned to the elements that are placed on the screen. The last transformation step of visual layout transformation creates a visual interface by placing one or more visual layouts on the screen. The main aspect of the reference model is that each of the layers can be manipulated by the underlying knowledge models that represent the usage behavior in relation to data. To illustrate the adaptive character of our model, we subdivided the measures models into volatile and persistent models, whereas volatile models are just valid for a momentary point of time, e.g. after a term was searched or an entity was focused. The adaptive characteristics of our model enable the adaptation of content on the data model layer and manipulate the data transformation. The layout adaptation is adapting the visual layouts based on the underlying models and manipulates the visual mapping transformation. The visual variable adaptation adapts visual variables, e.g. saturation, hue, transparency, shape etc. and manipulates the retinal mapping transformation, and visual interface adaptation creates the visual interface by choosing the visual layouts based on user preferences and manipulates the visual layout orchestration transformation.

The last section of this chapter introduced our visualization cockpit model for supporting the exploratory search. We first introduced two different views on the search process, the bottom-up search that starts with the formulation of a query and provide the result processing in an iterative manner of query refinement. In contrast to that the top-down search process starts with an overview on a knowledge domain and provides various interaction abilities to process the required detailed information. It is important in context of visual search to differentiate between these two search processes, thus the bottom-up search requires formulation ability and the top-down search relies more on the human recognition ability. Based on these assumptions, we introduced our visualization cockpit model that makes use of the visual layout orchestration to provide various views on the same or different data for exploratory and more complex tasks. Overall, we identified six different views that interconnect visual layouts and data with each other or disconnect them. The approach enables different perspectives or the same view on different data or the same data. We thereby differentiated in our model 'data' as a data-set of the same data-source and from different data-sources. The identified six views on data as visualization cockpits were described and illustrated exemplary.

We illustrated in this part of the thesis the entire conceptual model, introduced algorithms and models, and described in some cases based on example how they are used. Beside the different models, approaches, and algorithms the interplay of these were illustrated to provide a holistic picture of the conceptual model.

References

1. S.K. Card, J.D. Mackinlay, B. Shneiderman, Readings in Information Visualization: Using Vision to Think, 1st edn. (Morgan Kaufmann, 1999). http://www.amazon.com/exec/obidos/redirect?tag=citeulike07-20&path=ASIN/1558605339
2. C. Bizer, T. Heath, K. Idehen, T. Berners-Lee, in *Proceedings of the 17th International Conference on World Wide Web, WWW 2008*, Beijing, China, 21–25 Apr 2008 (2008), pp. 1265–1266
3. C. Bizer, T. Heath, T. Berners-Lee, Int. J. Semantic Web Inf. Syst. **5**(3), 1 (2009)
4. T. Heath, C. Bizer, Linked data—evolving the web into a global data space, in *Synthesis Lectures on the Semantic Web: Theory and Technology* (Morgan & Claypool Publishers, 2011)
5. K. Nazemi, M. Breyer, J. Forster, D. Burkhardt, A. Kuijper, *Human Interface and the Management of Information. Interacting with Information* (2011), pp. 239–248
6. G. Marchionini, Commun. ACM **49**(4), 41 (2006). doi:10.1145/1121949.1121979. http://doi.acm.org/10.1145/1121949.1121979
7. K. Nazemi, M. Breyer, C. Hornung, in *Universal Access in Human-Computer Interaction. Applications and Services*. Lecture Notes in Computer Science, vol. 5616, ed. by C. Stephanidis (Springer, Berlin, Heidelberg, 2009), pp. 83–91. doi:10.1007/978-3-642-02713-0_9. http://dx.doi.org/10.1007/978-3-642-02713-0_9
8. K. Nazemi, M. Breyer, C. Stab, D. Burkhardt, D.W. Fellner, in *International Conference on Education and New Learning Technologies. Proceedings [CD-ROM]* (International Association of Technology, Education and Development (IATED), IATED, Valencia, 2010), pp. 006,476–006,484. http://bibcd.igd.fraunhofer.de/bibcd/INI_Science/papers/2010/10p054.pdf
9. Freebase Consortium, Freebase—a community-curated database of well-known people, places, and things (2013). http://www.freebase.com/. Accessed Aug 2013
10. Freebase Consortium. Freebase api. build intelligent apps with freebase data (2013). https://developers.google.com/freebase/. Accessed Aug 2013
11. J. Forster, K. Nazemi (Supervisor), D.W. Fellner (Supervisor), Semantische visualisierung von suchergebnissen auf der basis von linked open data (2011). http://bibcd.igd.fraunhofer.de/bibcd/INI_Science/theses/2011/Forster_Bachelor.pdf. 77 S

12. K. Nazemi, R. Retz, J. Bernard, J. Kohlhammer, D. Fellner, in *Advances in Visual Computing*. Lecture Notes in Computer Science, vol. 8034, ed. by G. Bebis, R. Boyle, B. Parvin, D. Koracin, B. Li, F. Porikli, V. Zordan, J. Klosowski, S. Coquillart, X. Luo, M. Chen, D. Gotz (Springer Berlin Heidelberg, 2013), pp. 13–24. doi:10.1007/978-3-642-41939-3_2. http://dx.doi.org/10.1007/978-3-642-41939-3_2

13. B. Shneiderman, in *VL* (1996), pp. 336–343

14. J. Mackinlay, ACM Trans. Graph. **5**, 110 (1986). http://doi.acm.org/10.1145/22949.2295

15. K. Nazemi, A. Kuijper, M. Hutter, J. Kohlhammer, D.W. Fellner, in *I-KNOW 2014: Proceedings of the 14th International Conference on Knowledge Management and Knowledge Technologies* (ACM DL, 2014) (to appear)

16. D.M. Hilbert, D.F. Redmiles, A.C.M. Comput. Surv. **32**(4), 384 (2000). doi:10.1145/371578.371593. http://doi.acm.org/10.1145/371578.371593

17. K. Nazemi, C. Stab, D.W. Fellner, in *Advanced Intelligent Computing Theories and Applications*. Lecture Notes in Computer Science, vol. 6215, ed. by Huang, D. et al. (Springer, Berlin, Heidelberg, 2010), pp. 362–371. doi:10.1007/978-3-642-14922-1_45. http://dx.doi.org/10.1007/978-3-642-14922-1_45

18. C. Stab, K. Nazemi (Supervisor), D.W. Fellner (Supervisor), Interaktionsanalyse für adaptive benutzerschnittstellen. Diploma thesis at the Technische Universität Darmstat. Supervised by K. Nazemi (2009). http://bibcd.igd.fraunhofer.de/bibcd/INI_Science/theses/2009/Stab.pdf. 161 S

19. K. Nazemi, C. Stab, D.W. Fellner, in *IEEE International Conference on Intelligent Computing and Intelligent Systems. Proceedings* (IEEE Press, New York, 2010), pp. 607–612. doi:10.1109/ICICISYS.2010.5658514. http://bibcd.igd.fraunhofer.de/bibcd/INI_Science/papers/2010/10p127.pdf

20. K. Nazemi, C. Stab, A. Kuijper, in *Human-Computer Interaction. Design and Development Approaches*. Lecture Notes in Computer Science, vol. 6761, ed. by J. Jacko (Springer, Berlin, Heidelberg, 2011), pp. 480–489. doi:10.1007/978-3-642-21602-2_52. http://dx.doi.org/10.1007/978-3-642-21602-2_52

21. K. Nazemi, W. Retz, J. Kohlhammer, A. Kuijper, in *Human Interface and the Management of Information*. LNCS, vol. 8521, ed. by S. Yamamoto (Springer International Publishing, 2014), pp. 64–75

22. W. Retz, K. Nazemi (Supervisor), D.W. Fellner (Supervisor), Integration heterogener methoden der empfehlungssysteme zur visualisierungsadaption (2013). 115 S

23. C.R. Anderson, P. Domingos, D.S. Weld, in *KDD'02: Proceedings of the Eighth ACM SIGKDD International Conference on Knowledge Discovery and Data Mining* (ACM, New York, NY, USA, 2002), pp. 143–152. http://doi.acm.org/10.1145/775047.775068

24. M. Guzdial, Deriving software usage patterns from log files. Technical report (Georgia Institute of Technology, 1993)

25. D. Gotz, M. Zhou, in *IEEE Symposium on Visual Analytics Science and Technology, 2008. VAST'08* (2008), pp. 123–130. doi:10.1109/VAST.2008.4677365

26. D. Gotz, Z. Wen, in *Proceedings of the 14th International Conference on Intelligent User Interfaces, IUI'09* (ACM, New York, NY, USA, 2009), pp. 315–324. doi:10.1145/1502650.1502695. http://doi.acm.org/10.1145/1502650.1502695

27. D. Gotz, Z. When, J. Lu, P. Kissa, N. Cao, W.H. Qian, S.X. Liu, M.X. Zhou, in *Proceedings of the First International Workshop on Intelligent Visual Interfaces for Text Analysis, IVITA'10* (ACM, New York, NY, USA, 2010), pp. 1–4. doi:10.1145/2002353.2002355. http://doi.acm.org/10.1145/2002353.2002355

28. A. Künzer, F. Ohmann, L. Schmidt, MMI-Interaktiv **7**, 61 (2004)

29. W. Hacker, Allgemeine Arbeits- und Ingenieurpsychologie: psychische Struktur und Regulation von Arbeitstätigkeiten. Schriften zur Arbeitspsychologie (H. Huber, 1978). http://books.google.de/books?id=HnBZAAAAYAAJ

30. R. Sun, in *Sequence Learning*. Lecture Notes in Computer Science, vol. 1828, ed. by R. Sun, C. Giles (Springer, Berlin, Heidelberg, 2001), pp. 1–10. doi:10.1007/3-540-44565-X_1. http://dx.doi.org/10.1007/3-540-44565-X_1

31. G.I. Webb, M.J. Pazzani, D. Billsus, User Model. User-Adap. Inter. **11**(1–2), 19 (2001). doi:10. 1023/A:1011117102175. http://dx.doi.org/10.1023/A:1011117102175
32. D. Gotz, M.X. Zhou, An empirical study of user interaction behavior during visual analysis. Technical report (IBM Research Division, NY, 2008)
33. S. Greenberg, Using unix: collected traces of 168 users. Research report 88/333/45. Technical report (Department of Computer Science, University of Calgary, Calgary, Canada, 1988)
34. D. Sleeman, Int. J. Man-Mach. Stud. **23**(1), 71 (1985). doi:10.1016/S0020-7373(85)80025-0. http://dx.doi.org/10.1016/S0020-7373(85)80025-0
35. P. Brusilovsky, E. Millán, User models for adaptive hypermedia and adaptive educational systems, in *The Adaptive Web*, ed. by P. Brusilovsky, A. Kobsa, W. Nejdl (Springer, Berlin, Heidelberg, 2007), pp. 3–53. http://dl.acm.org/citation.cfm?id=1768197.1768199
36. K. Nazemi, D. Burkhardt, M. Breyer, A. Kuijper, in *Proceedings of the 6th International Conference on Universal Access in Human-Computer Interaction: Users Diversity, UAHCI'11*, Part II (Springer, Berlin, Heidelberg, 2011), pp. 88–97. http://dl.acm.org/citation.cfm?id=2027376. 2027387
37. A. Kobsa, Benutzermodellierung in Dialogsystemen. Informatik-Fachberichte (Springer, 1985). http://books.google.de/books?id=YzC7AAAAIAAJ
38. A. Kobsa, in Workshop im Rahmen der 17. Fachtagung für Künstliche Intelligenz (KI 93), Humboldt-Universität zu Berlin, ed. by 17. Fachtagung KI (Springer, Berlin, 1993)
39. H. Luo, C. Niu, R. Shen, C. Ullrich, Mach. Learn. **72**(3), 231 (2008). http://dx.doi.org/10. 1007/s10994-008-5068-4
40. S. Gong, J. Softw. **5**(7), 745 (2010). http://ojs.academypublisher.com/index.php/jsw/article/ view/0507745752
41. L. Guo, Q. Peng, in *Proceedings of the 2nd International Conference on Computer Science and Electronics Engineering (ICCSEE 2013)*. Advances in Intelligent Systems Research (Atlantis Press, 2013), pp. 1921–1924
42. P. Brusilovsky, J. wook Ahn, T. Dumitriu, M. Yudelson, in *Tenth International Conference on Information Visualization, IV 2006* (2006), pp. 142–150. doi:10.1109/IV.2006.16
43. P. Symeonidis, A. Nanopoulos, Y. Manolopoulos, in *Proceedings of the 11th International Conference on User Modeling, UM'07* (Springer, Berlin, Heidelberg, 2007), pp. 97–106. doi:10. 1007/978-3-540-73078-1_13. http://dx.doi.org/10.1007/978-3-540-73078-1_13
44. D. Oelke, H. Janetzko, S. Simon, K. Neuhaus, D.A. Keim, in *Proceedings of the 13th Eurographics/IEEE—VGTC Conference on Visualization, EuroVis'11* (Eurographics Association, Aire-la-Ville, Switzerland, Switzerland, 2011), pp. 871–880. doi:10.1111/j.1467-8659. 2011.01936.x. http://dx.doi.org/10.1111/j.1467-8659.2011.01936.x
45. A.M. Treisman, J. Souther, J. Exp. Psychol.: Human Percept. Perform. **12**, 107 (1986)
46. A.M. Treisman, S. Gormican, Psychol. Rev. **95**(1), 15 (1988)
47. J.M. Wolfe, W. Gray (eds.), *Integrated Models of Cognitive Systems* (2007), pp. 99–119
48. J. Bertin, *Semiology of Graphics* (University of Wisconsin Press, 1983)
49. K. Nazemi, D. Burkhardt, A. Praetorius, M. Breyer, A. Kuijper, in *Human Centered Design*. Lecture Notes in Computer Science, vol. 6776, ed. by M. Kurosu (Springer, Berlin, Heidelberg, 2011), pp. 566–575. doi:10.1007/978-3-642-21753-1_63. http://dx.doi.org/10.1007/978-3-642-21753-1_63
50. K. Nazemi, Zusammenfassender Sachbericht: CTC-WP5-Innovative Benutzerschnittstellen und Visualisierung. Final Report on the THESEUS Innovative User Interfaces and Visualization. Deliverable, Technical Report to BMWi (2012)
51. K. Nazemi, J. Kohlhammer, in *1st International Workshop on User-Adaptive Visualization (WUAV 2013). Extended Proceedings of UMAP 2013, CEUR Workshop Proceedings*, vol. 997 (2013). ISSN: 1613–0073
52. K. Nazemi, M. Breyer, D. Burkhardt, C. Stab, J. Kohlhammer, in *Towards the Internet of Services: The Theseus Program*, ed. by W.e.a. Wahlster (Springer, 2014), pp. 191–202
53. K. Nazemi, M. Breyer, D. Burkhardt, D.W. Fellner, Int. J. Adv. Corp. Learn. **3**(4), 26 (2010). doi:10.3991/ijac.v3i4.1473

54. K. Nazemi, C. Stab, M. Breyer, D. Burkhardt, T. May, T. von Landesberger, THESEUS CTC-WP5-Innovative Benutzerschnittstellen und Visualisierungen: Alleinstellungsmerkmale der CTC-WP5 Technologien (THESEUS Core Technology Cluster for Innovative User Interfaces and Visualizations: Unique Features of the CTC-WP5 Technologies). THESEUS Programme Deliverable of the Fraunhofer Institute for Computer Graphics Research (IGD), ed. by Nazemi K. (2010)
55. D.A. Keim, F. Mansmann, J. Schneidewind, H. Ziegler, J. Thomas, in *Visual Data Mining: Theory. Techniques and Tools for Visual Analytics*. Lecture Notes in Computer Science (LNCS) (Springer, 2008)
56. G. Ryle, *The Concept of Mind* (60th Anniversary Edition), 60th edn. (Routledge Taylor and Francis, London and New York, 2009 First published 1949). http://www.lightforcenetwork.com/sites/default/files/Gilbert%20Ryle%20-%20The%20Concept%20of%20Mind.pdf. Accessed Feb 2014
57. A.M. Treisman, G. Gelade, Cogn. Psychol. **12**(1), 97 (1980)
58. P. Baumgartner, S. Payr, *Lernen mit Software (Learning with Software)* (Studienverlag, 1999)
59. C. Ware, *Information Visualization Perception for Design* (Morgan Kaufmann (Elsevier), 2013)
60. R.A. Rensink, Annu. Rev. Psychol. **53**, 245 (2002)
61. R.A. Rensink, Vis. Cogn. **7**, 17 (2000)
62. K. Nazemi, D. Burkhardt, M. Breyer, C. Stab, D.W. Fellner, in *ICL 2010 Proceedings. International Association of Online Engineering (IAOE)* (University Press, Kassel, 2010), pp. 163–173
63. S.G. Kobourov, arXiv:1201.3011 (2012)
64. B. Shneiderman, ACM Trans. Graph. **11**(1), 92 (1992). doi:10.1145/102377.115768. http://doi.acm.org/10.1145/102377.115768
65. B. Shneiderman, C. Plaisant, Treemaps for space-constrained visualization of hierarchies (University of Maryland, 2009). http://www.cs.umd.edu/hcil/treemap-history/index.shtml. Accessed July 2013
66. C. Stab, M. Breyer, K. Nazemi, D. Burkhardt, C. Hofmann, D. Fellner, in *Proceedings of World Conference on Educational Multimedia, Hypermedia and Telecommunications 2010* (AACE, Toronto, Canada, 2010), pp. 911–919. http://www.editlib.org/p/34743
67. D. Burkhardt, K. Nazemi, M. Breyer, C. Stab, A. Kuijper, in *Human Centered Design*. Lecture Notes Computer Science, vol. 6776, ed. by M. Kurosu (Springer, Berlin, Heidelberg, 2011), pp. 491–499
68. N. Bhatti, in *Proceedings of ED-Media 2008* (Association for the Advancement of Computing in Education (AACE), 2008), pp. 312–317. http://bibcd.igd.fraunhofer.de/bibcd/INI_Science/papers/2008/08p129.pdf
69. C. Stab, K. Nazemi, M. Breyer, D. Burkhardt, J. Kohlhammer, in *The Semantic Web: Research and Applications*. Lecture Notes in Computer Science, vol. 7295, ed. by E. Simperl, P. Cimiano, A. Polleres, O. Corcho, V. Presutti (Springer, Berlin, Heidelberg, 2012), pp. 633–646. doi:10.1007/978-3-642-30284-8_49. http://dx.doi.org/10.1007/978-3-642-30284-8_49
70. C. Stab, D. Burkhardt, M. Breyer, K. Nazemi, in *Semantic Models for Adaptive Interactive Systems*. Human Computer Interaction Series, ed. by T. Hussein, H. Paulheim, S. Lukosch, J. Ziegler, G. Calvary (Springer, London, 2013), pp. 133–149. doi:10.1007/978-1-4471-5301-6_7. http://dx.doi.org/10.1007/978-1-4471-5301-6_7
71. C. Stab, K. Nazemi, M. Breyer, D. Burkhardt, A. Kuijper, in *Human-Computer Interaction. Users and Applications*. Lecture Notes in Computer Science, vol. 6764, ed. by J. Jacko (Springer, Berlin, Heidelberg, 2011), pp. 520–529. doi:10.1007/978-3-642-21619-0_64. http://dx.doi.org/10.1007/978-3-642-21619-0_64
72. C. Stab, K. Nazemi, D. Fellner, in *Advances in Visual Computing*. Lecture Notes in Computer Science, vol. 6455, ed. by G. Bebis, R. Boyle, B. Parvin, D. Koracin, R. Chung, R. Hammound, M. Hussain, T. Kar-Han, R. Crawfis, D. Thalmann, D. Kao, L. Avila (Springer, Berlin, Heidelberg, 2010), pp. 514–523. doi:10.1007/978-3-642-17277-9_53. http://dx.doi.org/10.1007/978-3-642-17277-9_53
73. C. Stab, M. Breyer, D. Burkhardt, K. Nazemi, J. Kohlhammer, in *Proceedings of SIGRAD 2012*. *SIGRAD*, Linköping Electronic Conference Proceedings; 81 (Linköping University Electronic

Press, Linköping, 2012), pp. 83–86. http://bibcd.igd.fraunhofer.de/bibcd/INI_Science/papers/2012/12p121.pdf

74. R. Schäfer, T. Becker, C. Burghart, K. Nazemi, P. Ndjiki, T. Riegel, Basistechnologien für das Internet der Dienste, in *Internet der Dienste* (Springer, Berlin, Heidelberg, New York, 2011), pp. 19–40. acatech DISKUTIERT. http://bibcd.igd.fraunhofer.de/bibcd/INI_Science/papers/2011/11p061.pdf

75. A. Neumann, Recommender systems for scientific and technical information providers. Ph.D. thesis, University of Karlsruhe (2008)

76. J.M. Wolfe, J. Exp. Psychol. Human Percept. Perform. **15**(3), 419 (1989)

77. B.S. Bloom, *Taxonomy of Educational Objectives* (David McKay Co., Inc., NY, New York, 1956)

78. R.W. White, R.A. Roth, *Exploratory Search: Beyond the Query-Response Paradigm, Synthesis Lectures on Information Concepts, Retrieval, and Services*, vol. 1, ed. by G. Marchionini (Morgan & Claypool Publishers, 2009). doi:10.2200/s00174ed1v01y200901icr003. http://dx.doi.org/10.2200/s00174ed1v01y200901icr003

79. K. Nazemi, O. Christ, in *Advances in Affective and Pleasurable Design, Advances in Human Factors and Ergonomics Series* (Taylor & Francis, 2012). http://books.google.de/books?id=WHtwWU7C_vYC

80. M.A. Hearst, *Search User Interfaces* (Cambridge University Press, Cambridge, UK, 2009)

81. G. Marchionini, *Information Seeking in Electronic Environments* (Cambridge University Press, Cambridge, 1995)

82. F. van Ham, A. Perer, IEEE Trans. Vis. Comput. Graph. **15**, 953 (2009)

83. K. Nazemi, Statusbericht zum verwertungs-workshop. THESEUS Status Report of the Fraunhofer Institute for Computer Graphics Research (IGD). Internal Document (2009)

84. D. Burkhardt, K. Nazemi, C. Stab, M. Steiger, A. Kuijper, J. Kohlhammer, in *Advances in Visual Computing. 9th International Symposium, ISVC 2013*. Lecture Notes in Computer Science (LNCS), vol. 8034 (Springer, Berlin, Heidelberg, New York, 2013), pp. 86–97. doi:10.1007/978-3-642-41939-3_9

Part III
Proof of the Conceptual Model

Chapter 7
SemaVis: An Adaptive Semantics Visualization Technology

The last two chapters introduced first the general idea of our model in an abstract level to enable an overview of the conceptual model with our high-level design. This was followed by a detailed description of the main applied approaches, algorithms, and models. The detailed description aimed at providing the replication of each integrated component for the interested audience. The described models and approaches were implemented in an iterative manner with the main purpose to provide a proof of our conceptual model and the real application of the designed approaches for various scenarios. We named our technology that integrates the conceptual design for adaptive semantics visualization *SemaVis*. SemaVis was originally designed as a modular framework for semantic visualization, editing and annotation of semantic content [1, 2]. We enhanced SemaVis with respect to an adaptive behavior and focused in particular on visualization and visual adaptation. SemaVis as a visualization technology enables visualizing various data-types, adapting to various influencing factors, and provides more functionalities than described in our conceptual model. SemaVis is implemented as a client-server technology, but it can also be used as a client application or compiled as desktop application with limited functionalities. A detailed description of implementation of SemaVis, even of those parts that were introduced in this work, would blast the length of this thesis enormously. As the main aspects were already introduced in the previous chapters, we illustrate in this section the general architecture of SemaVis and refer the interested audience to the appeared publications, book chapters, and articles for further readings. The general architecture aims at providing the technical interplay of the introduced approaches, algorithms, and models. It gives an overview of the implementation strategy and enables a mapping to the already introduced high-level design. Besides the general architecture of SemaVis three exemplary application scenarios will be introduced. The main goal of the application scenarios is to provide a proof of feasibility and an insight in the usage of the system in different application scenarios.

© Springer International Publishing Switzerland 2016
K. Nazemi, *Adaptive Semantics Visualization*, Studies in Computational
Intelligence 646, DOI 10.1007/978-3-319-30816-6_7

7.1   General Architecture of SemaVis

SemaVis is a client-server technology that can be best described in an abstract Model-View-Controller (MVC) [3] design pattern. Although, the implementation makes use of various further design and architectural patterns and methods, the MVC patterns enables a more comprehensible view. Figure 7.1 illustrates the abstract SemaVis architecture using the MVC layers. In general SemaVis with its client-server architecture uses distributed computing techniques. A dedicated server for SemaVis provides the controller layer and parts of the model layer. The data that are visualized are commonly not stored in a data-base of the SemaVis server. Instead of storing the data, SemaVis uses different sources on Web with their own data-bases and servers. With this main characteristic, SemaVis distributes the computing to various systems, whereas the user experiences a single application.

The SemaVis client is designed as fat-client that requires the Adobe Flash-Player. The client-technology was designed to enable the usage without the need of the SemaVis server. It is even possible to use the client as a single technology without a server thus it enables accessing semantic data from its own directories. In some application scenarios, e.g. for visualizing the structure of an ontology without any search capabilities, the single-client system may make sense. However, this way of using SemaVis is supported but not recommended and even scarcely used. A second way to use the SemaVis client without the SemaVis server, is the non-adaptive way of visualizing semantics. Therefore SemaVis accesses data from a data-base and may use server, but does not access the own server that contains the main controller and models for adaptation. This way of using SemaVis is commonly recommended for those application scenarios or institutions that restrict using other server technologies commonly for security reasons. In this second case of using SemaVis only as client and accessing another server without its own server, the SVML with the description of visual layouts and the configuration has to be part of the client. SemaVis enables to compile the data parser and the SVML-parser (Fig. 7.1: model layer) as parts of the client and support a single-client technology without the need of the SemaVis server.

The introduced usage of SemaVis as only client technology is not supporting the adaptation functionality of SemaVis and is consequently not the focus of this work. The introduced approaches in the previous chapters are implemented as a client-server solution with a dedicated SemaVis server. It should be outlined that some parts and components of our conceptual model were implemented just for proving the concept and the main idea and are not part of SemaVis. Whereas many implemented functionalities of SemaVis are not described in this thesis. The main goal is to prove the concept by evaluating the main idea of the thesis.

SemaVis supports the access to various data-bases, sources, and servers to visualize heterogeneous data from different data-bases. The client in this architectural-model is only the view-layer according to Fig. 7.1. Although, only the visual layers of our reference model are part of this layer, the client is a fat-client, due to the following reasons: (1) all the visual layouts are part of the client, (2) the horizontal transformations of our reference model, e.g. visual mapping or visual layout orchestration

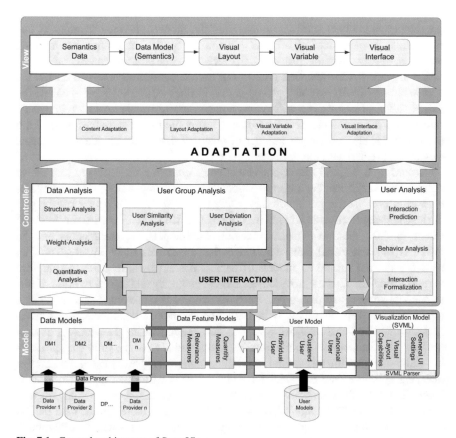

Fig. 7.1 General architecture of SemaVis

are performed in the client, and (3) the interactions are gathered in the client and transformed to the introduced formal representation. The major computations on the client are the different transformation steps that lead to visualizing semantics in a proper manner. The architectural choice that the client is responsible for the visual transformations can be argued with the need of single-client solutions as described. Only if the client is able to transform the data into a representation, its single usage is supported. Although, the client is designed as a fat-client, the response and computation time is performant enough so that users are scarcely perceive the delay caused by the various computations.

The approaches and models described in the previous chapters are partially realized as server components to enable a faster computation and to provide persistent storing for the different models. As already mentioned this architectural design has advantages and disadvantages. The main disadvantage is that the adaptivity of the system cannot be used without the server component. Most of the computations and storing mechanisms are deployed on a dedicated server. It is of course possible to provide a single-client solution with storing the models on users' computers and computing the introduced algorithms in the client application using the computing resources of users. But this procedure may lead to some lacks of security and usage of the unpopular *Cookies*. One goal of our architectural design was to enable the user to choose if information is stored or not. Further we did not want to enable the identification of users'. The main advantages of our distributed solution is that no information are stored on users' computers, less computation resources of users are needed that leads to a greater number of potential users, and the identity of users cannot be revealed with our architectural design.

According to Fig. 7.1 the layers *Model* and *Controller* are deployed to a server application. The Model layer hosts all models that are computed, generated, or gathered from other servers or data-bases. Only the SVML and its parser can be part of the client-application as already described. The Model layer consists of four different 'models', the data models, the data feature models, the user models, and the visual layout and visual interface models that are annotated in SVML. The data models store the structured models of data as described in Sect. 6.1.1. Therefore various implemented data providers enable the gathering and transformation of different data-sources and data-types. Each data-source is represented as a single data model and uniquely identified by its URI. The number of data models in SemaVis is not limited and enables storing various and heterogeneous data-sources as data models. The data models are the baseline and foundation of the following processing and transformation steps. First of all the structure analysis by using the iterative querying approach (see Sects. 6.1.1.1 and 6.1.1.2) is performed in the Controller-layer. Thereafter the quantitative analysis is performed based on the data structure (see Sect. 6.1.2.1). The result of the structural analysis is stored in the data model, whereas the results of the quantitative analysis are stored in the data feature model. The structured representation of the data model is then sent to the visualization models (SVML). The SVML enables a comparison of data structure in the data model with the capabilities of the visual layouts (see Sect. 6.2.3) and annotates the visual layouts that are able to visualize the underlying data model. This information is sent from the SVML modules to the user model module to enable choosing and determining adequate visual layouts. Further the information about the general visual interface (UI) settings is sent to the user model. Thereby all required information from SVML is stored in the user model. To summarize the described steps in a comprehensible way, we use a short example: Let us assume that a certain new user has searched for a term in different data-bases on the visual interface of SemaVis. Her interaction is first formalized (see Sect. 6.1.3.1). The formalized interaction of the search query is then sent to the data model. The data model starts an iterative querying in the data sources and providers. Further the gathered information is sent to the data analysis module that responds with structural information that are stored in the data model and quantitative information that are stored in the data feature model. The visual layouts, which are able to visualize the data are provided by the SVML and the information entity is then stored in the user model. Let us further assume that the user is a new user, so that the canonical user model is applied. The preferences, prior-knowledge and further modeled user information (see Sect. 6.1.3) in relation to the data are sent to the adaptation controller that adapts to the various layers. The adapted measures for each layer is sent to the view (client) that transforms the entire visual layers to an adapted user or visual interface and visualizes the results of the particular search term.

The described example illustrates only the initial step of a search. Both, the adaptation and the results are in the initial step strongly related to the data model. In the following steps, while the user interacts with SemaVis, the role of the user model, the interaction analysis, and further weighting measures increase and influence the adaptation effect. Each user interaction is first formalized according to our description of a relational formalization (see Sect. 6.1.3.1) and influences various modules of the Controller. Thereby the user interaction with the system is one of the most influencing modules in the Controller layer. The *User Analysis* module determines first the interaction behavior as described in Sect. 6.1.3.2. Each interaction is registered and trains the user model based on the given data and the interaction with visual layouts. Further the next possible action of the user is determined as described in Sect. 6.1.3.3. This information is volatile and leads to predict the next action for querying data and reduce the loading and measuring time for the loaded data. Consequently, this information is sent from the user model to the data model that starts with querying the data. The data are not visualized at this stage, thus a real interaction was not performed. Another volatile model is measured with each user's interaction in the *Data Analysis* module: the weight analysis as illustrated in Sect. 6.1.2.2. This measures the contextual relevance of a certain semantic instance and sends it directly to the *Adaptation* module. The results of the measurements of the algorithms are not stored persistently in the user model, but are stored for each session in the Data Feature Model. The user interaction is further affecting directly the data model for querying new data or semantic neighbors of a new selected data entity. With this step the data structure might be changed and lead to a new measurement of all the described initial steps. One further aspect of the user interaction is the analysis of similarity and deviation as described in Sect. 6.2. Each user interaction leads to train at least the canonical user model. If the user is logged-in as an individual user, the individual model is trained too, and if she belongs to a user group, the average user group model is trained too. With the analysis of similarities the individual user interactions can be compared with canonical user and other users. If a similar behavior is detected, the user is assigned to a user group that is a specification of the canonical user model. But if the system detects anomalies in user' interaction by a differing behavior from the canonical user model, the information are stored in the user model and only the individual user model is applied. The analysis of users' interaction and the described measurements and effects to the various models of the Model layer and modules of the Controller layer is performed permanently.

The main module of the Controller layer is the *Adaptation* module. It gets information about the data, the structure of the data, quantity information about the data structure, and the contextual weighting information from the Data Analysis module. The *User Group Analysis* module sends information about users' similarity or deviation to the Adaptation module and for modeling users to the user models. Further the Adaptation module gets the user model or user models, depending if the user is logged-in as an individual user and belongs to a group. The volatile information about the next possible users' action (interaction prediction) is sent from the User Analysis module to the Adaptation module. Based on this information the adaptation module starts the process of adaptation as described in Sect. 6.2.3. According to our reference model of adaptation (see Sect. 6.3) the information for adapting the various layers are processed in this module that consists of four main components: The content adaptation affects based on the user model and data analysis the amount of presented information and the focused entity. The layout adaptation determines based on the user models and the underlying data structure, visual layouts that are appropriate for the user and the underlying data. The visual variable adaptation makes use of the user history in the user model and the volatile models to adapt visual variables and the visual interface adaptation determines the number of visual layouts and their arrangement on screen. The adaptation criteria are the baseline for the entire adaptation that is performed on the SemaVis client, which is illustrated as the View layer of our architecture in Fig. 7.1. The SemaVis client consists of the visual layer of our reference model described in Sect. 6.3. Beside the adaptation criteria, the layer adapts to the horizontal transformations of data transformation, visual mapping, retinal variable mapping, and visual layout orchestration (see Fig. 6.36). The visual interface is thereby created during the interaction of users on the SemaVis client and further changes and adaptations are dependent on the user models, data characteristics, and users' behavior.

7.2 User Interface Design of SemaVis

SemaVis enables the use of a variety of visual layouts to visualize information. These visual layouts can be arranged and composed either by the user or by the system in a user interface. Further the visual layouts can be used as stand-alone visualization in Web-environments. This section introduces the user interface design that was used for the purpose of this work and applied for some application scenarios. With the adaptive and adaptable character of SemaVis, changing the user interface of SemaVis can be performed in an easy way. The description of the user interface in this section describes only one design that were used for the application scenarios and should be seen as a general description that introduces some terms for a more comprehensible way of describing the application scenarios. Thus our user interface design in all three application scenarios are similar, we start with the description of our user interface design. It should be outlined that the user interface design can be changed or adapted with our SVML. Special features of the user interface design will be therefore described in the illustration of the application scenarios.

The user interface of the SemaVis client as applied for our application scenarios is designed very simple and consists of two main areas (see Fig. 7.2). The areas are separated visually through a lightly varying color. The light-green area provides

Fig. 7.2 SemaVis user interface design for the application scenarios with the light-*green* UI-area for general functionalities of the user interface and the *white* VI-area for playing the visual layouts

general functionalities of the user interface and the white area is the visual interface for placing the visual layouts (see Fig. 7.2). For a comprehensible way of describing the user interface, we use the terms UI-area for the light-green area of the user interface with the general functionalities and VI-area for the visual interface area, where the visual layouts are placed and composed.

The UI-area can be further separated into three areas of *search interface*, *user-login area*, and the *visual recommendation area*. The search interface and the user-login area are placed on the top and the visual recommendation on the right bar of the user interface. Further on the top-left, the logo of SemaVis is placed that links to the website of SemaVis.

The search interface on top of the user interface provides the search functionality in various data-bases and different ways. It consists of a simple input field for searching terms, the names or logos of the data-bases (Fig. 7.3), and may further include special search functionalities, for instance *comparative search* (see Fig. 7.3b). Users are able to type their search term in the search field and send them through the established methods: pressing the Enter button or clicking on the search button next to the search field. The search can be performed in one or more data-bases. As Fig. 7.3 illustrates the user interface may include more than one data-base. The names or logos of the data-bases are placed as buttons next to the search field. Commonly the results of the first data-base are visualized initially. This behavior can be changed through the SVML to visualize the results of more than one data-base as initial search results. The user is able at any point to click on one of the data-base buttons and include new visual layouts. These visual layouts will visualize the results of the selected data-base. The special functionality of comparative search can be added to the UI-area for comparing different search terms from the same data-base as described as part of our

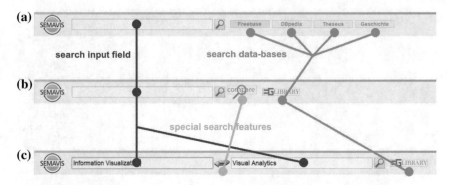

Fig. 7.3 Search interface of SemaVis: *blue circles* illustrates the input fields, *green circles* the buttons for the different available data-bases, and the *turquoise circles* some special search functions; **a** illustrates a simple search interface with four different data-bases, **b** a search interface with one data-base and the comparative search function, and **c** illustrates **b** with two search field after clicking on the compare button

visualization cockpit model (see Sect. 6.4.2). Thereby instead one search field, two search fields appear (Fig. 7.3c) for entering two different terms. Thereafter the results for both terms are visualized in a comparative way. The search interface on top of the user interface provides the search functionality in various data-bases and different ways. It consists of a simple input field for searching terms, the names or logos of the data-bases (Fig. 7.3), and may further include special search functionalities, for instance *comparative search* (see Fig. 7.3b). Users are able to type their search term in the search field and send them through the established methods: pressing the Enter button or clicking on the search button next to the search field. The search can be performed in one or various data-bases. As Fig. 7.3 illustrates the user interface may include more than one data-base. The names or logos of the data-bases are placed as buttons next to the search field. Commonly the results of the first data-base are visualized initially. This behavior can be changed through the SVML to visualize the results of more than one data-base as initial search results. The user is able at any point to click on one of the data-base buttons and include new visual layouts. These visual layouts will visualize the results of the selected data-base. The special functionality of comparative search can be added to the UI-area for comparing different search terms from the same data-base as described as part of our visualization cockpit model (see Sect. 6.4.2). Thereby instead one search field, two search fields appear (Fig. 7.3c) for entering two different terms. Thereafter the results for both terms are visualized in a comparative way.

The user-login area of the UI-area enables to login as individual user and get personalized adaptive visualizations. The user-login area is positioned on top-right of the user interface. At the initial start of the visualization system the user is not logged-in, thus we store no information on users computers. For a personalized or individual view on the visualized results, the user clicks on the login button and enters his acronym as illustrated in Fig. 7.4. The system indicates that an individual

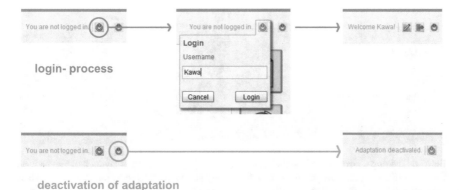

login- process

deactivation of adaptation

Fig. 7.4 User login area of SemaVis: the *upper* process illustrates the individual user login, the *bottom* process illustrates how to deactivate the adaptation of the visualization

user is logged-in by the text "*welcome 'user'!*", whereas 'user' is always the chosen acronym of the user. A further button appears next to the login button to enable a log out of the individual user. The process is illustrated in Fig. 7.4 under login-process. The user login-in area provides further the function to deactivate the adaptation of the system. Thus the user should have the control if the system's adaptivity is turned on or not, we included a button that provides this function. To deactivate the adaptation, even if no user is logged-in and the system adapts to the general user behavior based on the canonical user model, just one click on the adaptation deactivation button is required. The system indicates that the adaptation is deactivated next to the buttons as illustrated in Fig. 7.4: deactivation of adaptation. The user is able to login again and activate thereby the adaptation.

The last component of the UI is the visual recommendation area. It illustrates all available visualizations with icons. After a search, the system chooses automatically a set of appropriate visual layouts and places them into the VI-area. Besides this the visual recommendation recommends visual layouts for the user as described in Sect. 6.3.4 and illustrated abstractly in Fig. 6.55. The recommendation of visualizations leads to rank the visual layouts according to the user model and the data and provide the best suited visual layouts for the user and data. The user is able to put the visual layouts on the visual interface and compose an individualized visual interface. The system behaves as follows: if the user just clicks on a visual layout, the visual layout is placed in the bottom of the visual interface. The procedure and number of visual layouts is not limited but can be limited based on the data, user model, and the application scenario. Further the user is able to drag a visual layout from the visual recommendation area and drop it into the visual interface (VI-area). This way enables to place the visual layouts on a preferred area in the visual interface. Thereby the system behaves as follows: the user drags a visual layout from the recommendation area, the mouse cursor is enhanced by the icon of the selected visual layout. If the user moves the mouse cursor on the visual interface, a white area appears and indicates

(a)   (b)

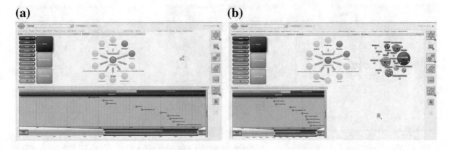

Fig. 7.5 Adding a visual layout on the screen: **a** the user dragged a visual layout and moved it on the visual interface, the area gets *white* and the already placed visual layouts rearrange themselves automatically and **b** the user dropped the first visual layout and dragged another one on bottom of the visual interface

the place of the visual layout, if the user drops it. Through this procedure the user gets an idea of how the interface would look like before he drops the visual layout on the visual interface. Figure 7.5 illustrates the described drag and drop procedure with the white space and rearrangement of visual layouts on screen.

The visualization environment does not only provide to drag and drop visual layouts from the visual recommendation area to the interface, but also to remove and rearrange them by drag and drop. The user can drag each visual layout on its top light-green area that indicates with its color to be a part of the user interface and drop it somewhere else. If the user drags the visual layout, the visual layout disappears from the screen and mouse cursor changes to the icon of the visual layout. Moving the mouse over the visual interface rearranges the placed visual layout and indicates with a white area the space if the visual layout is dropped. It is the same system behavior as dragging a visual layout from the recommendation area. If the mouse cursor is moved outside the visual interface the icon of the visual layout and thereby the mouse cursor changes to a visual layout icon with a red x. If the visual layout is dropped outside the visual interface, it disappears from the interface as illustrates in Fig. 7.6. User are further able to dismiss each visual layout by a 'x' in the top-right of each layout, according to the close metaphor of Windows operated systems.

The visual interface contains an arrangement of visual layouts that can be rearranged. Further each visual layout can be dismissed from the interface and new ones can be added. These are main functionalities of the visual interface. Beside these functionalities, the visual interface provides an interactive interaction history at the top. The interactive interaction history illustrates the user interactions with data representation on the visual interface. Thereby the interaction path of the user is illustrated for each session. The interaction path is interactive, so that the user is able to click on one previous interaction. In this case the data item is selected and visualized on screen. This function is illustrated in Fig. 7.6, where the interaction history is enlarged.

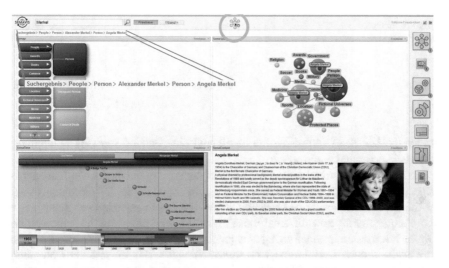

Fig. 7.6 Dismissing a visual layout from the screen: the SemaGraph layout from Fig. 7.5 (*top-center*) is in this figure dragged and moved outside the visual interface. The visual interface is rearranged and the mouse cursor (highlighted in turquoise) has been changed to the icon with a *red* 'x'. The figure further illustrates the interactive interaction history in an enlarged way

7.3   Selected Application Scenarios

We introduced in the previous section the general architecture of our conceptual model as a distributed system with a client-server solution. The main purpose was to illustrate how the conceptual models were applied at an abstract level. The introduced architecture of SemaVis that applied the conceptual models and algorithms was illustrated more as an overview to illustrate the realization as a distributed system and the interplay of the different components. We outlined that SemaVis can be used as a single client application without a server application, as a client that only accesses data-bases without adaptation, and as a client-server system that accesses the SemaVis server and enables adaptation and the visualization of different data-bases or computing resources. We further outlined that some of the components of our conceptual model was implemented to prove our conceptual model and are not part of the standard SemaVis technology, whereas many components that were designed and implemented are not described, due to the focus of our thesis. The main goal of SemaVis in context of this work is to prove our concept and the affects of visual adaptations.

SemaVis was already applied in different real-world application in several variations. It was applied by different enterprises, research institutions, and projects to visualize different kind of semantics starting from flat statistical data [4] to formal ontologies [1]. This section introduces three application scenarios of the adaptive SemaVis. We aim to provide not only a view on how SemaVis is applied in different

Fig. 7.7 Selected
application scenarios

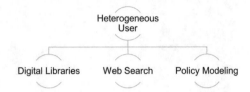

scenarios, but also to illustrate the distributed character of SemaVis by applying different server and data-sources.

To illustrate in particular the advantages of our conceptual model, we chose application scenarios with heterogeneous users as illustrated in Fig. 7.7. Commonly most of the application scenarios of SemaVis involve heterogeneous user groups. It is more a question of how different those users are, who are using the same system. Digital libraries for instance do not only involve users with different knowledge backgrounds, aptitudes, and interests, who are searching for scientific resources. Members of program committees and program chairs may search for adequate reviewers or institutions may search for adequate persons for working and research positions.

The second application scenario that we chose for illustrating SemaVis is Web-search. Searching in Web data-bases for gathering information is used by the entire spectrum of users. The differences in skills, prior knowledge, and preferences are enormous, due to the number of potential users. The third chosen application scenario is the visualization of social media in policy making. The stakeholders in policy modeling come from different disciplines and have various intentions [5–7]. They differ in their skills, roles, and prior knowledge.

The here described application scenarios use the same user interface design. The user interfaces of our application provides general functionalities like searching, enabling or disabling adaptation, or selecting and dismissing visual layouts as illustrated in Sect. 7.2.

7.3.1 SemaVis in Digital Libraries

Digital publications of journals, conference proceedings, or book chapter are more and more used for research and information acquisition tasks. Although, the existing digital libraries that host and provide information are established search data-bases, the visualization of those libraries are not established, due to the heterogeneous users, tasks, and intentions. The visualization of digital libraries is a broad area with many efficient and useful techniques. There exist already a huge number of visualizations that provide useful visual representations. The *BiblioVis* system by Shen et al. [8], the *Citation Map* of *Web of Knowledge* [9], the *PaperVis* system [10], the exploratory content-based layout by Bernard et al. [11], the *PaperCube* [12], or a methodology based on *Power Graphs* [13] may serve as examples. Thereby multiple linked views and a variety of graph-based visualizations were applied to provide additional value of bibliographic entries to the user. However, there exist already a huge number of visualizing digital content and metadata, adaptive visualization methods were not applied in the domain of digital libraries despite the heterogeneity of users interacting with these systems.

This section introduces one instantiation of our conceptual model that was applied in the domain of digital libraries. The main goal is to provide an insight of how the system behaves and which modules or components enhanced the conceptual model to integrate non-semantic metadata in a useful way. This section was partially published in [14]. We first give a brief introduction to the data integration and interpretation step of the system based on the chosen application scenario. Subsequently, we report on data-specific and user-dependent implications that influence the adaptive behavior of the visualization. Furthermore, we detail in how the adaptive capability of the system affects the automatic adaptation of our conceptual model.

In this application scenario SemaVis was applied as a client-server technology that uses the dedicated SemaVis server for adaptation and another server with its data-base for gathering the metadata that are not semantically annotated. It should be outlined that only one data-base is applied and the data-base is not deployed on the SemaVis server. We used the server of the *Eurographics Association* (EG) with its existing application programming interfaces (API) to access the Eurographics BibTeX entries. The search was performed in the EG-Library for results in *title*, *keywords* and *authors* with a HTTP-request and our iterative querying approach as described in Sect. 6.1.1.2 for enriching the data with semantics that are returned as plain-text BibTeX entries. We applied our data querying routines and the bag of words approach that enable the generation of semantics on the fly. Further a uniformed letter-comparison was applied to transforms all characters into the standard Latin character-set (8859-1). Since a query on the EG data-base was applied on the three categories *title*, *keywords* and *authors*, the result set might contain duplicates. We removed these redundant result entries by a duplicate detection routine. In a preliminary test case, we identified the need to disambiguate the author's names, because the same authors were stored in the data-base with different writings. To overcome this problem, we applied a rule-based algorithm that compares the last name, the first name, the first character of the first name, and the coauthors to disambiguate the author's name [14]. Figure 7.8 illustrates exemplary the system's disambiguation of the name 'Fellner' in the EG-digital library.

After removing the duplicate entries in the result set and the disambiguation of authors' names, we applied our iterative querying approach to create a semantic schema based on the respective BibTeX metadata. The applied routine detects relations and stores them as a schema on the SemaVis server. For example, we formalized authors of the same publication as 'co-authors of' and enhanced it with the relation 'written by' to papers and their corresponding authors, and assigned the relation 'author of', in return. This simple but efficient schema subsequently enables the adjustment of visualizations between different metadata attributes. The ACM Computing Classification System (CSS) were used to create categories that provide a lightweight hierarchy for the publications. This additional information, encoded as

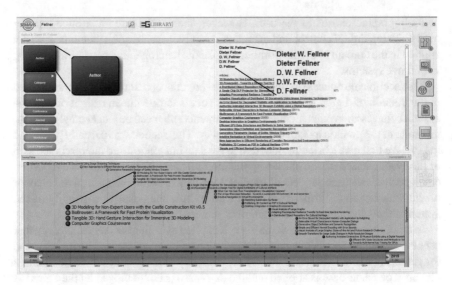

Fig. 7.8 Disambiguation of terms in SemaVis: illustration of the search result for the term "Fellner" generated automatically with the conceptual model for adaptation

a hierarchical data structure, is visually presented in the system with hierarchical aggregation metaphors. Since the data processing is executed in real-time, additional persistence layers were not necessary. Rather, the system is capable of immediate changes in the EG-digital library. For the client and thereby the visualization of the results, we applied the introduced user interface, whereas the number of the integrated visual layouts that are responsible for the placement and arrangement of data-objects on the screen, were limited to seven.

For the adaptation of SemaVis, we applied major aspects of our conceptual model. Thereby, the system starts always with the introduced empty user interface as illustrated in Fig. 7.2. The user is then able to search for a term and gets in return an adapted interface based on the search result and the canonical user model as described in Sect. 6.1.3.4. Thereby the capability of each visual layout is one indicator to recommend and automate the adaptation of the most appropriate visual layout. Another indicator is the users' interaction with visualizations. The user interactions on visualizations placed on the screen and the choice of alternative visualizations or their movement from the screen are used to derive the introduced canonical user model. Each of the transformations of our reference model was applied to adapt the visualization. The initial search of the user first detects in which search space (or class) the results were found. Therefore a search space definition (like the authors category) is not needed, thus the system recognizes automatically the search space. As Fig. 7.8 illustrates the term "Fellner" was detected in the class of authors. Therefore, this class is selected initially. Based on previous interactions and the data model a set of visual layouts is selected and orchestrated in the way how users did it previously for this kind of resulted data. The search for the term "Fellner" is visualized by three different visual layouts (see Fig. 7.8): on top-left, the SeMap layout provides a general and categorical view. Thereby the concept 'Author' is selected. On the top-left, the content-viewer (SemaContent) is used to illustrate the different writings, associated articles, and coauthors of the author of interest. On the bottom, a timeline (SemaTime) illustrates associated articles of the author based on the time-stamp and therefore the temporal spread of his publications. The entire adaptation is based on the canonical user model that was trained by real users of the system with different prior knowledge and aptitudes over a time period of more than one year. The canonical user model is therefore sufficient for using the visualization and provides the views that were commonly chosen by users. The orchestration of the visual layouts on the visual interface is based on previous interactions and thereby on the canonical user model too.

Let us introduce a further example to illustrate the adaptive behavior: The search for the term "Schneider" leads to a slightly different view of the results as illustrated in Fig. 7.9a. Thereby the term was found in the class of authors but the authors names in the result set are not the same person. Therefore the visual interface is enhanced with another visual layout that enables choosing an author. This view lists all authors with the name "Schneider" that are recognized as different identities. Thereby the author with the most publications is listed on top. In contrast to that, if the search is more precise and includes for instance the first name of the author like "David Schneider", SemaVis visualizes only the publications of the searched author. Thereby the placed visual layout might be different based on the canonical user model (see Fig. 7.9b).

As already mentioned, the entire pipeline of our reference model is adapted in this application scenario based on the canonical user model and the data. We illustrated the initial adaptation of the visual layouts and their orchestration on the visual interface. Further the visual variables are adapted in this application scenario. The adaptive visual variables are color (hue), saturation, transparency, color value (brightness), size, and order. The visual variables indicate the importance of the visualized data-entities (content adaptation) and guide the attention to certain entities. The number

(a)   **(b)**

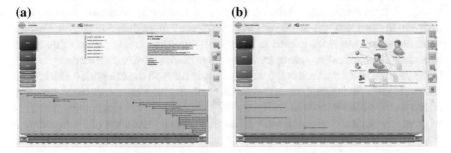

Fig. 7.9 Different views based on resulted data: **a** illustrates the results of the search term "Schneider", in top-center there appears a further list-view and visualizes interactively all authors with the name; **b** illustrates the search for the term "David Schneider" with other visual layouts and the searched author appear in the center of the graph layout

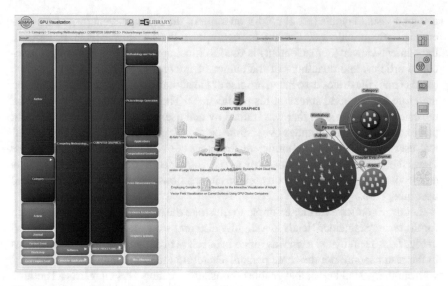

Fig. 7.10 Adaptation of visual variables: size indicates the amount of entities while color, brightness, and order indicate the user preferences according to the canonical user model

of results is used as an indicator for adapting the visual variable *size*. For example, SemaVis visualizes the icons of authors with the most publications slightly greater and the categories with the most publications are adapted in their size too. For content recommendation, the visual variables of order, brightness, transparency, and color hue were applied. Relevant content or generated classes are visualized in a deep blue color and the less a data entity is important, based on the user model, the more transparent and bright it is. Further the color hue changes from blue to a light-violet. Figure 7.10 illustrates the adaptation of visual variables. The user searched in this case for the term "GPU visualization". SemaVis provides two visual layouts and indicate that the term was found in the search space of *Articles*. Further in the left visual layout (SeMap) the classes are illustrated and indicates with their size that, although the search term was found in Articles, the most entries are in the class of *Author*. This is due to the fact that one article might be written by more than one person. Further there are more categories related to the search term than articles, this is due to the same fact that publications are commonly assigned to more than one category (ACM CSS). In Fig. 7.10 the user has started to navigate through the categories. During the interaction, the SemaGraph visual layout as the third visual layout was automatically added on the visual interface. The user navigates through the most dominant classes of categories that includes the most publications to the last class of *Picture/Image Generation*. The SeMap visual layout indicates with the order that users commonly chose the same path, except the last class of *Methodology and Techniques* that is on the top but has fewer articles than the chosen one. The SemaGraph visual layout provides a detailed view on the articles of the chosen class, whereas the SemaSpace on right of screen illustrates the entire relationships. Thereby the class of authors illustrates the number of all authors with publications in "GPU Visualization". The categories illustrate aside from the hierarchy, the same articles within the graph and in the class Articles the sum of articles. Further the class of authors illustrates all authors, whereas there are two dominant authors that have published more article than others.

The canonical user model does not require personal information about the user because the model itself provides a general data-dependent "initialization". As described in our conceptual model the over-generalization of visualization adaptation is faced by individual user models. The user of our application is able to login in the login-area as individual user. The process of individual user adaptation starts from this point as described in the Sect. 6.2. If the user does not yet have a user model, the user model of the canonical user is applied and the deviation and similarity measurements start. The interaction of the user, while she is logged-in as individual, changes during his interaction the individual user model and enhances it with individual behavior patterns. The visual layouts, their composition on the interface, the content, and the visual variables might be changed based on the individual user model. Therefore, the history of user's interactions is analyzed with a subsumption of the hierarchy of the schema he interacted with. For example if the user is more interested in user interfaces and visualizations (based on his previous interactions) and searches for the term *Interaction Design*, our application presents the 'categories-of-interest' visually highlighted (see Fig. 7.11b). In contrary, the canonical user model applies the average interest of users in the particular domain of interest to adapt the visualization (see Fig. 7.11a). SemaVis enables thereby not only a general adaptation but also an individual adaptation that may affect only the visual variables or the entire adaptation transformations of our reference model. The result might be a slightly different or even a completely different view on the same data. Figure 7.11 illustrates a slightly different view on the returned results for the term *Interaction Design*. Thereby the adaptation based on the canonical user model visualizes the information with three visual layouts, whereas the visual variables indicate the general interest in the selected categories by color and brightness (Fig. 7.11a). In contrast to that, the individual user model illustrates clearly that the particular user is more

(a)   **(b)**

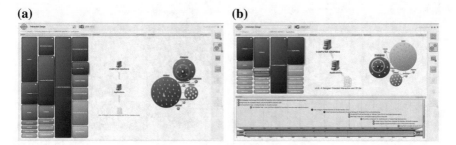

Fig. 7.11 Canonical versus individual user model adaptation: **a** illustrates the search results and the interaction path to the category *Applications* based on the canonical user model and **b** illustrates the same search with the same interaction path based on an individual user model

interested in applications and software and not so much interested in authors. Further the visual interface is changed slightly by visualizing a fourth visual layout for temporal information. It seems that the user has selected in previous similar searches the temporal view on articles (Fig. 7.11b).

If the data type changes within the exploration work flow, the system automatically adapts the set of provided visual layouts. For example, the user may request all publications of a specific author on demand. In this case the system automatically adapts the visualization for the specific results during the interaction with a visual layout. Changes in provided visualizations are performed as unobtrusively as possible in order to not confuse the user. This is performed by automatically suggesting the most similar visualization type (e.g. aim to apply a graph-based visualization when replacing another graph-based visualization). The information about the similarities of the visual layouts is gathered from the SVML as described in Sect. 6.2.3. Visual layouts that are not applicable anymore for currently analyzed data are temporarily excluded from the set of user-selectable visual layouts in the recommendation area.

The described application scenario illustrates how visualization can be adapted to various influencing factors, e.g. data and users. An initial design for the visual adaptation is not required anymore, thus the canonical user model provides a self-learning approach and improves the visualizations with each user for all users. But of course SemaVis provides the functionality to deactivate the adaptation. In this mode the results are illustrated with two predefined visual layouts (SeMap and SemaGraph) and all visual layouts are selectable in the recommendation area. The recommendation is then just recommending the visual layouts based on the data. No user information is considered anymore. Figure 7.12a illustrates the returned results of the term *GPU Visualization* without adaptation. Thereby the class of *Author* is selected by the user. We further included based on the same search some more visual layouts to provide a view on a not adaptive visual interface. Thereby the categories were interacted to illustrate the way of how visual variables affect the view (see Fig. 7.12b).

This section introduced one instantiation of our conceptual model as an adaptive visualization application for bibliographic entries in digital libraries. The application

(a)   **(b)**

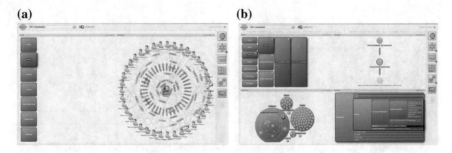

Fig. 7.12 Canonical versus individual user model adaptation: **a** illustrates the search results and the interaction path to the category *Applications* based on the canonical user model and **b** illustrates the same search with the same interaction path based on an individual user model

scenario aimed at providing a view on how the conceptual model was applied for visualizing bibliographic entries from digital libraries. We outlined that the underlying metadata were not semantic. Aside from an enrichment with semantic through our iterative querying approach, we cleaned the data from duplicated results, transform the characters into standard Latin, and implement a routine for disambiguating author's names. SemaVis was used in this application scenario with its dedicated server. The Eurographics server was used as data-provider, thereby the search was performed on the Eurographics server with their API and all the measurements on the SemaVis server. We illustrated in this application scenario in particular the role of the canonical user model. With the integration of the canonical user model as the average usage behavior there is no further need to train an adaptive system by experts. SemaVis learns from all interactions with the system regardless who the user is. Furthermore we illustrated some main adaptation effects on both, the canonical user model and the individual user model. We concluded the section by illustrating how the system behaves, if the adaptation functionality is deactivated. The main contribution and goal of this section was the proof of feasibility of our conceptual model. It should be shown that the complex measurements and adaptations are applicable.

7.3.2 SemaVis in Web Search

The introduced application scenario of SemaVis in digital libraries provided a sufficient insight into the feasibility of our conceptual model. The application scenario applied all aspects of our model and adapted according to our reference model based on both canonical and individual user. Although, the users of a digital library are very heterogeneous and intent different tasks to be solved, the heterogeneity of users, who search for information on Web are enormous. Almost every one searches the Web for information. The differences between the people, who are interacting relies not only on their prior knowledge, interests, education, visual abilities, or aptitudes, the users differ in their cultural and demographic background too. A search on Web databases might be performed by a fourteen years old girl from India for her research on school or a fifty years old professor for computer science in the United States for his research on search interfaces. However, the main aspect is that the application scenario of Web search has commonly the most heterogeneous users. The here described software was accessed by users all over the world. We registered access from China over Iran to the United States of America. Although the most of the Web accesses came from Europe and overseas, the heterogeneity of users is given and this fact affects the way of visualizing information enormously.

In our opinion, the heterogeneity of users in this application scenario is the main reason that visual representations of search results could not find their way to a regular usage in Web search. The common way of searching information on the Web is still the list-based textual representation of search results. Although, information visualization and visual analytics experienced enormous enhancements and developments, the techniques are still just used by special groups of users for special tasks. Although, we do not expect that SemaVis will be established as visual search environment that is regularly used and is part of the daily searching tasks, we think that the idea of adaptive visualization would enable this idea and SemaVis could be the first step to a visual search for everyone. The here introduced application scenario was published in [15].

SemaVis uses in this application scenario two slightly different data-bases with their search capabilities and own servers. On the one hand the *DBPedia* data-base with the structured Linked-Data [16] and on the other hand the *Freebase* data-base a Linked-Open-Data base of *Google* [17–19]. Similar to the application scenario for digital libraries, the search process is bottom-up (see Sect. 6.4.1) by means that the user starts the search process with a query. The main difference is that the process of data-cleaning and term-disambiguation is not necessary for these data-bases. Further the searched term is queried on both data-bases simultaneously that leads to results from two different data-bases and provides a more complex visualization process. The search results from the semantic data-bases are commonly instances without further semantic relations or contextual information. The returned instances have commonly a weighting-value, how good the queried term matches to the results. These resulting instances are our foundation to create a visual semantics and provide contextual information. Therefore, we apply our iterative querying approach (see Sect. 6.1.1.1) to generate a categorical hierarchy and a contextual semantics. Figure 7.13 illustrates a test environment with the results and their weighting for the term *Kabul* on left, the

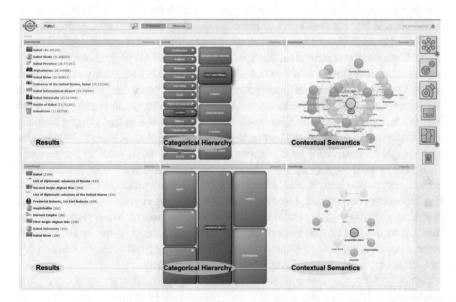

Fig. 7.13 Inclusion of semantics from Web data-bases: the returned results are enriched with our iterative querying approach with the semantics. On *left* the results for the term *Kabul* are illustrated, in the *center* the categorical hierarchy is determined, and on *right* the contextual semantic is illustrated as an arbitrary graph

categorical hierarchy in the center, and the contextual semantics on right. Thereby the upper visual layouts visualizes the results of Freebase and the lower the results of the DBPedia data-base.

The iterative querying approach enables to gather and visualize the semantic structure of the result set and provides an interactive picture of the searched term. This process is the foundation of visualizing the semantic structure. In this application scenario, we enhanced our approach based on the users' search intentions [20]. We determine based on the searched term and the weight-values of the data-bases, if a search is focused or more exploratory. Therefore the search terms are compared to the weighting values of the data-base. With the assumption that if a user searches for a specific fact, she defines more precise search terms, we implemented an algorithm that make use of the data-base weightings in relation to the search terms. If one result returns based on a specific search that may contain more than one search term is weighted significantly higher than other results, SemaVis visualizes the entire set of results but selects the result with the highest value initially. The process of exploratory search is thereby not limited. Although, the user searched for a very specific combination of terms, the entire set of results is visualized. The main difference is just that SemaVis already selected already the path to the result that is significantly high compared to the other resulted instances. This functionality can be best explained with an example: Let us assume that the user searched for the term *Obama*. As non-exploratory search engines would prefer and illustrate result on the president of the United States in their first pages, SemaVis visualizes all categories and semantic relations found for this term and provide an exploratory navigation (see Fig. 7.14a). This is due to the unspecific search. There is also a city in Japan named *Obama*. If SemaVis would just visualizes terms that are related to the president of United States, the user would not find the city of Obama easily. But vice versa, if the user searches for the term *Barack Obama*, it is obvious that the user wants to get information about the president. In this case, SemaVis visualizes all results of

(a)   **(b)**

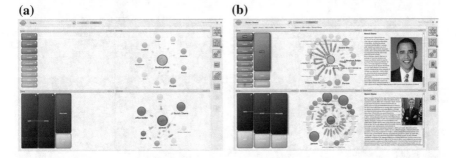

Fig. 7.14 Adaptation based on search term: SemaVis adapts in this application scenario based on the searched term. In **a** the user entered the term *Obama* for search, the results are giving in categories and hierarchies on both data-bases. In **b** the user entered the more specific search terms *Barack Obama*. In this case SemaVis visualizes all results, but selects the most appropriate result based on weighing measure of the data-bases

the query *Barack Obama* but selects the instance *Barack Obama* with based on the highest values (see Fig. 7.14b). Figure 7.14 illustrates this functionality of SemaVis in this application scenario.

Another aspect that is relied on the data-bases and supported by our iterative querying approach is the initial selection of concepts after a performed search. In many cases, the DBPedia data-base provides a concept-hierarchy consisting of one sub-class. In such cases the interaction costs of the user increases due to interacting through single concepts and getting at last stage either a separation of concepts or further just one concept with a set of instances. Regardless of the visual layout, the common procedure would be to select each concept and navigate through them. To reduce the interaction cost, we implemented a routine based on our quantitative measurements (see Sect. 6.1.2.1) that detects if a concept has just one sub-concept and navigate automatically through the concept hierarchy until either there are more than one sub-concepts and the user can choose one or there are no concepts anymore and the user can interact with the instances directly. We kept the concept-hierarchy to provide the hierarchical information for the user. During users' interaction, new data may be loaded on demand and new concepts may complement the hierarchical structure. Figure 7.15 illustrates this functionality. Thereby the search results in Fig. 7.15a just provided a single sub-concept hierarchy. SemaVis selects automatically the lowest level and visualized the related instances. In Fig. 7.15b the hierarchy was selected until more than one sub-concept appeared. The user is able to select a further level of hierarchy or interact with the related instances.

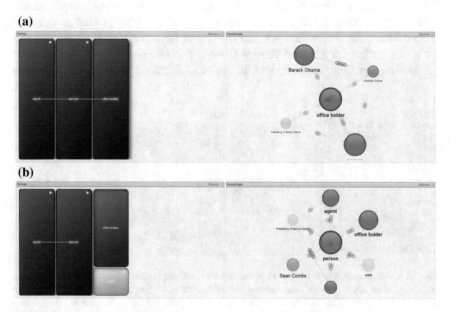

Fig. 7.15 Automatic selection of sub-concepts: sub-concepts of the semantic data are selected automatically by SemaVis until a kind of decision can be performed by the user

The adaptation in this application scenario follows our conceptual model. All major aspects of the conceptual model could be implemented in this scenario, thus real semantic data are accessed from the two mentioned data-bases. The application starts similar to the introduced application scenario of digital libraries with a blank user interface as illustrated in Fig. 7.2. The user interface is the same as already described, with its several areas for login, search, visual recommendation, and visual interface. The application scenario includes a set of eight visual layouts. These visual layouts were the most used ones since the first version was online accessible. The first adaptive version of SemaVis for Web search with limited functionalities was released in 2012 on the Web and is free accessible without restrictions. Thereby the canonical user model was integrated one year later and is trained since 2013 by various and very heterogeneous users.

SemaVis starts in this version similar to the digital library one, with a canonical user that adapts the entire visual interface based on the queried data and the user model. The main difference is that the search is performed simultaneously on two different data-bases. Therefore the visual variables that indicate relevance values and guide the users' attention differ in their color hue. This is to enable a differentiation between the results of the two data-bases. Beside this, each visual layout on interface is annotated with the corresponding data-base. The adaptation based on the canonical user model is as described in Sect. 6.1.3 is based on the interaction behavior of the user in relation to the data. So the results of the different data-bases may initially be visualized with different visual layouts. This is due to the different structure and content of the result data. Figure 7.16 illustrates the changed views on the slightly

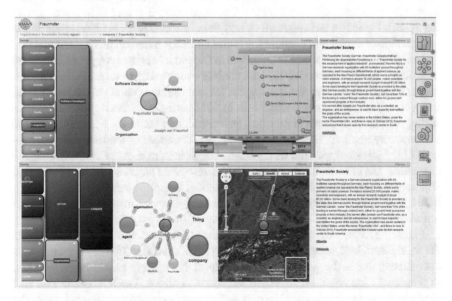

Fig. 7.16 Simultaneous visualization of different data-bases: on *top* the results of the Freebase data-base is visualized colored in *orange*, in *bottom* the results of DBPedia are visualized in *green*

differing data. Thereby a search for the term *Fraunhofer* was performed and navigated to the Fraunhofer Society. The similar visual layouts are due to the similar data, whereas the DBPedia data-base provides additional geographical information and Freebase temporal information. The visual interface is thereby adapted based on the user model and the underlying data. Further the visual variables that make use of color hue (Freebase: deep orange to a light green and DBPedia: deep green to a light turquoise), saturation, size, and order. The visual layouts are recommended on the recommendation area for each data-base separately. Further the user interaction history above the visual layout is visualized in the dominant color of the particular data-base. Figure 7.16 illustrates that with both data-bases were interacted. This can be seen from the different colors of the user interaction history.

The canonical user model is the main user model of this application scenario and therefore, beside the data structure, the foundation of adaptation. The workflow in the adaptive SemaVis is similar to the SemaVis version for digital libraries. Due to the very different data that are returned from the data-base, the adaptation effects are bigger and changes during the interaction with the system appear more often. A main aspect is that during the interaction, data may be loaded from the underlying data-bases on demand. The changed data structure in combination with the user model effects the visual interface immediately and enhances the interface with new visual layouts. Thereby the automatic dismissal of placed visual layouts are only then performed, if not data for that particular visual layout exist or the user starts a new search that returns other data with other data-structure. The canonical user model is in this scenario like in the other scenarios too, the average usage behavior of all users, who interacted with the system. Those users, who are logged-in as individual users are considered too. That means that regardless, if a user is interacting with the system logged-in as individual or not, SemaVis considers his interaction in the canonical user model and changes the entire behavior based on this user model.

In contrast to that, if a user is logged-in as individual and has not yet an individual user model or his user model does not contain enough data to determine his prefer- ences and behavior, SemaVis investigates for that user the canonical user model and trains simultaneously the individual one. In this case the introduced approach of mea- suring deviations and user similarities as illustrated in the Sects. 6.2.1 and 6.2.2 are continuously applied. Thereby the individual preferences of the user are measured and if his individual user model contains enough information for an individualized adaptation, the canonical user model is not investigated anymore for adaptation (but still trained further) and the individual user model is applied for adaptation. To illus- trate how individual user may change in their behavior, we illustrated in Fig. 7.17 the initial results of the term *Albert Einstein* of two differing users.

Figure 7.17 illustrates clearly that the deviated user models of the two user affects the visual variables, visual layout, and the visual interface. Thereby the content is not adapted, due to the very specific search term that leads to an automatic selection of a uniquely identified search result as described above. The left user in Fig. 7.17 seems to use more simple visual layouts and is more interested in temporal issues. The right user in contrast to that seems to have preferences on more complex graph-based structures, whereas the temporal aspects seem not to be interesting for him. Based

Fig. 7.17 Visual adaptation for differing user: the result of the search term *Albert Einstein* visualized by SemaVis based on two differing individual user models

on the visual recommendation area (right bar) the most preferred visual layouts in the context of the chosen data are illustrated. The right left user (Freebase as data-base is chosen) prefers in particular SeMap, SemaGraph, and SemaTime (see Sect. 6.3.3.2) whereas the right user (DBPedia is selected) seems to be more interested in geographical aspects and graph-based layouts. The fact that in both cases the content is illustrated is due to the general behavior of users (based on the canonical user model) that if an instance is referring directly to a Web-resource, this should be visualized. The SemaVis learns from the individual user behavior. If a user is permanently dismissing the view on content, this would even not be visualized, if an instance would refer to a Web-resource.

Let us take a look in contrast to the individualized visualizations of the term *Albert Einstein* to the visual representation based on the canonical user model for the same term. Figure 7.18 illustrates the results of the term, if a user is not logged-in. We illustrated in this figure two slightly different versions of the visual representation for the canonical user model. The left one illustrates the standard view, while the right one has turned on an iconic view. Thereby the figures from instances are loaded on demand and replace the default icons. The figure illustrates clearly that the left user from Fig. 7.17 is more similar to the canonical user and thereby the average user. Aside from his interest on temporal and somehow geographical issues, the visual layout orchestration seems to be the same. The right user by contrast differs from the canonical user much more. Besides the view on content there are no identical visual layouts placed on screen.

The adaptation process in SemaVis changes during users' interaction the visual interface and recommends other visual layouts. This effect is always coupled with the users' interaction to perceive the changes on screen as a reaction of the system. Major changes appear if the user searches for other terms, his interaction with data is loading a set of new data that has another structure and content, and in particular if the user is logged-in during the search and exploration process. Thereby the already placed visual layouts still remains, expect a new search is performed. During the interaction without searching for another term only new visual layouts appear without dismissing the placed ones. Figure 7.19 illustrates a scenario in which the user started a search for the term *Einstein* and navigated in both data-bases to the city

Fig. 7.18 Visual adaptation based on the canonical user model: the result of the search term *Albert Einstein* visualized by SemaVis based on the canonical user model. The *right* figure uses the icons the associated figure from the data-bases as icons

(a)   **(b)**

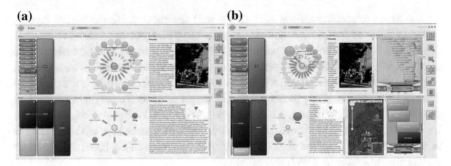

Fig. 7.19 Visual adaptation during the interaction process: **a** illustrates the visual presentation based on the canonical user model and **b** illustrates the same view after an individual user logged-in. The already placed visual layouts still remain, but the most preferred visual layouts of the user are added on the interface. Further the visual recommendation area and visual variables changes

of Princeton (Fig. 7.19a). Thereafter he logged-in as individual and the visual interface and the visual recommendation changed their appearance. The visual interface added new visual layouts, while the visual recommendation changed the order of the recommended visual layouts (Fig. 7.19).

Another mentionable aspect of the adaptation is the use of our prediction algorithm (see Sect. 6.1.3.3) in context of adaptation. Commonly SemaVis makes use of the algorithm to load data on demand prior to users' interaction. This procedure reduced the loading and measurement time for the adaptations. In some cases it might be useful for the user to see what is predicted by the system. Commonly this functionality is turned off (in all scenarios), but the user can turn this on to get recommendations for interactions. Thereby commonly the underlying user model is investigated to measure the predicted action. The predicted action may lead the users' attention and guide them in his exploration process. Figure 7.20 illustrates a scenario in which the user searched for the term *Fraunhofer*, thereby the interaction prediction is turned on and visualizes the recommended related instance or concept by a deep red color.

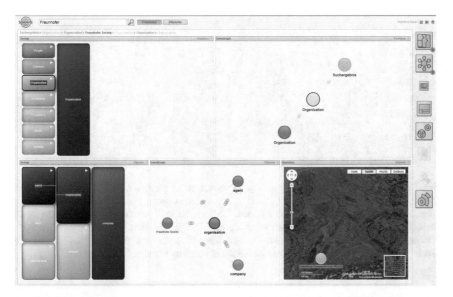

Fig. 7.20 Interaction prediction for guiding users: the predicted and thereby recommended entities are colored *red*. In the *upper* visual layouts, the concept *Organization* is colored *red*, in the *lower*, the instance *Fraunhofer Society*

The content adaptation of SemaVis is based on the underlying user model. While not logged-in users get a kind of content recommendation based on the overall average interaction behavior, the individual users' content is recommended based on their own user model. For the content adaptation in particular the users' interests are considered and a data-base independent model is generated as illustrated in Sect. 6.1.3.4 (Eqs. 6.13 and 6.14 and Table 6.4) that are part of the user model. The steady state vector on concept level enables a data-base independent recommendation of data. This is in particular in our Web search scenario of great interest, thus the user searches and interacts simultaneously on two different data-bases. Although, the data-bases have different data-structure and differing content, on an abstract level the interactions and interests can be determined for both data-bases. However, not all concepts use the same terms and can consequently not matched. Commonly the content is recommended by using the visual variable layer of our reference model and changes the items of interest in their size, color, order, and brightness. Figure 7.21 illustrates the search results for the term *Kabul* for two different users with different interests. We can see clearly that the user in Fig. 7.21a is more interested in events. In both data-bases the concept event can be found on the highest level of hierarchy and is clearly highlighted in contrast to all other items. In contrast to that in Fig. 7.21b the user seems to be more interested in places, locations and countries. In this context just one common concept can be found, the concept of *Country*. Although this concept is in DBPedia in the third level of hierarchy and in Freebase in the second level, the user model is able to determine based on the introduced algorithms the relevance of

(a)   **(b)**

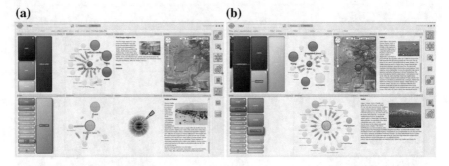

Fig. 7.21 Content recommendation in adaptive visualization: the figure illustrates the search result for the term *Kabul* based on the user model of two different users. The user in the *left* figure **a** is more interested in "events" and related issues. The term "event" occurs in both data-bases on the highest concept-level and is recommended by the use of visual variables. The user in the *right* figure **b** is more interested in places and related issues. Here the term "country" was found in different levels of concept-hierarchy and recommended for the user by the visual variables

this concept for the user and highlights it regardless from which data-base they are resulted.

The application scenario of Web-search with two different data-bases implements the major aspects of or conceptual model and the underlying reference model of adaptation. The adaptation is based on the combined model of user behavior, data structure and content. It is further enhanced with recommendation functionalities and prediction of recommended data that can be turned on, if the user wants such guidance. All other functionalities like adding, dismissing, and rearranging visual layouts can be further used in this application scenario. Each interaction of the user trains the user model and if the user is logged-in his individual user model too. But our application can be used as a visualization environment without the adaptive functionalities. The user has always the choice to use the adaptive version or the non-adaptive version that neither adapting the different visual layer nor recommending any visual layouts for the user. Figure 7.22 illustrates SemaVis for Web search in a non-adaptive manner. All adaptation functionalities were turned-off. Even the visual recommendation in the right bar does not recommend any visual layouts. All icons have the same size and the order is a default set order. The differing colors in the visual layouts are associated to the different data-bases and indicate if the entities are instances or concepts.

We introduced in this section the application scenario of Web-search with the full-adaptive SemaVis. This application scenario makes use of real semantic data provides thereby enhanced visual adaptation on different data-bases. We could illustrate that the major aspects of our conceptual model can even be applied to data from different data-bases with masses amount of data. The application scenario of Web search demonstrated clearly that there is no need to train the systems by experts. With our approach of the canonical user model, the system is continuously trained by real users. This version of SemaVis is free online accessible for interested audience. This

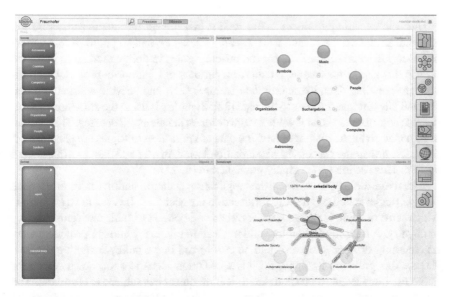

Fig. 7.22 SemaVis in a non-adaptive manner: the adaptation functionalities are turned off, the visual recommendations on the right bar are not recommending visual layouts

audience trained the major aspects of the user models by using the system similar to the application scenario for digital libraries.

The main difference between the introduced applications scenarios of Web-search and digital libraries were the underlying data-bases. While the digital library scenario provided non-semantic data that had to be first enriched with semantics, we used in this application scenario two well-known semantic data-bases. Further the application scenario of digital library just used on data-base in contrast to this application scenario. In the next section, we will introduce our ongoing work on the visual adaptation in context of policy modeling. Although, the work on this topic is ongoing, we introduce the application scenario due to its differing character of search. While the application scenarios for digital library and Web-search made use of a more traditional bottom-up search, we will illustrate in the next section that SemaVis is capable to support a top-down search too.

7.3.3 SemaVis in Policy Modeling

The involvement of citizens with their opinions and discussions in the policy modeling or creation process plays an increasing role. The Web provides vast amounts of social data, which can be used to identify problems and consider citizens' opinions in the policy creation process. The masses of information are difficult to handle. Everyday new opinions, discussions etc. and thereby new data are available. The application scenario introduced in this section investigates the adaptive visualization of social media data, e.g. data from *Facebook*, *Twitter*, and so on, to enable a view on emerging trends and topics for the different roles in policy modeling. SemaVis uses in this application scenario data from partners of a European project. Our partners are crawling data, extracting data features [21], and provide a structured and semantically enriched access to the data [22]. SemaVis is using the structured data to provide appropriate visual layouts for the different stages and processes. This section aims not at giving an insight of how the data are processes or topics are generated, it focuses more on the part of visual adaptation based on user and data as a distributed system. This section was partially published in [6, 23].

The main difference between this application scenario and the introduced ones is in particular relied on the way information are searched. It is not always obvious which term or topic is of interest for citizens in context of their environment, city or country. A bottom-up search would not lead to efficient results or provide a view on emerging or interesting topics to be considered in the policy modeling process. To face this problem, we apply for this application scenario a top-down search for gathering knowledge in problem identification on social (subjective) level etc. The top-down approach integrates the introduced search interface as well as a temporal overview of topics. The temporal overview provides a kind of faceting the search space to reduce the information amount on relevant aspects. On the visualization level "details-on-demand" and graph-based visualizations provide a comprehensible view on the information relationships. With the integrated visual layouts of SemaVis, the level of detail may reveal fine granular or textual information. Thereby we follow the visual information seeking mantra proposed by Shneiderman [24]. Figure 7.23 illustrates the temporal overview of SemaVis with the temporal spread of the topics. The topics are chosen from the left bar, in which a list illustrates all emerging topics. Topics of interest can be selected to view their temporal spread. The visualized data are extracted terms from Croatian social media resources.

The temporal overview on extracted topics is kept simple. It is not combined with the visual interface of SemaVis to provide a clear and comprehensible picture of the topics' evolutions. Only the SemaRiver visual layout (see Sect. 6.3.3.2) is placed to provide such a temporal view on the topics. This view on the data is not adaptive. Although the selection interactions of users are stored in the canonical user model, an adaptation is not provided due to the simple way of visualization. The work on this application scenario is ongoing. It is intended to enhance this view on categorical and geographical overviews [6]. With these enhancements the application of our adaptive model would make sense. However the simple overview on data enables the user to choose one of the visualized topics to get detailed information about the topic. Therefore the selected topic is searched in a semantic data-base provided by our project partners [22]. Similar to the other application scenarios, the data is not stored in the SemaVis server. It is again a distributed client-server solution that makes use of more than one server. The formalization of the crawled social Web data is provided in this application scenario as a light-weight semantic representation. The technologies provide feature extraction based on statistical models [21]. The

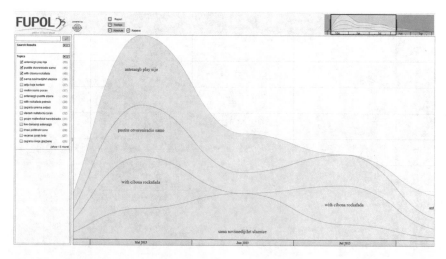

Fig. 7.23 Temporal overview of political topics: in the center (main) area the extracted topics from social media are visualized based on their temporal spread. On *left*, a selectable list of all available topics enables to choose certain topics of interest

extracted features are then formalized in a semantic relationship model, based on *SIOC* and *FOAF*, whereas project-specific classes are enhancing the formalization.

After a user has chosen a topic of interest, he gets with a double click on the SemaVis user interface as introduced in the previous application scenarios. Thereby the selected topic is visualized in more detailed manner. In particular the *channel* that refers to source of the topic, the *actors* that refer to the persons or institutions that published their position, and *post* that include the text that was published, is visualized with their semantic relations. In this application scenario the adaptation is limited to the visual layouts and their composition on screen based on the canonical user model. Commonly the size of icons or graphical representations of data refers to the amount of related or included postings. The average usage of the visual layouts and their placements are stored to enable an adequate view on the underlying data. Thus in this context commonly the topic, channel, or actor with the most postings is of interest for the policy makers, the visual variables only refers to the amount of relations and postings. Figure 7.24 illustrates the canonical adapted visual interface after the choice of a topic in the overview visualization. Thereby four visual layouts give detailed information about the chosen topic. We can see at a glance that all postings about the chosen topic were published via Twitter, due to its iconic logo. Further one actor seems to post more about the certain topic, due to the greater size of the icon on the SemaSpace visual layout. The temporal visual layout illustrates the amount of topics in a more detailed way. The figure illustrates the topic postings on a daily interval.

From this point the process of adaptation starts as described in our conceptual model and in the previous application scenarios. The main difference is that the

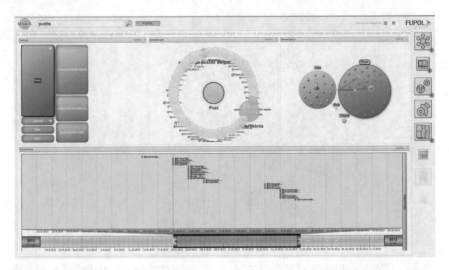

Fig. 7.24 Detailed view on a chosen topic: after selecting a topic, SemaVis illustrates the details about the topic in semantic manner. The visual layouts and their composition are based on a canonical user model

canonical user model just adapts the visual layout and their composition. The visual variables are dedicated for relevance measures of the project partners that provide a kind of weighting based on the various factors, e.g. amount of previous postings or relation to other relevant persons. Thereby only the size as visual variable is used. The interactions of users are enhancing the canonical user model and based on this the entire system behavior changes. Similar to our other application scenarios, the work flow of the users leads to immediate changes on the visual layout layer.

However, the visual variables are not used for the canonical user model; the user in this application scenario is still able to login as individual. Thereby all the functionalities of our conceptual model are applied including the adaptation of the visual variables based on user's interest and interaction history. Figure 7.25 illustrates a view on the data of another topic, which includes news articles. Thereby the user is logged-in as individual. The composition of the visual layouts, the choice of visual layouts, and the visual variables are adapted to the individual user model. The colors are the same as in the digital library application scenario to provide a coherent recognition effect, if the user works with different versions of SemaVis.

The work on this application scenario is still ongoing. During the investigation, we identified different roles in the modeling process of policies that should be considered in the adaptation process. Further we identified the necessity of simple one-dimensional visual layouts for quantitative data, thus the social opinions and their relationships would just fulfill a small part of the full set of requirements. Although, an enhanced version of SemaVis (see Fig. 7.26) [4] was applied for statistical data-bases, e.g. *Eurostat*, the requirements of the stakeholder goes beyond this data and visual layouts. The users in the domain of policy modeling differ in their prior knowledge,

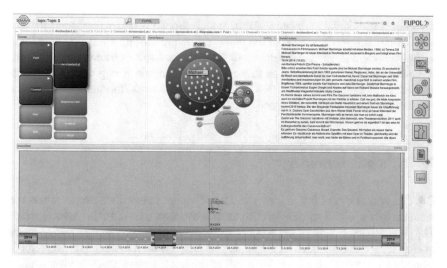

Fig. 7.25 Adaptation based on individual user model: visual variables, visual layouts, and their orchestration are adapted based on an individual user model

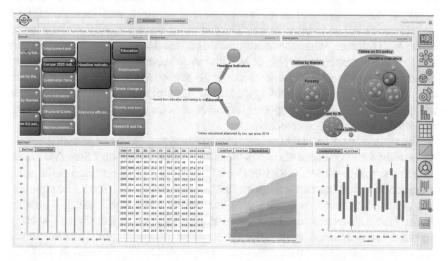

Fig. 7.26 Combined visualization of semantics and statistical data from the Eurostat data-base

interests, and intentions. But the main aspect is that they have further different political perspectives on the same issue. The goal should be to provide a more goal-oriented view on the facts and apply for instance process-oriented adaptation methods.

This section introduced SemaVis in the application domain of policy modeling. The main goal was to illustrate that the adaptive concepts can be applied even if further measurements affect the visual presentation. The application scenario illustrated further how the top-down exploratory information seeking is applied in SemaVis.

Therefore in particular the canonical user model is applied to adapt the visual interface and the visual layouts. Users' of the policy modeling domain are still able to login as individuals and get personalized views on topics from different social media channels. Although, the work on this application scenario is ongoing and requires further adaptation capabilities that are not focus our work, the main idea and the proof of feasibility were illustrated.

7.4   Chapter Summary

Our conceptual model introduced a variety of models, algorithms, approaches, and methods for the visual adaptation of semantics based on user and data characteristics. The detailed description of the conceptual model aimed at providing the replication of each integrated approach or model. In this chapter, we introduced the implementation of our conceptual model. We baptized the implemented adaptive visualization environment *SemaVis*, which was developed in an iterative manner. Our main goal in this chapter was to provide the proof of feasibility of our conceptual model based on SemaVis. We therefore introduced first the general architecture of SemaVis with the main focus on our conceptual model. The general architecture was described based on the MVC-pattern, whereas we outlined that this architectural design was chosen due to a more simple way of describing the main functionalities. The description of the entire architecture of SemaVis would go beyond the focus of this thesis. The description aimed more at providing the technical interplay of the various models and introduced the distributed computing character of SemaVis. SemaVis was designed as a distributed client-server technology.

The main adaptation functionalities are deployed to a dedicated SemaVis server, whereas the SemaVis client visualizes the result as Web-application. SemaVis can be used as a single-client technology or as a client-server technology without the dedicated server. In both cases the adaptive functionalities cannot be used. The common way of deploying SemaVis on Web is a client-server solution with multiple servers. Thereby the data are provided by other servers or data-bases and the SemaVis server measures the adaptation values. For the proof of feasibility, the SemaVis server plays the most important role. The main computations are performed in the Controller layer and stored in the Model layer. All components of the Controller layer and major parts of the Model layer are deployed on the SemaVis server. Based on this architectural design, we introduced three application scenarios with different approaches and goals. Therefore first the user interface design that was used in the application scenarios was introduced.

The UI-design of SemaVis was kept as simple as possible. The main area is the visual interface that enables the placement, rearrangement, and dismissing of visual layouts. A search area provides beside the search field, the choice of data-bases and special search features. The user login area enables users to login as individuals or deactivate the visual adaptation, and the visual recommendation area illustrates via icons various visual layouts arrange them based on the user model and the data characteristics. The visual layouts from this area can be dropped onto the visual interface.

After describing the user interface of the SemaVis client, we introduced our three application scenarios. The different application scenarios aimed at providing not only the proof of feasibility of our model, they further targeted to illustrate, in which contexts adaptive visualizations would make sense. One main application scenario was the use of SemaVis in context of digital libraries. We illustrated that not semantic data are enriched with semantics and adapted based on the canonical and individual user models. We could illustrate that major parts of our conceptual model were implemented and applied in this scenario that provided a full-adaptive visualization environment based on our reference model. The second main application scenario was Web-search. This application scenario makes use of two different data-bases simultaneously and applies major aspects of our conceptual model too. The third scenario was located in policy modeling domain and illustrated our ongoing work on visual adaptation. The main goal was to illustrate the top-down search in contrast to the full-adaptive scenarios of digital library and Web-search that use a bottom-up search approach.

With the introduced architecture and the different application scenarios, we could clearly prove the feasibility of our introduced conceptual model. Thereby a proof of the benefits of our conceptual model was not provided. In the following chapter, we will investigate the added values of our conceptual model and perform an empirical evaluation of the adaptive system. Thereby, SemaVis for digital libraries will be used for the main evaluation.

References

1. K. Nazemi, M. Breyer, D. Burkhardt, C. Stab, J. Kohlhammer, in *Towards the Internet of Services: The Theseus Program*, ed. by W.E.A. Wahlster (Springer, 2014), pp. 191–202
2. T. Becker, C. Burghart, K. Nazemi, P. Ndjiki-Nya, T. Riegel, R. Schäfer, T. Sporer, V. Tresp, J. Wissmann, in *Towards the Internet of Services: The Theseus Program*, ed. by W.E.A. Wahlster (Springer, 2014), pp. 59–88
3. F. Buschmann, R. Meunier, H. Rohnert, P. Sommerlad, M. Stal, *Pattern-Oriented Software Architecture: A System of Patterns*, vol. 1 (Wiley, 1996)
4. D. Burkhardt, K. Nazemi, C. Stab, M. Steiger, A. Kuijper, J. Kohlhammer, in *9th International Symposium on Advances in Visual Computing, ISVC 2013*. Lecture Notes in Computer Science (LNCS), vol. 8034 (Springer, Heidelberg, 2013), pp. 86–97. doi:10.1007/978-3-642-41939-3_9
5. P. Sonntagbauer, K. Nazemi, S. Sonntagbauer, G. Prister, D. Burkhardt (eds.), *Handbook of Research on Advanced ICT Integration for Governance and Policy Modeling* (Business Science Reference (IGI Global), Hershey PA, USA, 2014) (to appear)
6. K. Nazemi, M. Steiger, D. Burkhardt, J. Kohlhammer, in *Handbook of Research on Advanced ICT Integration for Governance and Policy Modeling*, ed. by P. Sonntagbauer, K. Nazemi, S. Sonntagbauer, G. Prister, D. Burkhardt (Business Science Reference (IGI Global), Hershey PA, USA, 2014) (to appear)
7. J. Kohlhammer, K. Nazemi, T. Ruppert, D. Burkhardt, IEEE Comput. Graphics Appl. **32**(5), 84 (2012). doi:10.1109/MCG.2012.107, http://bibcd.igd.fraunhofer.de/bibcd/INI_Science/papers/2012/12p079.pdf

8. Z. Shen, M. Ogawa, S.T. Teoh, K.L. Ma, in *Proceedings of the 2006 Asia-Pacific Symposium on Information Visualisation, APVis'06*, vol. 60 (Australian Computer Society Inc., Darlinghurst, Australia, 2006), pp. 93–102

9. T. Matthews, *Citation Map Visualizing Citation Data in the Web of Science* (Thomson Reuters, 2010). https://www.brainshark.com/thomsonscientific/citationmap/zCJzEJe9nz0z0

10. J.K. Chou, C.K. Yang, in *Proceedings of the 13th Eurographics / IEEE - VGTC conference on Visualization, EuroVis'11* (Eurographics Association, Aire-la-Ville, Switzerland, Switzerland, 2011), pp. 721–730. doi:10.1111/j.1467-8659.2011.01921.x, http://dx.doi.org/10.1111/j.1467-8659.2011.01921.x

11. J. Bernard, T. Ruppert, M. Scherer, J. Kohlhammer, T. Schreck, in *Proceedings of the 12th ACM/IEEE-CS Joint Conference on Digital Libraries, JCDL'12* (ACM, New York, NY, USA, 2012), pp. 139–148. doi:10.1145/2232817.2232844, http://doi.acm.org/10.1145/2232817.2232844

12. P. Bergstrom, D. Atkinson, in *Fourth International Conference on Digital Information Management, ICDIM 2009*, (2009), pp. 1–7

13. G. Tsatsaronis, I. Varlamis, S. Torge, M. Reimann, K. Nørvåg, M. Schroeder, M. Zschunke, in *Proceedings of the 15th international conference on Theory and practice of digital libraries, TPDL'11* (Springer, Heidelberg, 2011), pp. 15–26. http://dl.acm.org/citation.cfm?id=2042536.2042542

14. K. Nazemi, R. Retz, J. Bernard, J. Kohlhammer, D. Fellner, in *Advances in Visual Computing*, ed. by G. Bebis, R. Boyle, B. Parvin, D. Koracin, B. Li, F. Porikli, V. Zordan, J. Klosowski, S. Coquillart, X. Luo, M. Chen, D. Gotz Lecture Notes in Computer Science, vol. 8034 (Springer, Heidelberg, 2013), pp. 13–24. doi:10.1007/978-3-642-41939-3_2, http://dx.doi.org/10.1007/978-3-642-41939-3_2

15. K. Nazemi, D. Burkhardt, R. Retz, A. Kuijper, J. Kohlhammer, in *Advances in Visual Computing*, ed. by G. Bebis, et al. Lecture Notes in Computer Science (LNCS 8888) (Springer International Publishing, 2014), pp. 872–883. doi:10.1007/978-3-642-41939-3_2, http://dx.doi.org/10.1007/978-3-642-41939-3_2

16. P.N. Mendes, M. Jakob, C. Bizer, in *Proceedings of the Eight International Conference on Language Resources and Evaluation (LREC'12)* (Istanbul, Turkey, 2012)

17. Google Press Center. The Knowledge Graph (2013), http://www.google.com/intl/en/insidesearch/features/search/knowledge.html. Accessed Aug 2013

18. Freebase consortium. Freebase—a community-curated database of well-known people, places, and things (2013), http://www.freebase.com/. Accessed Aug 2013

19. Freebase consortium. Freebase api. build intelligent apps with freebase data (2013), https://developers.google.com/freebase/. Accessed Aug 2013

20. D. Burkhardt, M. Breyer, K. Nazemi, A. Kuijper, in *Universal Access in Human-Computer Interaction: Part I*. Lecture Notes in Computer Science (LNCS), vol. 6765, (Springer, Heidelberg, 2011), pp. 317–326. doi:10.1007/978-3-642-21672-5_35

21. G. Bouchard, S. Clinchant, W. Darling, in *Handbook of Research on Advanced ICT Integration for Governance and Policy Modeling*, ed. by P. Sonntagbauer, K. Nazemi, S. Sonntagbauer, G. Prister, D. Burkhardt (Business Science Reference (IGI Global), Hershey PA, USA, 2014) (to appear)

22. N. Rumm, B. Ortner, H. Löw, in *Handbook of Research on Advanced ICT Integration for Governance and Policy Modeling*, ed. by P. Sonntagbauer, K. Nazemi, S. Sonntagbauer, G. Prister, D. Burkhardt (Business Science Reference (IGI Global), Hershey PA, USA, 2014) (to appear)

23. K. Nazemi, D. Burkhardt, W. Retz, J. Kohlhammer, in *Advances in Visual Computing*, ed. by G. Bebis et al. Lecture Notes in Computer Science (LNCS 8887) (Springer International Publishing, 2014), pp. 333–344. doi:10.1007/978-3-642-41939-3_2, http://dx.doi.org/10.1007/978-3-642-41939-3_2

24. B. Shneiderman, in *VL* (1996), pp. 336–343

Chapter 8
Empirical User Study

The conceptual model, designed and implemented in context of this work, requires to be proved and validated. Furthermore, shortcomings and limitations of the approach should be enlightened to enable future work for the community on this topic. Thus, our approach is designed for use with heterogeneous users and provides a new way of human interaction with visual information the prove should be an empirical study on users. This chapter introduces the empirical study on our approach with an evaluation as a controlled experiment. We start with a general introduction into the topic of evaluation with a theoretical overview of the underlying psychological methods. In this context we will introduce some previously performed evaluations for similar applications. This should provide us with information to the right direction of evaluating our approach. Thereafter, a preliminary pilot study on evaluating only the effects of visual variables in context of information search will be introduced. The study was performed together with the psychological department of the Technische Universität Darmstadt. The main goal here was to find out, if the visual variables in terms of color and size have already an effect on search efficiency and enable us to identify appropriate questionnaires, limitations, and shortcomings.

The evaluation on SemaVis that focuses on our conceptual model will build the main part of this chapter. We will first introduce our assumptions on the benefits of the system and deduce hypotheses. Thereafter the evaluation method will be introduced. In this context we will describe our evaluation system that was developed in context of this work to provide an evaluation with full-automatic data-collection and reduced human intervention. This part will further introduce our group-design and Power Analysis to ensure that the number of participants is sophisticated for the evaluation. Thereafter, the collected data of the evaluation will be described together with the questionnaires and the entire procedure of the evaluation. The described method, procedure, and collected data should enable to prove the validation of the evaluation and enable to replicate the evaluation scenario.

© Springer International Publishing Switzerland 2016
K. Nazemi, *Adaptive Semantics Visualization*, Studies in Computational
Intelligence 646, DOI 10.1007/978-3-319-30816-6_8

338 8 Empirical User Study

The main evaluation was performed with the SemaVis application for digital libraries. This is due to the fact that the policy modeling scenario is ongoing work and does not include the entire spectrum of adaptation functionalities and the application scenario of Web-search makes use of two different data-bases. There exists no appropriate baseline for such an adaptive system. The results may be not confident enough.

One main part of the evaluation will be the introduction of results. This will include the number and demographic aspects of the participants, their visual limitations and of course the results of the evaluation in terms of measures for the different deduced hypotheses. This chapter will conclude with the discussion of the results in context of our hypotheses.

8.1 Foundations of Evaluating Adaptive Visualizations

The interactive visualization of information enables human access to increasing amount of data for solving a variety of informational tasks. As we worked out in this thesis, information visualization and in particular adaptive visualization is a prospering area of Human Computer Interaction and helps users retrieving and acquiring information and knowledge [1–4]. Beside aspects like design and implementation [5] or benchmarking techniques [6, 7], the evaluation of information and adaptive visualizations plays an increasing role in today's research [4, 8–12], due to the fact that information visualization builds the human interface to data, information, and knowledge. Komlodi and colleagues summarized the techniques on evaluating information visualization in the four main areas of *usability evaluation, controlled experiments comparing design elements, controlled experiments comparing two or more visualizations or tools*, and *case studies* [13, p. 2], [14]. Case studies, in which the users' task solving process is reported in their natural environment [13], are rarely used to evaluate visualizations. Thus, they are time-consuming and the results are not replicable [13]. Therefore, we will not investigate this type of evaluation in our thesis.

Usability testing or evaluation can be distinguished in two approaches depending on the prototypes' development progress: *formative testing* and *summative testing* [15]. Formative testing is used in early stages of the development to discover usability problems. Additionally, heuristics have been used in the past to evaluate information visualizations [16]. Heuristics comprise a set of usability guidelines, due to which deficits concerning usability can be detected. These problems can be a starting point for further formative usability evaluation [17]. In contrast to that *summative* testing aims at evaluating the application and reveal evidence for its goodness. The dominant method in summative testing are controlled experiments. This way confounding variables can be controlled by the experimental setting (e.g. [18]). However, well developed and reliable software is required [19]; otherwise the evaluation may be unsuited to provide proper results [20]. The conventional usability measures are effectiveness, efficiency, and satisfaction [21]. According to the ISO standard [21] *effectiveness* describes the accuracy of goal achievement, *efficiency* measures the relation of effort and effectiveness with respect to goal achievement, and *satisfaction* comprises perceived comfort as well as absence of discomfort. Aside from quantitative measures qualitative user data has been assessed in the past to discover patterns in users' behavior [22].

As already mentioned, formative testing aims to improve the usability of visualizations during the development phase, while summative testing compares different metrics of visualizations in controlled experiments [13, 15]. Komlodi et al. and Plaisant differentiate in the context of visualizations controlled experiments in comparing *design elements* and comparing *tools*. The comparison of design elements might include widgets or the mappings of data to certain visual layouts [13]. The comparison of tools is according to Plaisant the common type of study [13, p. 2]. It includes typically different visualizations and enables a summative evaluation on different visualizations. This way of testing goes beyond usability aspects and may involve metrics like perceptual speed, visual working memory, or verbal working memory [12].

Perceptual speed in context of visualization is the speed of encoding visual information. As we introduced in Sect. 2.2, commonly this is subdivided into preattentive information processing and attentive information processing [23–27]. One important metric for the evaluation of visualizations can be the speed in which certain information is perceived by users. Based on the fundamental works of Treisman [23, 28, 29], Wolfe [24, 30–32], Jun and colleagues introduced a visual information processing model for interactive visualization [33]. The model divides the processing of visual information into three stages: *feature extraction*, *pattern perception* and *goal directed processing*. Feature extraction is a low level process to extract main features from sensory information and is more related to the preattentive stage of visual information processing. If the information of interest is found (pattern perception), the following action is related to goal directed processing [33]. The model of Jun et al. is quite identical to that of Ware [25, 26] introduced in Sect. 2.2.2 and similar to that of Rensink [27, 34]. Another relevant concept for the perceptual speed is the *attentional weight*, a value assigned to each visual entity [35]. The attentional weight is associated with the strength of the sensory evidence and the context [35]. The similarity-choice theory suggests that the bigger the attentional weight of an object, the higher is the probability of choosing that object. Therefore homogeneous objects, as in a textual representation, have equal attentional weights. When using adaptive mechanism to visualize relevant objects differently, the probability of choosing them increases even more [36]. Within an application that uses visualization, the processing of visual information is faster and associated with less effort. Furthermore visualizations can make use of visual variables [37] like color, shape, size etc. to guide users' attention. Therefore, goal directed processing would takes less time.

Another important construct with respect to visual information processing is the visual working memory (*VWM*). As suggested by the biased competition hypothesis, VWM guides the allocation of attention [38]. It has been shown, that VWM is capable of storing an item during a task without influencing the task itself [38]. Other results show, that information present in the working memory is also possible to interfere with attention and lead to a lower performance on the task [39]. Lavie and Fockert also evidenced that the working memory interferes with the guidance of visual selective attention [40]. They discovered in their fMRI-study, that single distractors influence the performance in visual search tasks. Somervell et al. discovered that visualizations with high density require more attentional capacity than those with low density. The performance in a simple game was higher, when the visual search tasks were performed with respect to a visualization with low density [41]. Prior research has shown that features like position and presence are encoded differently from color. The visual memory stores those features automatically, whereas features like surface attributes are associated with a higher amount of required attention [42]. An adaptive visualization provides more guiding features than a static visualization, thus a visual search task is done with lower cognitive load [4, 12]. The cognitive load can be assessed through various methods. One way to measure the cognitive load is to perform multiple tasks parallel [43]. Another more direct way to assess the cognitive load is through EEG [11].

Verbal working memory refers to a measure of storage and manipulation capacity of verbal information [12]. In context of information visualization this subset of the working memory plays a secondary role [12, 44, 45] and is often not considered in the evaluation process. With respect to the measurement of systems' intuitiveness, the verbalization ability and the verbal working memory play a role [46, 47].

In summary, it can be said that there are established methods for evaluating information visualizations and proper metrics for evaluating their effectiveness, efficiency, and acceptance. The most common way to evaluate information visualizations with an advanced developing progress are summative evaluations, in particular by comparing the visualization tools in controlled experiments. Perceptual speed and visual working memory can be evaluated through the completion time of tasks compared to a baseline [12, 44, 45] and refer to effectiveness and efficiency of the visualizations. Furthermore, the aspects of satisfaction or acceptance can be gathered and evaluated through appropriate questionnaires that may involve the verbal working memory [46, 47].

In contrast to that the evaluation of adaptive visualizations are more complicated, due to the fact that the system learns from the user. To conduct an evaluation of adaptive visualization the participants have to work in long-term studies with the visualization system to train the underlying knowledge model. The effect of this process is obvious: while the adaptive visualization systems learn from the participant, the participant learns by interacting with the system. It is therefore difficult to measure which way of "learning" had higher effects. As mentioned, the studies have to be designed as long-term studies to train the system. Ahn conducted a study, in which the participants had about 50 min time to train the system with their individual profile [4]. After the training three search session were started followed by questionnaires [4, 48]. One main question arising from this procedure is how much did the participants learn during the training phase and how much did the system learn? A counterbalancing can be achieved by providing the same training session for an adaptive system and a non-adaptive system, but in this case "only" the adaptivity of the entire system can be measured without the effects that may occur by different levels of adaptation. Olson and Chun claimed that another problem with adaptive interfaces is that the spatial arrangement of items changes over time due to the adaptation effect [49]. This leads to losing the context in particular in repeated visual search tasks [49].

The effect of adaptive visualizations was evidenced by some user studies that measured the perceptual speed, visual working memory and verbal working memory [12, 44, 45, 50–53]. Toker et al. evaluated for example the perceptual speed, visual working memory, and verbal working memory with two different but equivalent visualizations regarding their information content, a bar and a radar graph [50]. They performed their study in three main steps: a cognitive test containing questionnaires about the participants' working memory (visual and verbal) and perceptual speed, followed by search tasks, and concluded with post-questionnaire [50]. The task were subdivided into "common search task" and "complex search task" (double scenario tasks). A complex search task had more the characteristic of exploratory tasks based on the model of Marchionini [54] or White and Roth [55]. An example for their complex task was *"Find the courses in which Andrea is below the class average and Diana is above it?"* [50, p. 277]. Their study measured three main aspects, task *completion time*, *visualization preference*, and *ease-of-use*. The task-completion-time was recorded by their software, the visualization preference was asked with a five-item question (from like to dislike), and the ease-of-use with five-item questions about the "understanding" of each visualization (from easy to understand to difficult to understand) [50]. The results of their study showed that bar graphs lead to faster task-completion-time for simple tasks. In more complex tasks no significant difference between the visualizations could be observed [50]. Further, the results showed a high effect of perception speed on task-completion-time when dealing with simple tasks. They discovered an interaction between the task and the visualization regarding the relation between perception speed and task-completion-time [50]. Although bar graphs lead to quicker task completion in accordance with other studies [56, 57], the task-completion-time was more affected by perception speed when participants used radar graphs [50]. Expertise had no influence on the performance in the simple task condition. Further, a positive effect of perception speed on task-completion-time was also revealed in the context of complex tasks [50]. However, there was no interaction. In contrast to simple tasks, expertise had an influence on task-completion-time. This effect was also present, when the expertise was related to the other visualization [50].

Based on the introduced studies and theoretical foundations, it is obvious that the evaluation of adaptive visualizations should be performed as controlled experiments with different conditions, on design or tool level. The main question remains how to train the user model without providing a learning effect for the participants? Further, it is more than relevant to evaluate the different aspects of adaptation and their effect to task completion and acceptance. The different aspects or layers of visual adaptation can be differentiated as described in Sect. 6.3 into *semantics and content*, *visual layout*, *visual variables*, and *visual interface* adaptation. Thus commonly the evaluation of adaptive metrics is performed only on the visual layout level [12, 44, 45, 50]. Another main aspect is the measurement of the task completion with different level of task complexity according to the exploratory search definitions [54, 55]. Further it should be expected from adaptive visualizations compared to static ones that the *cognitive load* is reduced, due to the adaptive characteristics.

8.2 Preliminary Study

The introduced studies on research of adaptive visualizations evidence that the adaptation of the visual layout improves the task-completion-time for simple search tasks [50]. Evaluations and empirical studies on the visual variable level, e.g. color or size is rarely performed to evidence the effect of parallel information processing in adaptive visualizations. Although, there exist several studies that evidence the "guidance characteristics" of those variables [24, 29] a controlled experiment by comparing two equivalent visual interfaces with and without adaptive visual variables should reveal the evidence that the visual variables support the user in simple search task by guiding the attention to the main search result. We therefore propose the hypothesis that the use of adaptive visual variables, exemplary with size and color, improves the performance of simple search-related tasks. We expect performance metrics to profit from the visual variable adaptivity of a visual interface.

Hypothesis 1: *The adaptation of visual variables in visualization environments provides a simpler access to required information than without an adaptive behavior. We there expect that more tasks are completed successfully in the same time.*

Beside performance, the way of presenting information influences the satisfaction, perceived by the user. Visualization of search results leads to high user-satisfaction [58] as well as visualizations of blog archives [59]. This effect was also experimentally verified in the context of problem solving tasks [60]. It has been shown, that visualizations containing color-term highlighting yield to higher user satisfaction even when not increasing the performance [61]. Chittaro and Combi [62] evaluated a visualization for temporal information. Different types of visualization were compared and although there was no evidence for a better user-performance, one of the visualizations was significantly rated better [62].

Visual aesthetics of a web-based application are closely related to perceived satisfaction. Lavie and Tractinsky addressed the importance of aesthetics by developing a measurement [63]. Cawthon and Moere discovered that perceived aesthetics also yield to better performance metrics, like effectiveness, when dealing with a visualization application [64]. Visual complexity is related to perceived aesthetic, but familiarity has the biggest effect [65]. Based on those findings, we expect the perceived satisfaction to be higher among users working with an adaptive visualization tool.

Hypothesis 2: *The adaptation of visual variables in visualization environments leads to a higher satisfaction and acceptance than visualization environments without adaptive visual variables. Hence, we expect participants give higher ratings with respect to usability dimensions.*

The goal of our preliminary study was to determine whether the adaptive behavior in terms of adaptive visual variable (color, size etc.) improves the usability of the application. The adaptive behavior, described in [66, 67], is supposed to help users get to the information of interest. To evidence our hypotheses, we conducted a preliminary user study that just involved the adaptation visual variables in a static visual environment. The adaptation was performed with a canonical user model (see Sect. 6.1.3.4) that was trained by a number of users, who searched for the terms before the study was conducted. With this model the relevant information was highlighted within the visualization by color and size. The study was conducted as a between subject design with ground-truth data of Freebase [68], a data-base for Linked-Open-Data. The participants were randomly distributed over two experimental conditions: 1. *visualization environment with adaptive visual variables*, 2. *visualization environment without adaptive visual variables*.

8.2.1 Method

Group Design

The study was conducted as a between subjects design. The participants were randomly distributed over two experimental conditions. Although each participant had to answer the same questions, the conditions differed in the tools applied in order to fulfill the tasks. Each condition is described below.

Condition 1 Under this condition the participant used SemaVis as visualization while the adaptive behavior was completely disabled (see Fig. 8.1a). The application

(a)   **(b)**

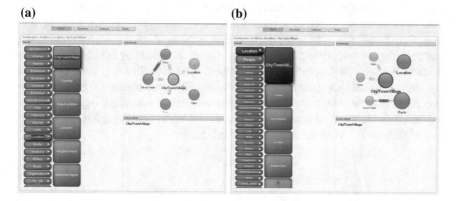

Fig. 8.1 Conditions of the preliminary user study. **a** SemaVis with fixed visualizations and without adaptive behavior. **b** SemaVis with adaptive behavior enabled

Table 8.1 Sample questions for solving the tasks in the preliminary study

Topic	Question
Paris	Where and when was Gustave Eiffel born?
Albert Einstein	What was Albert Einstein's cause of death and where did he die?
Vatican	Name three buidings in Vatican City
Plato	Name one person who died in Athens and two persons who were born there

contained of the three fixed visualizations *SeMap*, *SemaGraph*, and *SemaContent*. In this condition no further visualizations could be added nor could the presented visualizations be closed. However, the size of the single visualizations was editable. The top part of the application showed buttons for each topic of the tasks to provide a simpler access. In addition, this approach unified the initial search by contrast to using a common search field. The breadcrumbs navigation above the visualizations gave access to the recent history of the interaction.

Condition 2 The second group performed the same tasks by using the same constellation of visualizations, but with enabled adaptive behavior of the visual variables (see Fig. 8.1b). The adaptive visual variable used an initially trained user model. The user model was trained by about 20 persons who did not participate the study but were asked to solve the same questions. The visualizations just changed two visual variables with respect to the underlying canonical user model: color and size. All other parameters were exactly the same as in the first condition.

Tasks

The participants were asked to answer 20 questions. Each question belonged to one of the following topics: *Paris, Albert Einstein, Vatican,* and *Plato*. Table 8.1 shows a sample question for each topic. The tasks were constructed such that they were answerable in each condition. Some tasks could be accomplished within a single visualization, other required additional steps. In addition to that the tasks differed in the number of questions, e.g. the sample task related to the keyword Paris shown in Table 8.1 contained of two questions, whereas the sample task related to *Vatican* was answerable with only one question. Due to tasks performance metrics were assessed, the effectiveness could be measured as the ratio between correct and wrong answers [69]. In this study effectiveness was operationalized by the absolute amount of correctly answered questions in predefined time slot. Since all participants had the same time slot for answering the questions, the measurement was focused on correctly answered questions. The efficiency of a search tasks can be operationalized by the mean duration of a search. Measures related to task-completion-time were extracted from the log-files described in Sect. 8.2.2.

8.2.2 Collected Data

In order to validate the hypothesis two main aspects were collected. The answered questions were collected as a printed form to be filled out while interacting with the system and two questionnaires. The printed form consisted of the questions to be answered and empty areas for the responses. The participants were asked to read the question on paper, search and interact with the system, and write down the answer in the printed form. The number of answered questions and the number of correctly answered questions was assessed to measure the efficiency of the conditions. Further, the interactions of the participants were logged by the system to get the amount of interactions, find out which visualizations were used by the participants, and in particular if the answers were given based on a real search and interaction with the system. The content of the generated log-files is shown in Table 8.2. By analyzing the interaction additional variables (time spent to complete a certain task, derivation from optimal path etc.) were extracted. Further potential problems were detected, that could have influenced the performance of participants (e.g. long loading delays).

Hypothesis 2 predicts a higher level of satisfaction perceived by users in the adaptive visualization condition, hence satisfaction had to be operationalized and assessed. Satisfaction was measured through self-assessment methods. Prior research has shown the impact of computer abilities on the performance, therefore the level of computer literacy was assessed as a confounding variable [50, 70]. After the tasks were solved within a predefined time, the participants were asked to fill out two questionnaires, one gathered information about the intuitiveness of the system and the other the self-assessed computer-literacy.

Intuitiveness

Intuition is a process with the following characteristics: It is fast and effortless, unconscious, based on gut feelings and not explainable [46]. Ullrich and Diefenbach constructed based on prior definitions of intuitiveness from the areas of decision making, Human-Computer Interaction, usability, and user experience, the INTUI-questionnaire that measures the intuitiveness of products and applications [46, 47]. The questionnaire contains four relevant *subcomponents* (sub-scales): *Effortlessness*, *Gut Feeling*, *Magical Experience* and *Verbalizability*. The sub-scales are surveyed with a set of 16 seven-point semantic differential items [46]. The questionnaire contains a further seven-point semantics differential item for the global sub-scale

Table 8.2 Logfiles containing the user interactions

Part. no.	Timestamp	Interaction	Visualization	Content
1_DR13	13824142891	Mouse.button.left.click	SemaVis.SeMap.Concept	Location.CityTown
1_MK31	13824144701	Mouse.button.left.click	SemaVis.SemaGraph.Individual	Location.CityTown.Paris
2_OW14	13824150223	Keyboard.type	Sema Vis.searchbar	Einstein
2_AA12	13824178163	Mouse.button.left.click	SemaVis.SeMap.Concept	People.Person

Intuitiveness. The first version of the INTUI-questionnaire [46] included the sub-scale *Attention*, which was added in their newer version [47] to the sub-scale *Effortlessness*.

Effortlessness measures the amount of cognitive effort needed during the interaction with products or applications. This dimension is closely related to attentional resources. In the development process of the INTUI questionnaire, the initial sub-scales Effortlessness and Attention were combined to Effortlessness. The authors argue that intuitive responses are reached with little apparent effort and typically without conscious awareness ([71] in [47, p. 802]). Sample items for this sub-scale are *Using the product—was difficult / ...was easy* or *...required my close attention / ...ran smoothly*. Overall 5 out of 17 items refer to the sub-scale *Effortlessness*. **Gut Feeling** describes whether the decisions made by the product's user are based on intuitiveness or on reasoning. Gut Feeling refers thereby to the process of decision making that is performed more visceral without thinking about an issue. Sample items for this sub-scale are *While using the product... ...I was guided by reason / ...I was guided by feelings* or *......I acted deliberately / ...I acted on impulse*. Overall 4 out of 17 items refer to the sub-scale *Gut Feeling*. **Verbalizability** assesses how accurate the interaction with the product can be described after the interaction. Originally, this dimension has its foundation in the theory of intuitive decision making [46]. According to this, intuitive decisions go along with a reduced amount of Verbalizability. Sample items for this sub-scale are *In retrospect...—...it is hard for me to describe the individual operating steps / ...I have no problem describing the individual operating steps* or *......I can easily recall the operating steps / ...it is difficult for me to remember how the product is operated*. Overall 3 out of 17 items refer to the sub-scale *Verbalizability*. **Magical Experience** assesses the emotional component of the interaction arising from the difference between the users' expectations and the true interaction. Magical experience refers according to Ullrich and Diefenbach resulting not expected feeling that makes the use of a product exceptional [47]. Sample items for this sub-scale are *Using the product—was inspiring / ... insignificant* or *......was trivial / ...carried me away*. Overall 4 out of 17 items refer to the sub-scale *Magical Experience*. The sub-scale **Intuitiveness** is measured by one main item (out of 17) that asks the user directly, if the usage of the product was intuitive on not.

The questionnaire contains 17 seven-point semantic differential items and shows good reliability [46, 47]. Correlations between the four factors of INTUI and the free factors of the self-assessment *Manikin* suggest that Magical Experience and Gut Feeling correspond to positive emotions, whereas Effortlessness and Verbalizability correspond to missing negative emotions [47]. The factor Effortlessness is more important to devices that are used on the daily bases for their functionality (usability). Gut Feeling shows higher values if the fun aspect of the products is focused. Verbalizability is positive correlated to Effortlessness and negative correlated to Magical Experience [47].

Computer Literacy Chen and Yu [72] performed a meta-analysis to examine the influence of users' cognitive abilities on efficiency and accuracy when dealing with visualizations. High abilities lead to better performance. They therefore recommend assessing cognitive abilities when performing an evaluation. The INCOBI [73] is a questionnaire for computer literacy and attitudes towards computers [73]. For this study only the sub-scale *Fragebogen zur inhaltlich differenzierten Erfassung computerbezogener Einstellungen* (Questionnaire for content-differentiated gathering of computer-related attitudes—FIDEC) was assessed. *FIDEC* contains three dichotomous constructs and measures computer-related attitudes [70, 73].

8.2.3 Procedure

The study took place at a laboratory of the Department of Psychology of the "Technische Universität Darmstadt" (University of Darmstadt). All participants performed the tasks on a 19" monitor using a mouse equipped with a mouse-wheel and a keyboard as interaction devices. The participants received an instruction paper; where they were ask to generate their ID for the study. The IDs were composed through the following algorithm: The second letter of the father's first name was taken as the first symbol. The second character was the first letter of the mother's maiden name. The two last symbols were taken from the participant's birthday—the two digits, that specify the day (see Table 8.2 first column).

In the next step the participants were asked to input the generated participant's ID into a form. After confirming their ID, the distribution to the two groups was done randomly using a random function in SemaVis. Subsequently, the participants were asked to turn to the page in the instruction paper specified by the number presented on the screen, where they received a brief introduction in the system. Once all participants had finished reading the instructions and confirmed this by upholding their hand, the experiment started. The participants had 15 min to perform the given tasks. The tasks contained four topics with five questions per topic. The answers were written down by the participants next to the questions presented on a printed form. After the 15 min had expired, the participants filled out the two mentioned questionnaires. Figure 8.2 illustrates the procedure of the preliminary study.

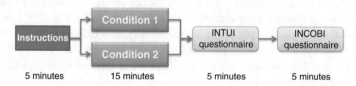

Fig. 8.2 Procedure of the preliminary user study

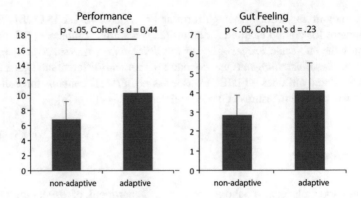

Fig. 8.3 Results of the preliminary study [67]

8.2.4 Results

A total number of 14 students (10 female) of the department for psychology participated in the study. The age ranged from 19 to 28 ($M = 23.9$, $SD = 2.7$). The average
usage of the computer was 21 hours per week ($SD = 28$) since 145 ($SD = 59$) months.
The mean number of programs used on a daily basis was 18 and ranged from 2 to
150. The INTUI questionnaire showed acceptable reliability ($\alpha = 0.95$).

In order to evaluate the hypothesis several t-Tests were computed. A t-Test is a
statistical procedure to compare two groups with regard to their means. The calculated
test-statistic follows a student's t-distribution. The requirements for the conduction
of t-Tests are normally distributed and interval scaled data. Both requirements are
fulfilled by the assessed data.

The effects of the condition are shown in Fig. 8.3. Hypothesis 2 was confirmed by
the data. The participants using the SemaVis with adaptive visual variables were able
to answer more question correctly than the participants using the SemaVis without
adaptive visual variables, ($p = .0385$). This result was expected, hence a one sided
test for significance was performed. Also participants in the condition SemaVis with
adaptive visual variable scored significantly higher on Gut Feeling ($p = .365$), as
expected, they perceived their actions more guided by gut feeling (Fig. 8.3: right).
The sub-scales Magical Experience, Verbalizability, and Effortlessness showed no
significant differences between the conditions. We assume that the small size of
participants and the changes in attention between a printed form and the visual
environment led to these factors.

8.2.5 Discussion and Limitations

Our preliminary study investigated the effect of visual variables adapted to the search
behavior of users regarding effectiveness and satisfaction. Although, the number of
participants was limited to just 14, an effect regarding task completion and thereby

to effectiveness could be observed. The first hypothesis could be validated. Adaptive visual variables lead to simpler access to information. The effect was expected due to the outcomes of the studies in visual perception, in particular in preattentive information processing. The second hypothesis that adaptive visual variables leads to higher satisfaction and acceptance was measured with the INTUI questionnaire. One of four dimensions could be confirmed with the preliminary study. While Effortlessness, Verbalizability, and Magical Experience showed no significant improvements, Gut Feeling was affected by the adaptive behavior. One reason could be the small sample size or the method of evaluation. Thus, the preliminary study had several limitations: The small sample size leads to big standard errors. In addition, smaller effects could not be investigated. Another limitation was the procedure of the evaluation. The answers were written on a printed form, therewith the attention of the participants changed from the computer-aided visualization to paper and vice versa. Further, no temporal information (e.g., task-completion-time for each task) was assessed and only simple search task were considered. But in general the main goal of the preliminary study was successful, thus it could be evidenced that visual variables like color and size has an effect on the effectiveness of task completion. The limitations of the study do not allow proposing this effect for general cases. Therefore, an empirical study was performed that made use of all adaptation effects, considering different task types, more conditions, no printed forms for questions, and with a textual search interface as baseline. The empirical study is described in the following Sect. 8.3.

8.3 Evaluation of SemaVis

The conceptual design and proposed models described in Chap. 6 require a more sophisticated evaluation that validates the proposed advantages. The preliminary user study already showed potentials by adapting the visual variables based on a previously modeled user in simple search tasks. But the study had various limitations and can therefore not be used for validating the conceptual model of SemaVis. In particular the aspect of data-based adaptation, the adaptation of the different layers of visualization, and exploratory tasks were not considered in the preliminary study. Beside these limitations, we could identify shortcomings in the design of the study. Further the small number of participants with relatively similar background led to small statistics power.

To evaluate the main propose of our conceptual model, namely that adaptive semantics visualizations as described in Chap. 6 lead to a more effective knowledge acquisition in terms of amplifying cognition, faster access to information, and more acceptance of a visualization environment by non-expert users, we conducted an evaluation that aimed to minimize the limitations of the preliminary user study and maximize the proof of our conceptual model. The goal of the evaluation is to evidence that adaptive semantic visualization simplifies the human information search process and in particular amplifies the higher cognitive search tasks as defined by [54]. Existing systems do not provide an adaptation of the entire spectrum of the visual layers. The main goal of the evaluation was therewith is to illustrate that the combination of adapting semantics and content, visual layouts, visual variables, and visual interfaces based on users' interaction behavior and data characteristics leads to solving lower and higher cognitive search tasks faster than without adaptation [54, 55, 74].

In general the evaluation of SemaVis should consider following main aspects to provide a valid proof of our model:

- The system should make use of the adaptation capabilities of the different visualization layers (content (semantics), visual layout, visual variables, and visual interface)
- The system should make use of different levels of task complexity as proposed by Marchionini [54]
- The system should make use of a well-trained user model regarding data and visual layouts adaptation as proposed in the conceptual model
- The system should make use of ground-truth data
- The evaluation method should make use of a computer-based evaluation tool to ensure that the participants' attention is predominantly on the screen
- The evaluation should involve a valid and higher representative number of participants
- The evaluation should avoid learning effects by applying an appropriate randomization method
- The evaluation should use a common textual interface and further non-adaptive visualizations to provide a proper baseline and proper conditions
- The evaluation should make use of automatic data collection to reduce human error

For evaluating the full-adaptive SemaVis, we used "ground-truth" data of the Eurographics Digital Library [75] and the associated application scenario of SemaVis for digital libraries (see Sect. 7.3.1). We chose this application scenario because it was the only one that used one data-base for visualizing search results with all the above mentioned adaptation criteria. SemaVis for digital libraries implements major aspects of our conceptual model and is consequently the appropriate application scenario to prove our conceptual model. We hypothesized that less cognitive resources are needed for information processing in working with adaptive semantics visualization compared to other forms of information presentations. The evaluation contained four different ways of information representation that were evaluated in a within-subject design randomized by a Latin Square [76] to avoid possible learning effects as far as possible. The baseline of our evaluation was a textual representation of the ground-truth digital library data with real-time access via a Web-interface. The textual interface of the digital library was modified to provide exactly the same information as in the conditions with visual interfaces. Further, an evaluation tool was developed to gather all the collected data automatically and provide the tasks on the same screen as the participants were asked to solve them. The evaluation tool further contained the used questionnaires. The data from the questionnaires were collected automatically too. To reduce the human interventions, we further included the instructions for evaluation in the tool. This was to ensure that all evaluation settings have at least a similar precondition for all participants. The calculation of the total sample size (number of participants) was performed with GPower 3 [77, 78] by Power Analysis.

This section describes the method, procedure, and results of our conducted evaluation. Further, the results and impacts of the results will be discussed. The goal is to provide a valid and replicable evaluation for proving our conceptual model and its added values. Therefore, first main research questions are formulated as hypotheses to be evaluated. In the following part the evaluation method and used system will be described. This includes the classification of tasks and the identification and illustration of the conditions. Thereafter, the procedure of the evaluation will be described followed by a discussion of the results.

8.3.1 Hypotheses

The main research goal of our work can be summarized as an adaptive semantics visualization that enables solving exploratory search tasks with less cognitive effort by using different methods and algorithms for data and user analyzing and adapting four layers of visual representation to the generated model. This simple and abstract research goal has two main points that should be worked out for the hypotheses. First, the use of data characteristics and data model in combination with users' behavior that affects the adaptation process, and second the combined adaptation of content and semantics, visual layouts, visual variables, and visual interface, as described in the conceptual model. We assume that if a visualization environment makes use of both kind of influencing factors, data and user, and adapts the visualization on the named four layers, the adaptation process leads to general lower level of cognitive effort, due to the adaptation effects. By lower cognitive effort, the human's effort is meant that occur during solving problems and tasks. We consider in that context two levels of tasks according to Marchionini [54] and Bloom [74]. The simple search tasks are characterized by low effort in general. They can be solved without comparing, investigating or learning. Further, we summarize the two levels of exploratory tasks as those tasks in which the aspect of comparison, investigation or learning occurs. This kind of tasks needs commonly more than one search iteration and are performed sequentially. The cognitive effort or cognitive load for solving these tasks is generally higher [12, 44, 50, 54, 55, 74].

In this context the dependent variables can be identified as the time that is needed (efficiency) for completing a task correctly (effectiveness) according to the main goals of visualization proposed by Card et al. that visualizations should enable human to interpret faster, distinct graphical entities, or make to fewer errors [3, p. 23] (see Sect. 2.1). Further, the use of auxiliaries (paper and pen) that are not components of the tasks or the system leads to the assumption that a task needed higher cognitive efforts. For example, if a visualization system is not able to provide the right visual representations for certain complex or comparative tasks, so that the cognitive load is higher that the working memory of the user, they will make use of paper and pen for remembering and noting interim results of the tasks. In other words the fluent solving of a complex task without using any auxiliaries leads to the assumption that the cognitive working memory load was lower [38, 74, 79].

Other dimension for evaluating the mentioned aspects are acceptance and satisfaction of users [21]. In this context two aspects plays a key-role: Intuitiveness and User Experience. The intuitive interaction with a system leads to the assumption that less cognitive effort and load was required for solving the task [46, 47]. Intuitiveness is therefore a key-factor for evaluating the acceptance of a visualization system. User Experience is according to Hassenzahl [80] more related to hedonic values of computational systems and indicates the "joy-of-use". In this context in particular the attractiveness of system plays a role that can be measured too [80].

We expect that our full-adaptive system will enable users to solve search both types of tasks *correctly* (simple and exploratory) *faster* than with textual representations (baseline), non-adaptive, and adaptive visualizations that just adapts the visual variables.

Therefore we assume that:

Hypothesis 1 [H1]: *The full-adaptive semantics visualization enables a more efficient way of solving tasks compared to other information representations. We there expect that the tasks are solved faster with the full-adaptive visualization compared to other information representations and partially adaptive visualizations.*

We further assume that the above hypothesis is valid for both kinds of tasks. Therefore we subdivide the hypothesis into the following two sub-hypotheses:

Hypothesis 1-1 [H1-1]: *The full-adaptive semantics visualization leads to smaller task-completion times of simple tasks compared to other information representations.*

Hypothesis 1-2 [H1-2]: *The full-adaptive semantics visualization also leads to smaller task-completion times of exploratory tasks compared to other information representations.*

Our second hypothesis addresses the correctness of solved tasks as proposed by Card et al. [3] that visualizations should lead to fewer errors:

Hypothesis 2 [H2]: *The full-adaptive semantics visualization enables a more effective way of solving tasks compared to other information representations. We there expect that more tasks are solved correctly compared to other information representations and partially adaptive visualizations.*

The correctness of tasks can be determined by two complementary measures: the number of correctly solved tasks and the number of omitted tasks. The number of omitted tasks provides information about those tasks that were either too complex to solve or were not understood correctly. We therefore assume that the hypothesis [H2] is valid, if the following two sub-hypotheses are verified:

Hypothesis 2-1 [H2-1]: *The full-adaptive semantics visualization leads to a higher percentage of correctly solved tasks (simple and exploratory) compared to other information representations.*

Hypothesis 2-2 [H2-2]: *The full-adaptive semantics visualization leads to a smaller percentage of omitted tasks (simple and exploratory) compared to other information representations.*

Another main purpose of our visual adaptation approach is to reduce the cognitive load of users by providing sufficient adaptations on the different visual layers. We therefore assume that:

Hypothesis 3 [H3]: *The full-adaptive semantics visualization consumes less cognitive resources while solving tasks compared to other information representations.*

The measurement of consumed cognitive resources can be determined in various ways. To enable an investigation from different perspectives, we enhance [H3] with the following sub-hypotheses:

Hypothesis 3-1 [H3-1]: *The process of tasks solving with our full-adaptive semantics visualization requires less use of further auxiliaries (pen and paper) than by using other ways of information representations. We there expect that less paper and pen auxiliaries are used for solving the same tasks with the full-adaptive semantics visualization than with other forms of information representations.*

The use of additional auxiliaries like paper and pen is one way to measure in particular the needed cognitive resource regarding the working memory, thus small information units and comprehensible views on data should lead to a seamless use of the visualization system without the need for further auxiliaries. A more formal way to measure the use cognitive resources and the usability of products is the sub-scale *Effortlessness* of the INTUI-questionnaire [46] (see Sect. 8.2.2). We therefore further assume that:

Hypothesis 3-2 [H3-2]: *Working with the full-adaptive semantics visualization is perceived as more effortless compared to other information representations. We therefore expect the ratings of the INTUI sub-scale Effortlessness to be higher with the full-adaptive semantics visualization than with other forms of information representations.*

Beside performance measures like the efficiency and effectiveness of our conceptual model and the use of cognitive resources, the aspects of users' satisfaction and acceptance is important for our evaluation. In general, we expect that the ratings for our conceptual model in terms of acceptance are higher. We therefore hypothesize that:

Hypothesis 4 [H4]: *The full-adaptive semantics visualization leads to a higher satisfaction and acceptance than other information representations. Hence, we expect participants give higher ratings with respect to acceptance and satisfaction dimensions.*

To measure the acceptance and satisfaction, we will use two different questionnaires that measure on the one hand the intuitiveness of a system and the attractiveness. We therefore sub-divide our hypothesis in the following three sub-hypotheses to ensure that the evaluation is replicable:

Hypothesis 4-1 [H4-1]: *The full-adaptive semantics visualization leads to a more intuitive usage than other information representations. Hence, we expect participants give higher ratings with respect to the dimension Intuitiveness of the INTUI questionnaire.*

Hypothesis 4-2 [H4-2]: *The full-adaptive semantics visualization is associated with a higher hedonic quality compared to other information representations. Hence, we expect participants give higher ratings with respect to the dimensions Hedonic Quality of the AttrakDiff questionnaire.*

Hypothesis 4-3 [H4-3]: *The full-adaptive semantics visualization is perceived as more attractive than other information representations. Hence, we expect participants give higher ratings with respect to the dimension Attractiveness of the AttrakDiff questionnaire.*

The introduced four hypotheses with their sub-hypotheses summarize our expectation of the main contributions of our model.

8.3.2   Method

The preliminary study described in Sect. 8.2 enlightened several shortcomings and limitations on the procedure of the study. Therefore it was more than important to provide a sophisticated and replicable evaluation on the full-adaptive visualization. This section describes the underlying method, tasks, and tools used to endure such a valid evaluation with less error-proneness.

8.3.2.1   Evaluation System

One of the main shortcomings of the preliminary study was that the participants had to fill out printed forms for the underlying tasks. This procedure may lead to human errors and to lack of attention, due to the change of human's attention between paper and the visual environment. To face this problem we developed an evaluation system that guides the user through the entire evaluation scenario and takes over the some part of the experiment leader role. The evaluation system was designed and implemented as a user interface framework that enabled logging every interaction of the user regardless which system is evaluated. This was required due to the baseline of our evaluation that is a textual representation similar to conventional search engine human interfaces. The visualizations, questionnaires, and instructions were completely embedded into the evaluation system. There was no need to switch between

(a)   **(b)**

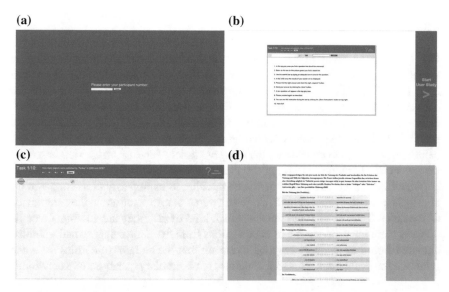

Fig. 8.4 Screenshots of our implemented evaluation system. **a** Entrance and participants number. **b** Exemplary instruction page. **c** Main panel with task (*top*) and information representation (*down*). **d** Exemplary questionnaire page

the experiment administration, tasks, and the visualization. The system comprises different parts, and the procedure can be defined with an underlying XML-file.

It contained in the evaluation instruction pages (see Fig. 8.4b), wherever necessary. The main window (see Fig. 8.4c) comprised two major parts: The upper panel contained the administration of the evaluation and the lower panel contains the system being evaluated. The administration unit contained the current task to be solved with a multiple choice of possible answers. Further the participants had the possibility to see whenever they want, the main instructions again. In this period of reading the instructions, the measured time was paused. In general every step of the evaluation was logged and the time for each step was saved together with the log-files. The participants had always the task and the search environment on the same screen. It was not necessary to switch the center of attention. However, due to the complex character of some selected tasks, we expected that some participants make use of paper and pen for solving the tasks and answering the questions.

Our evaluation system has advantages in comparison of providing the tasks as printed forms: First, additional performance measures (e.g., task-completion-time) can be assessed. Second, the participant's attention stays focused on the screen throughout the whole experiment. Third, the results can be processed automatically with respect to reducing both effort and error-proneness of the data assessment. Fourth, by handling almost all interaction with the participants through the framework confounding effects of the experiments administrator are minimized due to automation. The evaluation framework enables automated conduction of complex evaluations.

8.3.2.2 Tasks

One of the main goals of our study was to find out, if our approach of adaptive visualizations will affect the exploratory tasks as well as simple so called *Lookup* tasks [54]. Therefore we investigated the ground-truth data to identify ten tasks for each condition, five should be simple to solve and five should have a more exploratory character. We therefore categorized the tasks according to [37, 54, 55] in "explicit search tasks" and "exploratory search tasks". To have a more concrete view on the exploratory tasks, we considered the *compare* tasks of the *Learning* step based on the model of Marchionini [54] and the *analysis* from the *Investigate* stage [54, 74].

We differentiated in our system the tasks either as exploratory or as simple to be able to measure the effect of both. The users were not instructed about the complexity of the tasks. They even could not see which kind of task they are solving. Overall, 40 tasks were identified. Each of the questions from the same category had a similar amount of entities as query respond to ensure that the amount of entities does not influence the evaluation. Table 8.3 illustrates ten exemplary questions of a condition.

Table 8.3 Exemplary tasks of the evaluation

Exemplary tasks of the evaluation		
Question	Answers	Type
How many papers were published by Fellner in 2008 and 2009?	7, 5, 6, 4	Simple
In which year had Sadlo the first publication?	Picture/Image generation, applications, graphics systems, database applications	Exploratory
In which year had Sadlo the first publication?	2005, 2006, 2010, 2013	Simple
Which author has the most publications in the area of Visual Rendering?	Walker, Kohlhammer, Keim, Chalmers	Exploratory
How many papers about Semantic Visualization were published?	3, 4, 7, 1	Simple
Which of the named topics gained recently more interest in research?	Geometry rendering, 3D rendering, GPU rendering, Detail rendering	Exploratory
With whom had Kohlhammerin 2013 no publications?	Davey, Steiger, May, Ertl	Simple
With how many different co-authors had Schwenk publications?	6, 4, 7, 5	Exploratory
In which year had Ritschel the most publications?	2004, 2012, 2009, 2010	Simple
Which of these authors had after 2010 fewer publications than before 2010?	Fellner, Kohlhammer, Miksch, Kuijper	Exploratory

The entire tasks that were randomized over the conditions can be found in the Appendix D. We chose as respond choices four possible answers, whereas exactly one is correct. The evaluation system provided therefore a single-choice to indicate that only one respond is correct. Most tests make use of four to five respond options in such a multiple-choice scenario [81, 82]. Various studies have shown that three options are a convenient choice for evaluation. Previous studies discovered that the mean time necessary to answer the item increases with the number of distractors. There have been no significant differences in reliability between three, four or five options. *Baghaei* and *Amrahi* recommend three items per question due to the fact, that additional options increase the effort while not improving the test [81]. Other authors also suggest three alternatives to be sufficient (e.g. [82]). The evaluation system contained the information if a task is exploratory or simple to enable differentiated measurements of the effects. Further for each task the task-completion-time was assessed by the evaluation system and saved together with the information if the task was solved correctly or not in the log-files.

8.3.2.3 Group Design

The evaluation was conducted in a within-subjects design randomized by Latin Square. Overall, four different ways of information presentations were chosen for the bibliographic data. All user interfaces used the same library with the same data and the same way of querying. Each presentation varied with respect to the degree of visualization as well as the degree of adaptation. In all conditions, the participants completed two types of tasks as described in Sect. 8.3.2.2. Each participant finished all four conditions in the within-subject design, each containing tasks from the two categories. Overall 40 tasks were solved by each participant in four conditions, after each condition they were asked to fill out two questionnaires.

Conditions

The within-subject design of our evaluation necessitated that each participant solves the tasks with each condition. This procedure may lead to learning effects, due to the similar tasks and information representations. To control these learning effects the *Latin Square* group design was applied with four conditions (visual appearances) and two interventions (task-level). The tasks were grouped in four sets of tasks with each five simple and five exploratory tasks and randomized against the 4×4 Latin Square, which contained the conditions. The interactive user interfaces of all these conditions presented the same data of the Eurographics Digital Library. The evaluation was conducted within an implemented evaluation system (see Fig. 8.4). In the top part of the screen the tasks were presented. The tasks consisted of multiple choice questions with four possible answers. For each task there was exactly one correct answer. Depending on the condition, one of the four different visual representations was presented in the rest of the screen. Due to the integration of the task into the screen the participants did not need to switch their focus between the task and the tool. Further, in contrast to assessing the answers on printed forms, we collected performance measures, such as task-completion-time, automatically. The participants

(a)   **(b)**

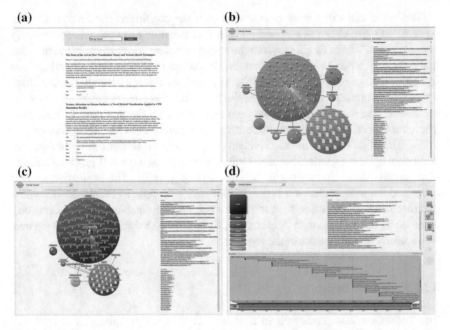

(c)   **(d)**

Fig. 8.5 The four conditions of the evaluation. **a** Textual representation (Baseline). **b** Non-adaptive visualization. **c** Adaptive visualization on visual variables level. **d** Full-adaptive Semantics Visualization

used a simple search field to enter a query based on their own knowledge and the given task. They further navigated through the visual or textual environment in order to retrieve relevant information and solve the given task.

The four different tools used in the four conditions are illustrated in Fig. 8.5. In the first condition, the textual representation (see Fig. 8.5a), the results were displayed as a simple list of bibliographic entries. This condition was our baseline and simulated the most common way of the information representation as results of search engine. In order to provide exactly the same information as in the visual environment, this condition was implemented by us. The standard representation of the Eurographics Library was not appropriate as baseline, due to its complicated result lists. Further we used the same querying to gather exactly the same amount and structure of data. So we applied our iterative querying as described in Sect. 6.1.1 for the textual representation too. In the second condition, the non-adaptive visualization, (see Fig. 8.5b), the participants used a static version of SemaVis, where the system-adaptive behavior was disabled. We used therefore the SemaSpace visualization, thus this visual interface provides all necessary information as a nested force-directed graph. Next to SemaSpace, the content visualization was place that enabled the textual view on the content. In the third condition, the adaptive visualization on visual variables level, (see Fig. 8.5c) exactly the same constellation of visualization were used as in the second condition with the difference that the adaptivity of the visual variables was enabled. This condition simulates our preliminary study with the main difference that there is now a baseline that is more known to the general users. Again the results were adapted with respect to their size and color according to the canonical user model. All other features were the same as in the second condition. Finally, in the fourth condition, the full-adaptive semantics visualization, (see Fig. 8.5d) all adaptation layers and functionalities were enabled as described in Sect. 7.3.1. The fourth condition adapted the content, visual variables, visual layout, and visual interface based on the canonical user model and represented therewith our main approach.

Canonical User Model

The canonical user model was trained in more than 10 000 search-sessions by different and very heterogeneous users in a period of one year (see Sect. 7.3.1). Although, our system was password protected, we provided in many workshops, conferences and visits, different passwords and user-names to the interested groups. Our system modeled based on the queries of these users the canonical user model anonymously. The interested groups were researcher from different areas, e.g. Eurographics community or User Modeling and Adaptive Systems community, further many students, librarian, and decision makers from industry and politics trained indirectly the canonical user model through their natural interaction with the system.

Latin Square Design

In a Latin Square design each participant completes all conditions in a predefined but randomly assigned order [83–85]. Figure 8.6 depicts the general setup for a 4 × 4 Latin Square. Instead of choosing the order in which the conditions are randomly performed, one order is chosen (In this case: A, followed by B, C, and D). The completion of the square in Fig. 8.6 is done by shifting the original order to left in every row. The participants are equally distributed to the four sequences. There are many ways of creating a Latin Square. The one used in this evaluation shows the conditions in ascending order in the first row and in the first column is sometimes referred to as standard Latin Square [76].

A Latin Square design cancels order effects by counterbalancing the number of participants completing a specific condition first [86]. In our study, we expected learning effects due to the similarity of the different visualizations and tasks. Thus, participants are more likely to achieve good results in later conditions due to knowledge gained in early conditions. As stated in [86], the approach fails its purpose when the condition and the order interact. This would be the case, for example, if Condition A would lead to a greater amount of learning and therefore to better results

Fig. 8.6 Example of a 4 × 4 Latin Square with conditions A, B, C, and D

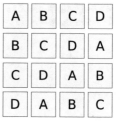

in subsequent conditions than Condition B. In the case of a non-zero interaction, the computed F-test statistics can be biased (e.g., [87]) and the results are affected with respect to their validity.

The Latin Square experimental design, originally applied in the context of agriculture, has proliferated over the last century (see e.g., [88] for an overview). Since then, it has been successfully applied in various disciplines, such as pharmacology [83–85], educational sciences [89–92] and perception research [93]. Prior research has also shown the Latin Square design to be a convenient choice for software-evaluation [94], human computer interaction [95], and the evaluation of adaptive visualizations [4]. The efficiency of the experimental design is an attractive feature. The Latin Square is suitable for providing evidence even with small sample sizes [76]. In our study we faced the problem of limited access of participants due to the lack of resources; therefore the Latin Square design was used.

To provide not only a proper randomization of the conditions, we randomized the tasks too. We created four task-sets with each ten tasks consisting of five simple and five exploratory tasks. The order of the task-set was set in the evaluation system as a fix order beginning with the "task-set 1" and ending with the "task-set 4". Thus, each sequence in our Latin Square design started with another condition and the order of the task-sets were fixed, the tasks were randomized over the conditions. Figure 8.7 illustrates the randomization design of our evaluation. For example "Sequence 1" starts with the Condition A and the Task-set 1 in the second sequence the first Condition is B but the task-set is "Task-set 1". This procedure randomizes all conditions with all task-sets.

8.3.2.4 Power Analysis

In order to provide a valid evaluation with a valid number of a sample we used Power Analysis to determine the number of participants needed to reach significant results. The computations were performed based on the designed study with within-subject design randomized by Latin Square, whereas the measurements of the results would be performed with repeated measures of analysis of variances (ANOVA). Based on this assumption we computed with GPower 3 [77, 78] a critical $F = 2.69939$ (see Fig. 8.8). In accordance with Ahn [4, 48], we expected an effect size between $F = 0.2$ and $F = 0.4$ corresponding to a medium and large effect, respectively [4].

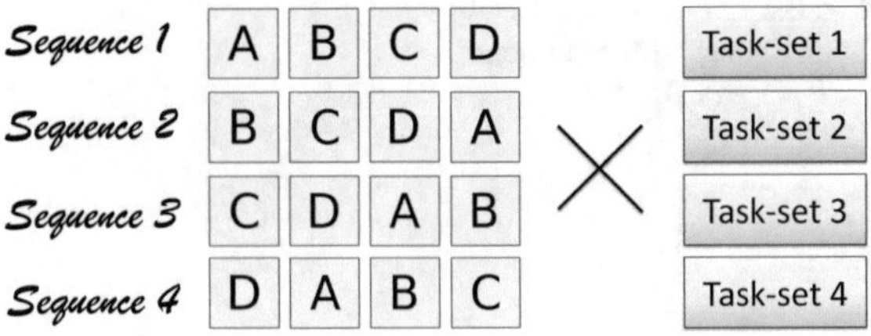

Fig. 8.7 Randomization design of the evaluation

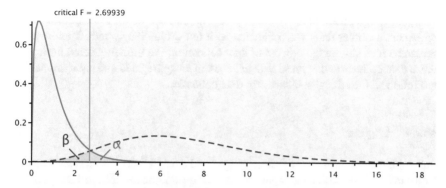

Fig. 8.8 Critical effect size F in the evaluation, computed with GPower 3 [77, 78]

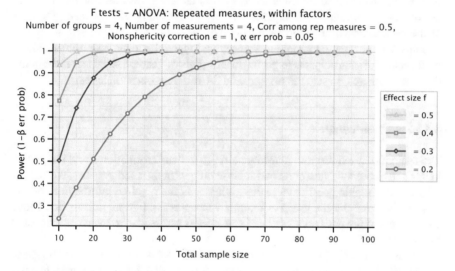

Fig. 8.9 Power analysis in correlation to total sample size (computed with GPower 3 [78])

Based on our study design and the repeated measures ANOVA the number of participants results in 36 (for $F = 0.2$) and 10 (for $F = 0.4$) for the main effects, respectively. We used a Type I error of 0.05 as well as a Type II error of 0.95. Since we did not have any prior information about the correlations between the within-factors, we used the default value of 0.5 in GPower 3. Figure 8.9 illustrates the sample size in correlation to the Power of the study with the values used in our evaluation.

8.3.3 Collected Data

In the entire evaluation different kinds of data were collected. This includes the data from the interactions with the system, different questionnaires and one printed form for the agreement of the photographic protocol. Beside the printed form for the agreement all other data were collected with our evaluation system. There was no human intervention in the process of data collection. We further ensured during the entire data collection that these data are used in an aggregated and anonymous way and followed the ethical standards for data collection.

8.3.3.1   Logfiles

Our implemented evaluation system logged and saved all the required information for the evaluation. Beside the inputs of the used questionnaires that will be described in the next section, the system logged the entire task-completion procedure of each participant. The logged information was saved as tables with the relevant item for evaluating the results of our experiments. This included beside the tasks, the type of the task, changes on responds during the interactions, task-completion-time in seconds, etc. Further, potential problems were logged that may have influenced the performance of a participants (e.g. long loading delays). Table 8.4 illustrates the items that were gathered, beside the responds on the different questionnaires.

8.3.3.2   Questionnaires

The evaluation used different questionnaires to gather information about the participants and about the interaction and their feeling with the used systems. We started with a demographic questionnaire that gathered all the required information for our context. In particular the usage of internet was assessed. Further information about any barriers of the participant's visual system was assessed, e.g. by asking for color blindness etc.

As opposed to the preliminary study the COMA sub-scale of the INCOBI questionnaire [73] was not used in the evaluation. This was due to the main reason that the adaptive visualization system was designed to fulfill the requirements of any kind of users, regardless if they are experienced or not. Thus the COMA sub-scale of INCOBI measures the self-assessed competence with computers [73]; this questionnaire was not of interest for our evaluation anymore.

In contrast to that an influential factor in the interaction with computer systems is whether users perceive the system as being intuitive and satisfying [46, 96]. Even if a system facilitates to solve tasks effectively it might not be designed intuitively, thus the interaction with the product is perceived as complicated, unforeseeable or inefficient. A lack of intuitive interaction might lead to dissatisfaction. As a result users may avoid using the system or may not feel comfortable in working with it. Therefore, it is very important to evaluate, how intuitive users perceive the interaction with a computer system [46, 47, 96]. For that reason, we used again the INTUI questionnaire [46, 47], a measurement tool to collect data on how intuitive the interaction with computer systems and software is. We assessed again the four

Table 8.4 Logged items of the evaluation

Logged items of the evaluation

Item	Description	Exemplary values
Participants' no.	Unique given number for each participants	160001
Time stamp	Computational time of the performed tasks	1386851532272
Answering	Indication if a question was answered	"answered" or "not answered"
Condition	Condition in which the task was solved	Evaluation SemaVisEG_EvalD
Task no.	Number of the task	Question 10
Task type	Information about the character of the task	"s" (for simple) and "e" (for exploratory)
Task	Textual representation of the task to be solved	How many papers were published by Fellner in 2008 and 2009?
Answer	Time time needed to answer the question in seconds	79
Respond changes	Changes in respond before sending the final respond	2
Task correctness	Information if the task was solved correctly	"incorrect" or "correct"
Visualization	Visualization interacting with (only conditions 2, 3, and 4)	SemaVis.SeMap
Search	Typed search term	Fellner
Results received	Indication that the result data were received from server and presented	Search result received
History	Used the navigation in the interaction history	History, back

sub-scales on *Effortlessness*, *Gut Feeling*, *Magical Experience*, and *Verbalization* ability (Verbalizability) as described in Sect. 8.2.2.

In addition, in order to measure the user experience, the AttrakDiff [80] was used. AttrakDiff is a questionnaire to measure perceived hedonic values and perceived pragmatic quality of interactive systems [80]. AttrakDiff provides the measurement of *Pragmatic Quality* (PQ), *Hedonic Quality* (HQ), and *Attractiveness* (ATT) [80] with a set of 28 seven-point semantic differential items as opposite adjectives [80]. Thereby, contrastive pairs of adjectives are presented to the participant. The participant gives ratings for the visualization with respect to the contrastive pairs. Ratings are assessed through a seven-point scale, where small values indicate preference for the adjective presented on the left side of the scale. High values indicate that the interaction with the visualization was described more accurate by the adjective on the right side of the scale. AttrakDiff provides the measurement of three main sub-scales: *Pragmatic Quality* (PQ), *Hedonic Quality* (HQ), and *Attractiveness* (ATT), whereas the sub-scale Hedonic Quality is further divided into the sub-scales *Hedonic Quality—Identity* (HQ-I) and *Hedonic Quality—Stimulation* (HQ-S) [80].

Pragmatic Quality reflects to what extent a product facilitates reaching objectives [97]. Similar to the INTUI sub-scale Effortlessness, the Pragmatic Quality refers the usability aspects of an interactive system and measure in particular how effective an interactive system is. The sub-scale Pragmatic Quality is measured by a set of 7 seven-point semantic differential items as opposite adjectives. Sample items for this sub-scale are *complicated—simple* or *technical—human* [80].

The sub-scale **Hedonic Quality** refers to the perceived hedonic values of an interactive system [80]. It measures the users' experience with an interactive system in terms of stimulating through visual appearance or perceived as cool, modern, or beautiful [80, 97]. This sub-scale comprises two distinct components, each measured through seven contrastive item pairs: Stimulation (HQ-S) and Identity (HQ-I). *Hedonic Quality—Stimulation* measures the product's ability to provide new experiences, stimulating humans' abilities and knowledge, or new and interesting features of an interactive system. Sample items for this sub-scale are *conventional—inventive* or *dull—captivating* [80]. *Hedonic Quality—Identity* measures the product's hedonic values in terms of evoking human's identity through the interactive system. It refers more to those hedonic values that refer to identity, creativity, or elegance [80, 97]. Sample items for this sub-scale are *tacky—stylish* or *conventional—inventive* [80]. Both sub-scales are commonly measured as *Hedonic Values* of an interactive system with together fourteen seven-point semantic differential items as opposite adjectives.

The sub-scale **Attractiveness** measures the global perceived attractiveness of an interactive system. It refers to the general values that refer to the subjective perceived liking of a system [80, 97]. The sub-scale Attractiveness is measured by a set of 7 seven-point semantic differential items as opposite adjectives. Sample items for this sub-scale are *disagreeable—likeable* or *repelling—appealing* [80].

8.3.4 Procedure

The general setting of the evaluation was as follows: The participants were guided into a laboratory in the Fraunhofer Institute for Computer Graphics Research (IGD). In each session only two participants could be evaluated, due to the limited number of equipment and rooms. Each participant had a Windows PC with equivalent performance: Intel I7 CPU, 1GB GPU, and 4GB RAM and exactly the same monitors, 27" full-HD. The participants agreed (or disagreed) with the photographic protocol. In case of disagreement no pictures were made from the session. Thereafter the participants were asked to sit in front of the monitor. The examiner then shortly explained the procedure and the goal of the evaluation and gave them their participant's number. Subsequently, participants read the instructions presented in the web-browser. Based on the participant's number the Latin Square sequence was computed through a Euclidean division. Figure 8.10 illustrated the entire procedure of the evaluation schematically.

Fig. 8.10 General procedure of the evaluation

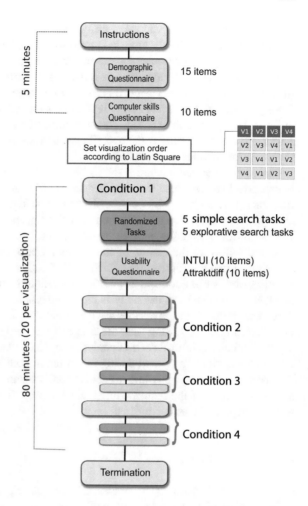

Overall, 40 different tasks were presented (10 per condition). Additional information about the participants' actions was stored in log-files for later analysis. In addition, self-assessment data were collected after each condition. Perceived intuitiveness and effortlessness was assessed through the INTUI-questionnaire [46]. Further participants were asked to fill out the AttrakDiff questionnaire [80] in order to measure the user experience.

8.3.5 Results

The data from the evaluation was analyzed using two different statistical methods: Analysis of Variance (ANOVA) and chi-square Tests. Analysis of Variance is a statistical method to compute mean differences across multiple groups. It therefore extends the classic t-Test for more than two groups. In the first step the total variance is split with respect to different sources of variance (e.g., different groups/evaluation conditions). These different variances can be used to compute a test-statistic, which follows an F-distribution. Hence, this F-value corresponds to a p-value. Subsequent to a significant ANOVA, pairwise t-Tests are conducted to determine which groups differ with respect to their mean. The advantage of using ANOVA is that pairwise t-Tests are computed only if the overall test suggests significant differences. Thus, accumulation of alpha errors is reduced.

The second applied method is the Chi-Square ($\mathscr{X}^2$) test. The method is applied in context of frequencies. The observed frequencies are compared to the expected frequencies. A term can be computed from these two quantities which approximately follows a Chi-Squared distribution. This distribution can be used to determine whether differences in frequencies are significant.

In the following the results of the evaluation are presented. First, the sample is described in a descriptive manner. Since different visualizations were evaluated, visual capability was a concern. In addition, the requirements for participation are outlined. Next, the results of the analysis regarding efficiency are given. Subsequently, differences between the conditions are investigated with respect to measures regarding effectiveness. Following this, the usage of auxiliaries (pen and Paper) as well as questionnaire data of INTUI is analyzed to determine differences regarding the cognitive resources consumed during the different conditions. Then, the results of the questionnaire data regarding acceptance and satisfaction are presented. Finally, a short summary of the main findings will be given. Thereby, connections to our research hypothesis are established.

8.3.5.1   Participants

Overall, $N = 53$ people (13 females, 24.5%) participated in the evaluation over a time period of two weeks. The participants' age ranged from 21 to 61 years ($M = 29.62, SD = 8.41$). Figure 8.11 illustrates the age distribution of the participants. One participant was excluded from the analysis due to incomplete data. Consequently each experimental order of the applied Latin-square was performed by 13 participants. Most of the participants (81.1%) had already completed their education. More than half of the participants (50.9%) showed visual impairments, but all were corrected to normal sight through either glasses or a comparable optical aid. Two participants suffered from dyschromatopsia (the disability to discriminate between red and green). One participant stated to have astigmatism.

All participants stated to use a computer on a daily basis. 43% stated that the most used computer is owned by them, 15.1% mostly use the computer that is provided at their workplace. The rest mostly use their parent's computer. On average, the participants have been using a computer for 16 years ($SD = 6$ years). The amount of time spent using a computer ranged from 5 to 90 hours per week ($M = 43.68, SD = 20.77$). The mean time spent using the internet was 31 hours per week

Fig. 8.11 Age distribution of the participants

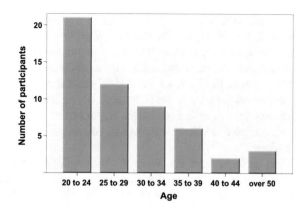

$(SD = 27)$. The proportion of participants with experiences in HTML programming was 62.3 %. 66.0 % had experiences in programming languages that go beyond HTML. Experiences with HTML-programming did not show this effect.

Participants were selected according two additional constraints: First, since the tasks as well as the multiple choice answers were presented in English, sufficient English skills were mandatory for participation. However, the psychological questionnaires were given in a German version. Therefore, only native German speakers or those with equivalent language abilities were recruited. The tasks, in particular the exploratory ones, were complex to solve. In order to ensure solvability only participants with higher education, at least students were recruited.

There were no significant differences in age $(F(3, 49) = 0.975, p = .412)$, computer literacy $(F(3, 49) = 0.785, p = .508)$, and gender $(\chi^2(3, 52) = 4.887, p = .180)$ among the different experimental orders of the Latin square. Therefore, these variables are not considered as confounding variables in the analysis.

8.3.5.2 Efficiency

Commonly the performance of interface visualizations and other information representation systems are measured through two main values: efficiency and effectiveness [3, 44, 98]. Thereby efficiency of a system commonly refers to the consumed time for solving tasks [44, 45]. Although, some studies use the interaction costs, by means of how many interactions were performed to solve a task [99], the value of time still remains the most adequate for measuring the efficiency of interactive systems [45]. We therefore used in our evaluation the task-completion-time for measuring the efficiency of the different conditions.

According to Hypothesis 1 [H1], the full-adaptive semantics visualization increases the efficiency of task solving. In this study, efficiency was assessed by measuring the mean task-completion-time for each condition. In addition to effectiveness, efficiency takes the effort involved for reaching a goal into account.

Consequently, task-completion-time is a convenient choice. Since efficiency is expected to be higher for the full-adaptive semantics visualization, we expected the mean task-completion-time to be significantly smaller in this condition. Task-completion-time was measured automatically by the implemented evaluation system and stored in the log-files. For a single task it constitutes the time elapsed between the first appearance of the task on the screen and the participant's confirmation of an answer. The main interest for Hypothesis 1 is to investigate, whether the mean task-completion-time (over all tasks and all participants) differed across the visualizations.

In order to meet the normality assumptions of the applied statistical tests for significance, task-completion-time was transformed through the logarithm function prior to the analysis. A two-way within subjects ANOVA was performed to compare the effect of the visualization and the question type on the mean time needed to complete the tasks. There was a significant effect of the visualization (Wilks' Lambda $= 0.096$, $F(3, 48) = 149.95$, p $< .001$) as well as a significant effect of the question type (Wilks' Lambda $= 0.188$, $F(1, 50) = 215.99$, $p < .001$). Also, we found a significant interaction between the visualization and the question type, Wilks' Lambda $= 0.817$, $F(3, 48) = 3.58$, $p = .02$.

Since the main effects were significant, pairwise t-Tests were computed (Table 8.5). The results showed, that all differences between the different conditions were significant ($p < .001$). Figure 8.12a illustrates the findings. This was the case for both question types (Fig. 8.12b). The mean task-completion-time as well as the corresponding standard deviations can be found in Table 8.6. We therefore conclude that the full-adaptive visualization leads to smaller task-completion-times. The effect is present for both types of tasks, exploratory and simple tasks. The analysis of the task-completion-time reveals that Hypothesis 1-1 [H1-1] and Hypothesis 1-2 [H1-2] are confirmed. Both, visualization and adaptation seem to improve the efficiency. The non-adaptive visualization led to significantly smaller task-completion-times compared to the textual baseline. In addition, the significant difference between the non-adaptive visualization and the visual variable adaptive visualization suggests, that adaptive behavior of the visual variables is connected to efficiency as well. Overall, the analysis of the efficiency metrics confirmed Hypothesis 1 and its sub-hypothesis.

Table 8.5 Pairwise t-Tests of Task-Completion-Time

	Paired differences		t	df	Sig. (2-tailed)
	95 % Confidence interval of the difference				
	Lower	Upper			
C versus D	23.49490	43.14357	6.809	51	.000
B versus D	52.63880	72.06874	12.879	52	.000
A versus D	134.85204	166.69136	19.005	52	.000
B versus C	15.10181	40.49819	4.395	51	.000
A versus C	99.66908	135.74631	13.100	51	.000
A versus B	70.84527	105.99058	10.097	52	.000

Fig. 8.12 Efficiency of the full-adaptive visualizations. **a** Task-Completion-Time of each condition. **b** Task-Completion-Time of each condition and intervention (task type)

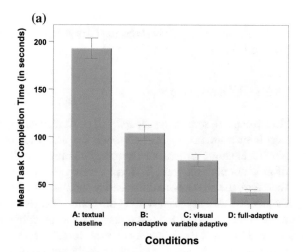

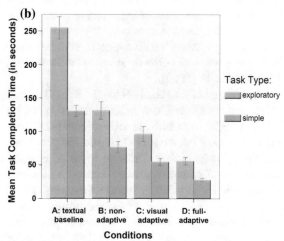

Table 8.6 Mean Task-Completion-Time of each condition in seconds

Condition	Simple tasks		Explorative tasks	
	M	SD	M	SD
A: textual-baseline	130.41	43.45	254.79	86.70
B: non-adaptive visualization	76.75	35.71	131.62	57.58
C: visual variable adaptation	54.51	22.99	96.42	52.16
D: full adaptation	27.84	13.48	55.82	32.64

The full-adaptive semantics visualization outperformed all other conditions in terms of efficiency. This aspect was observed for both types of tasks as well as for all tasks (see Fig. 8.12).

8.3.5.3 Effectiveness

Effectiveness describes the accuracy of goal achievement [21]. Therefore, it is convenient to measure effectiveness through the proportion of achieved goals. In the case of this evaluation, the goal was to solve tasks successfully using different the different information presentations. The tasks were presented as multiple choice questions. Therefore, the answers could be easily compared to the correct choices. By contrast to other task designs (e.g., writing the correct answer on a piece of paper), determining the correctness of the given answers was unambiguous. Each answer could be assigned to one of the following classes: correct, incorrect, omitted. The participants were encouraged to omit tasks they could not solve instead of picking an answer by chance. On the basis of these classes, two measures for effectiveness could be computed: the proportion of correctly answered tasks and the proportion of omitted tasks. Hypothesis 2 [H2] suggests, that the full-adaptive visualization leads to a higher proportion of correctly answered tasks [H2-1] as well as a smaller proportion of omitted tasks [H2-2].

Percentage of Correctly Solved Tasks: The computation of the percentage of solved tasks as a measure of effectiveness was straightforward. According to Hypothesis 2-1 [H2-1] the full-adaptive visualization was expected to facilitate a more effective way of interaction. Therefore, the percentage of correctly solved tasks was anticipated to be higher compared to the other visualizations. Figure 8.13a depicts the number of correctly answered questions for each condition. The full-adaptive Visualization lead to fewer incorrect answers than all other visualizations ($\mathcal{X}^2(1, N = 2099) = 174.95, p < .001$). The results show, that the full-adaptive SemaVis leads to greater effectiveness compared to both a reduced amount of adaptation and a lower degree of visualization. We therefore conclude, that Hypothesis 2-1 [H2-1] is verified. The differences among the other conditions were not significant ($p > .05$).

Percentage of Omitted Tasks: Regarding the proportion of omitted tasks, Hypothesis 2-2 [H2-2] states that the full-adaptive SemaVis leads to the smallest values of omitted tasks. Figure 8.13b depicts the number of omitted tasks. The analysis showed, that the number of omitted tasks was significantly smaller for the full-adaptive visualization than all other visualizations ($\mathcal{X}^2(1, N = 2099) = 94.23, p < .001$). In addition, less tasks were omitted in the visual adaptive condition compared to the non-adaptive visualization and the textual baseline ($\mathcal{X}^2(1, N = 1569) = 13.23, p < .001$). However, there was no significant difference between the textual baseline and the non-adaptive visualization ($p = .914$). We conclude, that Hypothesis 2-2 [H2-2] is confirmed.

The analysis of measures regarding effectiveness revealed, that more tasks could be solved correctly using the full-adaptive visualization compared to all other conditions. Also, fewer tasks were omitted. Overall, since all sub-hypothesis showed validity, the results confirm Hypothesis 2 [H2].

Fig. 8.13 Effectiveness of
the full-adaptive
visualizations. **a** Task
correctness. **b** Omitted tasks

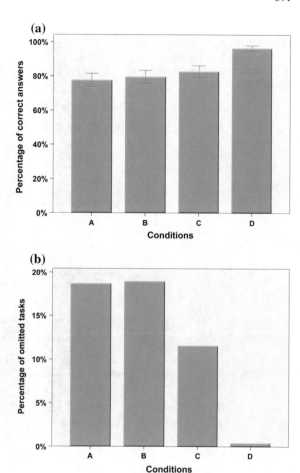

8.3.5.4 Cognitive Load and Perceived Effort

According to Hypothesis 3 [H3] less cognitive resources are consumed while working with the full-adaptive semantics visualization compared to other information
representations. The amount of cognitive resources needed for solving the tasks was
assessed by the use of auxiliaries (pen and paper) [H3-1], due to the assumption
that users write down notes, if their working store is not capable to remember the
intermediate results for complex tasks. The use of paper and pen is therefore an indicator for a higher cognitive load while solving tasks [12, 74]. Further a more formal
way for measuring the perceived effort was performed with the INTUI sub-scale
Effortlessness [H3.2]. Effortlessness refers similar to the use of further auxiliaries
the perceived ease to solve a task (Fig. 8.14).

Fig. 8.14 **a** Sub-scales of
the INTUI questionnaire:
Effortlessness. **b** Percentage
of participants taking notes
for each condition

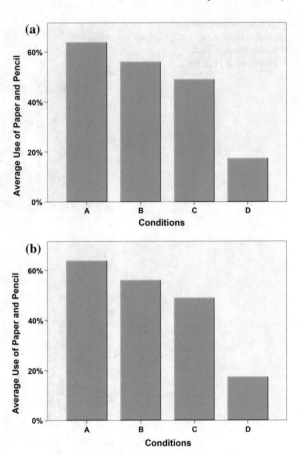

Auxiliaries: Use of paper and pen

The need for taking notes to aggregate visual data is correlated with a higher cognitive load. We hypothesized, that the full-adaptive visualization reduces the cognitive load needed to successfully complete the tasks. Therefore, we expected the participants to take fewer notes during the tasks. Figure 8.14b depicts the proportion of participants taking notes during the different evaluation conditions. Our results show, that the proportion of participants taking notes was significantly smaller when working with the full-adaptive visualization, $\mathcal{X}(1, N = 93) = 10.41, p < .001$. The differences between the other conditions were not significant ($p > .05$). Therefore, we conclude that Hypothesis 3-1 [H3-1] holds.

Effortlessness

In addition to the usage of auxiliaries, we assessed the perceived effort through the INTUI sub-scale *Effortlessness*. We investigated, whether the reduced need for cognitive resources when solving tasks with the full-adaptive visualization was perceived

Table 8.7 Pairwise *t*-Tests for the INTUI-sub-scale *Effortlessness*

	Effortlessness			t	df	Sig. (2-tailed)
	Paired differences					
	Mean	Std. deviation	Std. error mean			
C versus D	−0.87667	1.43246	0.20258	−4.328	49	.000
B versus D	−1.95353	1.57315	0.21816	−8.955	51	.000
A versus D	−1.74528	2.04003	0.28022	−6.228	52	.000
B versus C	−1.07500	1.10852	0.15677	−6.857	49	.000
A versus C	−0.88667	1.69330	0.23947	−3.703	49	.001
A versus B	0.23237	1.75434	0.24328	0.955	51	.344

by the participants as more effortless. According to Hypothesis 3-2 [H3-2], the full-adaptive visualization was expected to be perceived as more effortless compared to the other conditions. The analysis revealed, that participants rated the visualizations significantly different with respect to the dimension *Effortlessness*, Wilks' Lambda = 0.341, $F(3, 47) = 30.33$, $p < .001$. All differences shown in Fig. 8.14a were significant except between the conditions A and B (see Table 8.7 for the results of the pairwise *t*-Tests). The full-adaptive semantics visualization turned out to be usable with the least perceived effort. The difference to condition C was significant, $t(49) = −4.33$, $p < .001$. The results confirm Hypothesis 3-2 [H3-2].

In conclusion, Hypothesis 3 [H3] was confirmed. The full-adaptive visualization leads to significantly less usage of further auxiliaries [H3-1] (pen and paper). Further the perceived effort was significantly lower while working with the full-adaptive visualization in contrast to all other conditions [H3-2]. Thus the two sub-hypotheses were confirmed, we conclude that hypothesis 4 [H3] holds and the adaptive semantics visualizations consumes less cognitive effort and the work with it is perceived as more effortless.

8.3.5.5 Satisfaction and User Experience

According to Hypothesis 4 [H4] the full-adaptive visualization leads to higher satisfaction and user experience. Therefore, the increased positive perception of the full-adaptive visualization is expected to manifest in differences of ratings the INTUI sub-scale *Intuitiveness*, thus an intuitive interaction with systems leads to a more satisfied usage [46, 47]. Furthermore the AttrakDiff questionnaire should enable us to measure these values in a reliable manner. Thus the AttrakDiff questionnaire was designed to measure in particular heconic values [80, 97] of an interactive system beside usability aspects. The main sub-scales of AttrakDiff that refer to satisfaction and user experience are *Hedonic Quality* and *Attractiveness*. We hypothesized therefore that the full-adaptive SemaVis is perceived as more intuitive by higher ratings in the INTUI sub-scale Intuitiveness [H4-1], receives higher rating in the

Table 8.8 Pairwise *t*-Tests for the sub-scale intuitiveness of the INTUI questionnaire

Intuitiveness				t	df	Sig. (2-tailed)
Paired differences						
	Mean	Std. deviation	Std. error mean			
C versus D	−0.58974	1.27151	0.20360	−2.897	38	.006
B versus D	−1.36585	1.51255	0.23622	−5.782	40	.000
A versus D	−1.44118	2.29876	0.39423	−3.656	33	.001
B versus C	−0.90000	1.01519	0.14357	−6.269	49	.000
A versus C	−1.16279	2.24595	0.34250	−3.395	42	.002
A versus B	−0.26667	2.11488	0.31527	−0.846	44	.402

AttrakDiff sub-scale Hedonic Quality [H4-2], and the AttrakDiff sub-scale Attractiveness [H4-3].

Intuitiveness

There was a significant difference in how intuitive the different information representations and visualizations were perceived, Wilks' Lambda = 0.441, $F(3, 29) = 12.24$, $p < .001$ (see Fig. 8.15). Thus, pairwise *t*-Tests were performed (Table 8.8). The results show that all effects are significant on a five percent level except the difference between Condition A and Condition B. The full-adaptive semantics visualization outperforms all other visualizations. We therefore conclude that Hypothesis 2-1 [H2-1] holds.

Hedonic Quality

Beside the dimension Intuitiveness of the INTUI questionnaire, the perceived Hedonic Quality was assessed for each condition using the AttrakDiff questionnaire. The

Fig. 8.15 Mean intuitiveness of each condition measured with the INTUI questionnaire

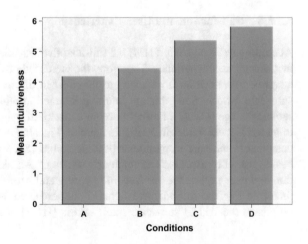

Table 8.9 Pairwise *t*-Tests of hedonic quality (total score)

	Hedonic quality—total score			T	df	Sig. (2-tailed)
	Paired differences					
	Mean	Std. deviation	Std. error mean			
C versus D	−0.26429	0.77974	0.11027	−2.397	49	.020
B versus D	−0.40408	0.91738	0.12722	−3.176	51	.003
A versus D	−2.26125	1.06956	0.14692	−15.392	52	.000
B versus C	−0.13738	0.54499	0.07707	−1.782	49	.081
A versus C	−2.03550	0.99696	0.14099	−14.437	49	.000
A versus B	−1.87044	1.04313	0.14466	−12.930	51	.000

Fig. 8.16 Mean total score on Hedonic Quality for each condition measured with the AttrakDiff questionnaire

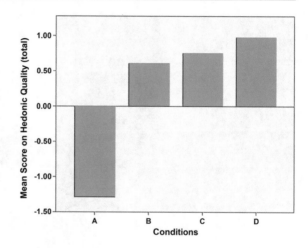

total score for Hedonic Quality comprises 14 seven-point semantic differential items with contrastive adjectives. We expected the full-adaptive SemaVis to show higher mean ratings than the other conditions (see Hypothesis 4-2). The analysis of the total hedonic quality confirms Hypothesis 4-2 [H4-2]. The mean scores for the full-adaptive visualization were significantly higher than for all other conditions (see Table 8.9). The results are depicted in Fig. 8.16. All differences are significant except between the conditions B and C. Hypothesis 4-2 [H4-2] was confirmed by the results. The full-adaptive visualization was rated best with respect to the overall Hedonic Quality.

Table 8.9 illustrates the results of the pairwise *t*-Tests of the AttrakDiff sub-scale Hedonic Quality. The full-adaptive visualization shows significant higher ratings on Hedonic Quality compared to all other conditions. The adaptation of visual variables showed no differences in contrast to the non-adaptive visualization. The textual baseline showed significant lower ratings compared to all other conditions.

The Hedonic Quality can be partitioned into two further sub-scales: Identity (HQ-I) and Stimulation (HQ-S) [80]. Both sub-scales are measured by seven items.

Fig. 8.17 Sub-scales of the
AttrakDiff questionnaire:
HQ-I and HQ-S. **a** Hedonic
Quality—Identity (HQ-I).
b Hedonic Quality—
Stimulation
(HQ-S)

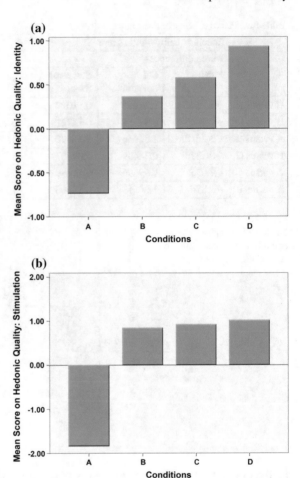

The analysis for the first sub-scale, Hedonic Quality—Identity, is given in Fig. 8.16a.
The overall test was significant, Wilks' Lambda $= 0.314$, $F(3, 47) = 34.20$,
$p < .001$. Thus, pairwise t-Tests were conducted (see Appendix Table E.4. All differences reached significance. Hedonic Quality—Stimulation also differed significantly
between the experimental conditions, Wilks' Lambda $= 0.142$, $F(3, 47) = 95.04$,
$p < .001$. The results of the pairwise t-Tests are given in Table E.5 in the Appendix
of this thesis. There were no significant differences between the visualizations (Condition A, B, and C). However, the ratings for the textual baseline were significantly
lower than for all other conditions. The results on the full-adaptive visualization for
the sub-scale Hedonic Quality-Identity showed significant higher ratings compared
to all other conditions, whereas the results for the sub-scale Stimulation did not reach
significance compared to non-adaptive or partially adapted visualization. Figure 8.17
illustrates the mean scores on the two partitioned sub-scales of Hedonic Quality.

Table 8.10 Pairwise *t*-Tests for the AttrakDiff-sub-scale *Attractiveness* pairwise *t*-Tests for the AttrakDiff-sub-scale

	Attractiveness			t	df	Sig. (2-tailed)
	Paired differences					
	Mean	Std. deviation	Std. error mean			
C versus D	−0.17519	0.54479	0.07704	−2.274	49	.027
B versus D	−0.40018	0.60403	0.08376	−4.777	51	.000
A versus D	−0.29784	0.63252	0.08688	−3.428	52	.001
B versus C	−0.22957	0.56622	0.08008	−2.867	49	.006
A versus C	−0.13290	0.71211	0.10071	−1.320	49	.193
A versus B	0.11584	0.59278	0.08220	1.409	51	.165

Fig. 8.18 Sub-scales of the AttrakDiff questionnaire: ratings on perceived Attractiveness

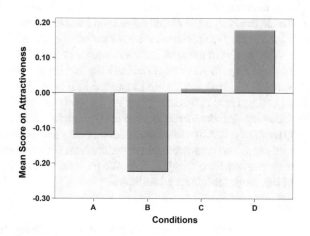

Attractiveness Finally, the full-adaptive visualization was expected to yield the highest ratings for attractiveness [H4-3]. The analysis revealed a significant main effect for attractiveness, Wilks' Lambda $= 0.681$, $F(3, 47) = 7.327$, $p < .001$. Pairwise *t*-Tests (see Table 8.10) illustrate the full-adaptive visualization to be significantly perceived as more attractive than all other visualizations. Therefore Hypothesis 4-3 [H4-3] is confirmed (Fig. 8.18).

Hypothesis 4 [H4] proposes the full-adaptive visualization to show better performance regarding Intuitiveness [H4-1], Hedonic Qualities [H4-2], and Attractiveness [H4-3]. In general all three hypotheses were confirmed, therefore hypothesis 4 [H4] holds. The partitioned sub-scale of Hedonic Quality—Stimulation (HQ-S) does not confirm that the full-adaptive visualization leads to higher rating in Hedonic Quality—Stimulation (Fig. 8.17). This sub-scale shows significant differences between all visualization (conditions B, C, and D) and the textual baseline. The adaptation effect seems not to have influences on the Stimulation factor of the Hedonic Quality. However, in the overall rating on the sub-scale Hedonic Quality, the full-adaptive semantics visualization outperforms all other conditions. Therefore Hypothesis 4 [H4] and all its sub-hypotheses hold.

8.3.5.6 Further Results

In this evaluation, the INTUI questionnaire as well as the AttrakDiff questionnaire was used to obtain measures for certain perceived quality dimensions of the different conditions. Our main hypotheses covered only a subset of the entire questionnaires to confirm our assumptions. We used the dimension *Effortlessness* of the INTUI questionnaire to measure the perceived effort and cognitive load in Hypothesis H3-2. Further, the main sub-scale of Intuitiveness of the INTUI-questionnaire was used to measure the perceived intuitiveness of our conditions in context of satisfaction and user experience [H4-1]. From the AttrakDiff questionnaire, we used all sub-scales beside the *Pragmatic Quality*. We used instead the real effectiveness of the systems by measuring the performance in terms of efficiency [H1] and effectiveness [H2].

The results of the remaining sub-scales of the INTUI questionnaire *Gut Feeling*, *Magical Experience*, and *Verbalizability* and the sub-scale *Pragmatic Quality* of the AttrakDiff questionnaire are illustrated in this section. There were no expectations regarding these sub-scales prior to the analysis, thus the effects presented in the following are considered to be entirely exploratory and complement the evaluation in terms of reliability. Interested audience may gather correlations from the illustrated measures. The measurement procedure here was the same as in the sub-scales that were relevant for our hypotheses. The pairwise t-Tests are illustrated in the appendix of this work for interested audience.

Gut Feeling

The INTUI sub-scale *Gut Feeling* is created by taking the mean over the ratings of four items. It assesses whether the interaction with the visualization was guided by reason or feelings. The scores differed significantly between the conditions, Wilks' Lambda $= 0.669$, $F(3, 47) = 7.77$, $p < .001$. *Pairwise t-Tests* (Table E.1) showed that the textual baseline yielded to significant smaller scores by comparison to all other visualizations. However, visualization B, C, and D did not differ significantly. Thus, the visualization component seems to have an impact on the way the interaction with the system is guided. The influence of the different levels of adaptation showed no effects. Figure 8.19 illustrates the outcomes of this sub-scale.

Magical Experience

The INTUI sub-scale *Magical Experience*, which was not part of the Hypotheses is measured with four seven-point items too. It determines whether the interaction with the system was perceived as a special experience. A sample item for this sub-scale is *nothing special* in contrast to *a magical experience*.

The main effect of the visualizations on the sub-scale *Magical Experience* was significant, Wilks' Lambda $= 0.205$, $F(3, 47) = 12.24$, $p < .001$. Figure 8.20 depicts the mean values of the sub-scale for each visualization. The results of the pairwise

Fig. 8.19 Mean scores on perceived Gut Feeling

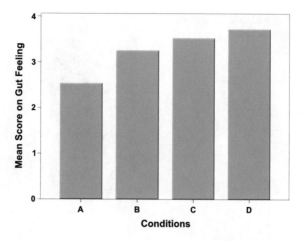

Fig. 8.20 Mean scores on perceived Magical Experience

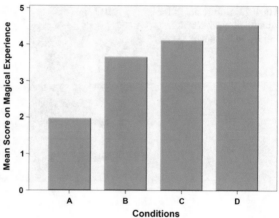

t-Tests revealed significant differences between all conditions (see Table E.2). The analysis reveals, that the interaction with a visualization tool is perceived as more magical compared to the textual baseline. In addition, the each layer of adaptation increases the scores on magical experience. The full-adaptive visualization reached the highest values and outperformed all other conditions.

Verbalizability

The sub-scale Verbalizability of the INTUI questionnaire was assessed using three seven-point items. Verbalizability is the extent of how the interaction can be described in retrospect. The better a system can be described, the less intuitive it is perceived according to Ullrich and Diefenbach [46, 47]. Repeated measures ANOVA revealed a significant difference between the visualizations with respect to Verbalizability, Wilks' Lambda $= 0.834$, $F(3, 49) = 3.05$, $p = .038$ (Fig. 8.21). The only significant difference was between Condition A and Condition B. The pairwise t-Test on this sub-scale is illustrated in Table E.3.

Fig. 8.21 Mean scores on
Verbalizability

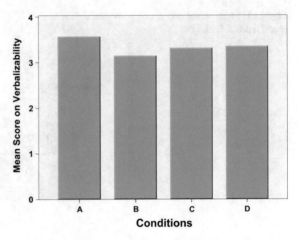

Fig. 8.22 Sub-scales of the
AttrakDiff questionnaire:
ratings on Pragmatic Quality

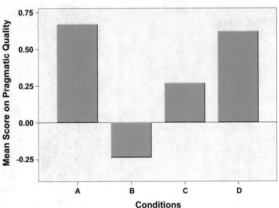

AttrakDiff Sub-scale Pragmatic Quality

The results of the AttrakDiff sub-scale *Pragmatc Quality* are illustrated in Fig. 8.22.
The overall test showed a significant effect, Wilks' Lambda = 0.371, $F(3, 47)$ =
26.517, $p < .001$. In order to examine the observed differences with respect to significance, pairwise t-Tests were computed. The results are given in Table E.6. However, no significant differences in the ratings were found between the full-adaptive
visualization and the textual baseline. This could be due to the participants' familiarity with the textual baseline. Significant effects could be observed between the
conditions with different degree of adaptation (see Table E.6).

8.3.6   Summary of Results and Discussion

The main goal and contribution of this thesis is to provide a model for adaptive
semantics visualization that enables a faster, more accurate and more accepted way
of exploratory search. To achieve this goal a variety of algorithms, models, and

approaches were designed (see Chap. 6) and implemented as SemaVis (see Chap. 7) to evidence the feasibility of our conceptual model. We referred in our conceptual model to the gaps and potentials of existing approaches and systems and tried to close the gaps and make use of the potential in each model or algorithm provided in this thesis. We aimed not only at providing enhances for the idea of adaptive visualizations, but also to provide a feasible, replicable, and verifiable approach. Thereby two main concepts should enhance the existing work on visual adaptation: a reference model that sub-divides the layers of interactive visual representation in more granular way based on studies and assumption of visual perception and a model that investigates user and data and is trained during the usage without the need of any experts. We proposed in our conceptual model, besides others, these two approaches with the main goal to provide a more effective, more efficient, and more accepted adaptive visualization approach for a broad range of heterogeneous users. Our reference model for visual adaptation enhanced the work of Card et al. [3] by adding further layer of visual representation. The transformation that was originally designed a horizontal transformation pipeline was enhanced by the influencing factors that affect each layer based on user and data model. In this context the visual layers of *content* or semantics, *visual layout*, *visual variable*, and *visual interface* were identified based on existing works on human perception and study outcomes on visual information processing (see Sect. 6.3). To train the self-learning adaptive system without the intervention of visualization experts, the model of canonical user was applied and enhanced with deviation and similarity measures.

Although, SemaVis enables an individual adaptation based on individual's user models, the necessity of training the adaptive behavior by expert is obsolete with the new approach. These models and further enhancements were proposed in our conceptual model that led to the development of SemaVis. We could illustrate that major aspects of our conceptual model were implemented as part and core of SemaVis and evidenced the feasibility of our conceptual model. Thereby, three application scenarios for adaptive visualizations were introduced (see Sect. 7.3), whereas two of them implemented major concepts our model and were adequate enough to prove our assumptions on improved efficiency, effectiveness, and satisfaction. For evaluating the conceptual model, we chose the application scenario of SemaVis in digital libraries (see Sect. 7.3.1), due to the following main reasons: This application scenario implements major parts of our conceptual model, it makes use of non-semantic data, the canonical user model is trained well by heterogeneous users, and it uses just one data-bases. The choice that the visual system should access just one data-base is important for the comparing the system with other conditions. Our visualization cockpit model enables the use of various data-bases simultaneously as described in Sect. 7.3.2 and adapts the visualization based on user and data model. Finding a baseline and further conditions for this kind of visualization with same the same data is a difficult and nearly impossible task. Evaluating the application scenario for Web-search would lead to results that are not reliable enough and to advantages for our approach. Thus, the work on the application scenario for policy modeling is ongoing and the users are not heterogeneous enough, we chose the application scenario for digital libraries to be evaluated. Thereby, some functionalities of the adaptive visualization were turned off to enable a valid evaluation. The deviation and similarity algorithms and the relevance measurements of contextual information were not investigated in our evaluation. We aimed to provide a coherent and replicable evaluation that can be performed within a given time-frame as described in the procedure (see Sect. 8.3.4) of the evaluation. The adaptation of SemaVis was based on a canonical user model that was trained over a period of about one year by real users of SemaVis. The users were asked in workshops, conferences, and other events to work with the system. Therefore the canonical user model in the evaluation reflects real users behavior modeled as a canonical user model.

The conditions chosen in this evaluation were oriented on our reference model. We used a textual representation as baseline for our study. We developed the textual representation to ensure that the same kind of data and same content is presented after a certain search. The baseline referred to the layer of *semantics* or *content* of our reference model. The first visual condition made use of all four layers of the reference model but did not adapt the visualization based on the reference model. We ensured that in this non-adaptive visualization, the same information is always available as in all other conditions too. The third condition referred to the adaptation of visual variables similar to our preliminary study. The same visual layouts from the second conditions were used with adaptive visual variables. We wanted to find out how much the full-adaptive SemaVis differs from a SemaVis version that just adapts the visual variables. Here, the canonical user model was used too. The last condition was our full-adaptive SemaVis as described in Sect. 7.3.1.

For evaluating the system, we deduced from existing literature four main hypotheses. According to Card et al. [3] and many other works on visualization [44, 52, 100–102], the aspect of performance plays an essential role for evaluating visualizations. We therefore investigated the performance measure of the full-adaptive visualization with two main metrics: efficiency and effectiveness [3, 45, 102]. The tasks to be solved were subdivided into exploratory and simple tasks according to previous works on exploratory search [54, 55, 74]. Based on these works, we hypothesized that solving tasks with our full-adaptive semantics visualization is faster and more effective. The efficiency was measured with the dimension task-completion-time [H1], whereas we further hypothesized that this is valid for both kind of tasks simple [H1-1] and exploratory [H1-2]. The effectiveness was measured through two main aspects, the number of correctly solved tasks [H2-1] and the number of omitted tasks [H2-2]. We asked the participants to omit a task, if they are not able to solve instead of choosing an answer by chance. Beside performance measures, we wanted to find out if the full-adaptive visualization leads to less cognitive load and effort [H3]. We therefore gave all participant pen and paper and asked to answer in a questionnaire, if they made use of these auxiliaries. We assumed that the need for using paper and pen is an indicator for an overload of the working memory [3-1] [38, 74, 79]. We also used a more subjective way to measure the cognitive effort with the sub-scale *Effortlessness* of the INTUI questionnaire [46, 47]. Effortlessness refers to the *effort of learning* according to DIN EN SIO 9241-110 and according to Hogarth "intuitive responses are reached with little appearing effort" ([71] in [47, p. 802]). We assumed in our hypothesis 3-2 [H3-2] that the full-adaptive visualization leads to lower perceived effort compared to the textual baseline or the non- or partially adaptive visualizations. Another main aspect for us was satisfaction and user experience. Full-adaptive semantics visualization may lead with the included machine learning approaches to a faster and more effective way of information acquisition. But is the full-adaptive visualization perceived as more attractive, more intuitive, and leads to higher perceived user experience, and consequently to more acceptance by users? To investigate this question, we used the AttrakDiff questionnaire [80, 97] that measures in particular the perceived *Hedonic Quality* and *Attractiveness* of interactive systems. According to Hassenzahl et al. these both dimensions gives a clear measure on the acceptance, satisfaction, and user experience of an interactive system that goes beyond but includes usability [80, 97]. We hypothesized that the full-adaptive visualization leads to higher satisfaction and user experience [H4]. This hypothesis was measured through the perceived *Intuitiveness* [H4-1], a sub-scale of the INTUI-questionnaire, the *Hedonic Quality*, a sub-scale of the AttrakDiff-questionnaire, and the perceived *Attractiveness* [H4-3] another sub-scale of the AttrakDiff questionnaire. We hypothesized that our full-adaptive SemaVis leads to better rankings in all the mentioned dimensions of satisfaction and user experience.

The **results of the evaluation** illustrate clearly that all our assumptions are confirmed. The full-adaptive semantics visualization is more efficient, more effective, leads to less cognitive load and effort and to higher satisfaction and user experience. In all our hypotheses, the full-adaptive SemaVis outperforms non-adaptive visualizations, partially adaptive visualizations, and the textual baseline. The results confirmed increased efficiency of the full-adaptive SemaVis in terms of task-completion-time [H1]. Both, simple tasks [H1-1] as well as exploratory tasks [H1-2] were solved significantly faster. With respect to effectiveness [H2] the analysis revealed that the full-adaptive visualization outperforms all information presentation interfaces. The percentage of correctly solved tasks was significantly higher [H2-1]. In addition, significantly fewer tasks were omitted [H2-2] while working with the full-adaptive SemaVis. Besides increased efficiency and effectiveness the full-adaptive SemaVis consumes less cognitive resources and effort [H3]. The usage of auxiliaries (taking notes during the tasks) was significantly less frequent [H3-1] and the interaction was perceived as more effortless by the means of the INTUI sub-scale *Effortlessness* [H3-2]. Finally, the full-adaptive SemaVis showed the best results with respect to user experience measures [H4]: First, the interaction was perceived as significantly more intuitive [H4-1]. Second, the overall hedonic quality assessed by the AttrakDiff questionnaire outperformed all other visualizations [H4-2]. Third, the users perceived the full-adaptive SemaVis as significantly more attractive by means of the sub-scale *Attractiveness* of the AttrakDiff-questionnaire [H4-3].

In summary, the evaluation clearly confirmed all hypotheses. Consequently, the conceptual model proposed in this work leads to significantly more efficient, effective, and accepted systems that reduce significantly cognitive load and effort and are perceived as more intuitive, more attractive, and are related to higher hedonic qualities.

Beside the sub-scales of the questionnaires INTUI and AttrakDiff that were used to confirm our hypotheses, there are sub-scales that are part of the questionnaire but were not considered in our evaluation. Although, they were not part of our evaluation scenario, we measured all items of the questionnaires and presented them as further results (see Sect. 8.3.5.6). One of these sub-scales was *Gut Feeling* of the INTUI questionnaire that refers more to an intuitive decision making [47]. In this sub-scale, there were no significant differences between the different levels of adaptation. However, a significant difference could be observed between all visual representations and the textual baseline. We think that the process of decision making is already sufficiently supported by visualizations and consequently the adaptation level has no effects. A similar observation was made for the INTUI sub-scale *Verbalizability*. The work with the non-adaptive visualization could be retrospectively verbalized significantly lower than the textual baseline. The lower verbalization ability of a system refers according to Ullrich and Diefenbach to higher degree of intuitiveness [46, 47]. The authors claim that the lower degree of verbalization ability after using a system is more characterized by intuitive decisions rather than intuitive interaction [46]. Thus, they were uncertain about this dimension, as they claimed "However, we still wanted to include this component in our questionnaire to find out whether any correlations to the other components or overall ratings could be revealed." [3, 46], we did not consider this dimension. In our opinion the correlation between retrospective verbalization ability and the intuitiveness cannot be confirmed in context of visualizations.

Another sub-scale of INTUI that was not considered in our evaluation was the dimension of *Magical Experience*. Although, the full-adaptive SemaVis outperformed in the dimension *Magical Experience* all other conditions, this dimension was not of interest for us. We already covered the aspect of user experience with the sub-scale *Hedonic Quality* that showed more reliability in previous studies. A similar situation was given by the sub-scale *Pragmatic Quality* of the AttrakDiff questionnaire that refers to the perceived quality of achieving the goals. Thus, we investigated in our evaluation the real achievements in terms of efficiency and effectiveness, this dimension was not considered. The results on this dimensions showed significant higher values according to the adaptation levels. The full-adaptive SemaVis outperformed all other visual representations. But compared to the textual baseline, there were no differences between the full-adaptive SemaVis and the textual baseline. We think that the perceived pragmatic quality of the textual baseline is due to users' prior experience.

8.4   Chapter Summary

In the previous Chap. 7, we described the implementation of our conceptual model as proof of feasibility. We could illustrate based on the architectural design and the different application scenarios that our conceptual model is implemented and in use in different contexts. While we evidenced in the last chapter the feasibility of our conceptual model with SemaVis, proof of the benefits of our model was not provided. This chapter aimed at providing a proof of the benefits and added values of our conceptual model and enlightening shortcoming and potentials that should be investigated in future. Thus our conceptual model on adaptive semantics visualization with its human centered design targets at providing benefits for users of the system, we chose an empirical 'a posterior' evaluation of our approach. To provide a valid and replicable evaluation of our model, we introduced first some theoretical foundations and performed empirical studies in context of adaptive visualizations. The main goal here was to provide a theoretical overview of the underlying psychological methods.

Thereafter a preliminary pilot study on evaluating only the effects of visual variables of our reference model in context of information search was introduced. The preliminary study was performed together with the psychological department of the Technische Universität Darmstadt in the laboratory of the Department of Psychology. A total of 14 persons participated the preliminary study with a between-subject group design. The main goal here was to find out, if the visual variables in terms of color and size have already an effect on search efficiency and enable us to identify appropriate questionnaires, limitations, and shortcomings. Although, the number of participants was limited to just 14, an effect to task completion and therewith to effectiveness could be observed. The effect was expected due to the outcomes of the studies in visual perception, in particular in preattentive information processing. Further we used the INTUI questionnaire with the four sub-scales of Effortlessness, Verbalizability, Gut Feeling and Magical Experience. Only the dimension Gut Feeling was affected by the adaptation of the visual variables. But the main outcomes from the preliminary were for us the identified shortcomings and limitations. The small sample size led to big standard errors. In addition, small effects could not be investigated. Another limitation was the procedure of the evaluation. The answers were written on a printed form, therewith the attention of the participants changed from the computer-aided visualization to paper and vice versa. Further no temporal information (e.g., task-completion-time for each task) was assessed and only simple search task were considered. But in general the main goal of the preliminary study was successful, thus it could be evidenced that visual variables like color and size has an effect on the effectiveness of task completion.

In the main evaluation, we tried to eliminate all identified shortcomings from the preliminary study. We therefore deduced our hypotheses first to have an idea what kind of evaluation should be performed. Overall, four hypotheses were deduced with a total number of nine sub-hypotheses. One main limitation of the preliminary study was that the participants had to move their attention from the screen to the printed forms for filling out the questionnaires and solve the tasks. To avoid this procedure, we developed an evaluation system that visualized the task on top of the screen. Further, all main instructions were provided by the developed evaluation system. The users were guided through the entire evaluation procedure without the need of paper forms to be filled out during the task-solving process. Further the evaluation systems enabled to measure the task-completion-time and reduce human errors, thus every answer and solved tasks was depicted and stored automatically. Another shortcoming of the preliminary study was the tasks. As we focused in our preliminary study only on simple tasks, the main evaluation contained well-defined simple and exploratory search tasks according to the previous works on exploratory search [54, 55, 74].

Further, we defined based on our reference model four conditions, instead of two. We defined a textual baseline to validate our conceptual approach in contrast to the most common way of information presentation. The textual representation of search results used the same data-base, the same approach for gathering the data, and illustrated exactly the same data as in all other conditions. To achieve this, the textual presentation of the data was developed by us. The second condition was a static version of SemaVis with exactly the same data as all other conditions. For the third condition, we applied the second condition and turned the visual variable adaptation on. In this condition only the visual variables were adaptive according to the canonical user model. The fourth and last condition was our full-adaptive SemaVis as described in Sect. 7.3.1. We applied the application scenario of digital libraries, thus this application scenario provided major components of our conceptual model and used just one data-base. To amplify the power of our evaluation, we decided to design the evaluation as a within-subject group. All participants had to solve similar tasks (see Tables D.1 and D.2) in all conditions. To reduce the learning effect, we applied a Latin-Square design with four conditions and two interventions (task-types). The randomization was performed by the evaluation system with an integrated Euclidean division to ensure that all sequences of the Latin-Square design were conducted by the same number of participants.

The limitation of the small sample size in the preliminary study was eliminated with a 'a priori' power analysis. We measured with how great our sample size should be to reach enough power for the evaluation and the settings. We used for the evaluation beside the INTUI questionnaire the AttrakDiff questionnaire that showed good reliability. All data including the task-completion-times were collected automatically with our self-developed evaluation system. The evaluation was conducted in a laboratory of Fraunhofer IGD as a controlled experiment. Overall 53 persons participated the evaluation in a time-period of two weeks. The results of the evaluation illustrated clearly that all our assumptions were confirmed. The full-adaptive semantics visualization was evaluated as more efficient, more effective, with consuming less cognitive load and effort and higher ratings on satisfaction and user experience. In all our hypotheses, the full-adaptive SemaVis outperformed non-adaptive visualizations, partially adaptive visualizations, and the textual baseline. The results confirmed increased efficiency of the full-adaptive SemaVis in terms of task-completion-time. Both, simple tasks as well as exploratory tasks were solved significantly faster. With respect to effectiveness the analysis revealed that the full-adaptive visualization outperforms all information presentation interfaces. The percentage of correctly solved tasks was significantly higher. In addition, significantly fewer tasks were omitted while working with the full-adaptive SemaVis. Besides increased efficiency and effectiveness the full-adaptive SemaVis consumed less cognitive resources and effort. The usage of auxiliaries (taking notes during the tasks) was significantly less frequent and the interaction was perceived as more effortless. Finally, the full-adaptive SemaVis showed the best results with respect to user experience measures, the interaction was perceived as significantly more intuitive, the overall hedonic quality outperformed all other visualizations, and SemaVis was perceived as significantly more attractive compared to all other conditions. Overall, the evaluation clearly confirmed that our conceptual model leads to significantly higher efficient, effective, and accepted systems that reduce cognitive load and effort and are perceived as more intuitive, more attractive, and are related to higher hedonic qualities.

References

1. J.D. Fekete, J. van Wijk, J. Stasko, C. North, Information Visualization pp. 1–18 (2008)
2. D. Keim, J. Kohlhammer, G. Ellis, F. Mansmann, *Matering the Information Age Solving Problems with Visual Analytics* (Eurographics Association, 2010)
3. S.K. Card, J.D. Mackinlay, B. Shneiderman, *Readings in Information Visualization: Using Vision to Think*, 1st edn. (Morgan Kaufmann, 1999). http://www.amazon.com/exec/obidos/redirect?tag=citeulike07-20&path=ASIN/1558605339
4. J.W. Ahn, Adaptive visualization for focused personalized information retrieval. Ph.D. thesis, School of Information Sciences, University of Pittsburgh (2010). http://etd.library.pitt.edu/ETD/available/etd-08262010-150850/
5. L.C. Koh, A. Slingsby, J. Dykes, T.S. Kam, in *2010 14th International Conference Information Visualisation 0* (2011), p. 90. http://doi.ieeecomputersociety.org/10.1109/IV.2011.32
6. G.G. Grinstein, P. Hoffman, R.M. Pickett, S.J. Laskowski, *Information Visualization in Data Mining and Knowledge Discovery*, pp. 129–176 (2002)
7. C. Plaisant, J.D. Fekete, G. Grinstein, IEEE Trans. Vis. Comput. Graph. **14**(1), 120 (2008). doi:10.1109/TVCG.2007.70412
8. B.S. Santos, position paper (2005). http://www.dis.uniroma1.it/beliv08/pospap/santos.pdf
9. J.J. Van Wijk, in *Visualization, VIS 05*, (IEEE, 2005), pp. 79–86
10. S. Carpendale, *Information Visualization*, pp. 19–45 (2008)
11. E.W. Anderson, K.C. Potter, L.E. Matzen, J.F. Shepherd, G.A. Preston, C.T. Silva, Comput. Graph. Forum **30**(3), 791 (2011). doi:10.1111/j.1467-8659.2011.01928.x
12. B. Steichen, G. Carenini, C. Conati, in *Proceedings of the 2013 International Conference on Intelligent User Interfaces* (ACM, New York, NY, USA, 2013), IUI '13, pp. 317–328. doi:10.1145/2449396.2449439. http://doi.acm.org
13. C. Plaisant, in *Proceedings of the Working Conference on Advanced Visual Interfaces*, pp. 109–116 (2004)
14. A. Komlodi, A. Sears, E. Stanziola, Information visualization evaluation review. Technical report, ISRC Tech. Report, Department of Information Systems, UMBC. UMBC-ISRC-2004-1 (2004)
15. J. Lazar, J. Feng, H. Hochheiser, *Research Methods in Human-Computer Interaction* (Wiley, Indianapolis, 2010)
16. T. Zuk, L. Schlesier, P. Neumann, M.S. Hancock, S. Carpendale, in Proceedings of the 2006 AVI Workshop on BEyond Time and Errors: Novel Evaluation Methods for Information Visualization, pp. 1–6 (2006)
17. J.L. Gabbard, K. Swartz, K. Richey, D. Hix, Simulation Series 31 (1999)
18. A. Falk, J.J. Heckman, Science **326**(5952), 535 (2009)
19. H. Lam, E. Bertini, P. Isenberg, C. Plaisant, S. Carpendale, Technical report 2011, University of Calgary (2011)
20. A. Veerasamy, N.J. Belkin, in *Proceedings of the 19th Annual International ACM SIGIR Conference on Research and Development in Information Retrieval*, pp. 85–92 (1996)
21. O.I.d. Normalizacion, *ISO Guide 34 General Requirements for the Competence of Reference Material Producers* (ISO, 2000)
22. D. Sprague, M. Tory, Motion in casual InfoVis and the interrelation between personality, performance, and preference. Technical report, University of Victoria. British Columbia, Canada (2009)

23. A.M. Treisman, G. Gelade, Cogn. Psychol. **12**(1), 97 (1980)
24. J.M. Wolfe, W. Gray (Ed.), Integr. Models Cognitive Syst. 99–119 (2007)
25. C. Ware, *Information Visualization Perception for Design* (Morgan Kaufmann Publishers, 2004)
26. C. Ware, *Information Visualization Perception for Design* (Morgan Kaufmann (Elsevier), 2013)
27. R.A. Rensink, Ann. Rev. Psychol. **53**, 245 (2002)
28. A.M. Treisman, Sci. Am. **255**, 106 (1986)
29. A.M. Treisman, S. Gormican, Psychol. Rev. **95**(1), 15 (1988)
30. J.M. Wolfe, J. Exp. Psychol. Hum. Percept. Perform. **15**(3), 419 (1989)
31. J.M. Wolfe, Psychon. Bull. Rev. **1**(2), 202 (1994)
32. J. Wolfe, T.S. Horowitz, Scholarpedia **3**(7), 3325 (2008)
33. E. Jun, S. Landry, G. Salvendy, Int. J. Human Comput. Inter. **27**(4), 348 (2011)
34. R.A. Rensink, Vis. Cogn. **7**, 17 (2000)
35. S. Kyllingsbàek, Behav. Res. Methods **38**(1), 123 (2006)
36. G.D. Logan, Ann. Rev. Psychol. **55**(1), 207 (2004). doi:10.1146/annurev.psych.55.090902. 141415
37. K. Nazemi, J. Kohlhammer, in 1st International Workshop on User-Adaptive Visualization (WUAV 2013). Extended Proceedings of UMAP 2013, CEUR Workshop Proceedings. ISSN 1613–0073, vol. 997 (2013), CEUR Workshop Proceedings. ISSN 1613–0073, vol. 997
38. P. Downing, C. Dodds, Visual Cogn. **11**(6), 689 (2004). doi:10.1080/13506280344000446
39. C. Lucas, J. Lauwereyns, Experimental Psychology (formerly Zeitschrift für Experimentelle Psychologie) **54**(4), 256 (2007). doi:10.1027/1618-3169.54.4.256
40. N. Lavie, Visual Cogn. **14**(4–8), 863 (2006). doi:10.1080/13506280500195953
41. J. Somervell, D.S. McCrickard, C. North, M. Shukla, in *Proceedings of the Symposium on Data Visualisation 2002*, pp. 211–216 (2002)
42. V. Aginsky, B.U. IV, M.J. Tarr, Visual Cogn. **7**(1–3), 147 (2000)
43. L. Matzen, L. McNamara, K. Cole, A. Bandlow, C. Dornburg, T. Bauer, in *BELIV'10: BEyond time and errors: novel evaLuation methods for Information Visualization*. Workshop of the ACM CHI 2010 (2010)
44. C. Conati, H. Maclaren, in *Proceedings of the Working Conference on Advanced Visual Interfaces* (ACM, New York, NY, USA, 2008), AVI '08, pp. 199–206. doi:10.1145/1385569. 1385602. http://doi.acm.org/10.1145/1385569.1385602
45. C. Conati, G. Carenini, M. Harati, D. Tocker, N. Fitzgerald, A. Flagg, in *AAAI Workshops* (2011). http://www.aaai.org/ocs/index.php/WS/AAAIW11/paper/view/3944
46. D. Ullrich, S. Diefenbach, in Mensch & Computer 2010: 10. Fachübergreifende Konferenz für interaktive und kooperative Medien. Interaktive Kulturen (Oldenbourg Wissenschaftsverlag, 2010), p. 251
47. D. Ullrich, S. Diefenbach, in *Proceedings of the 6th Nordic Conference on Human-Computer Interaction: Extending Boundaries* (ACM, 2010), pp. 801–804
48. J.W. Ahn, P. Brusilovsky, in *User Modeling, Adaptation, and Personalization, Lecture Notes in Computer Science*, ed. by P. Bra, A. Kobsa, D. Chin, vol. 6075 (Springer, Berlin, 2010), pp. 4–15. doi:10.1007/978-3-642-13470-8_3
49. I.R. Olson, M.M. Chun, Visual Cogn. **9**(3), 273 (2002). doi:10.1080/13506280042000162
50. D. Toker, C. Conati, G. Carenini, M. Haraty, in *User Modeling, Adaptation, and Personalization, Lecture Notes in Computer Science*, ed. by J. Masthoff, B. Mobasher, M. Desmarais, R. Nkambou, vol. 7379 (Springer, Berlin, 2012), pp. 274–285. doi:10.1007/978-3-642-31454-4_23
51. D. Toker, C. Conati, B. Steichen, G. Carenini, in *Proceedings of the SIGCHI Conference on Human Factors in Computing Systems* (ACM, New York, NY, USA, 2013), CHI '13, pp. 295–304. doi:10.1145/2470654.2470696. http://doi.acm.org/10.1145/2470654.2470696
52. B. Steichen, G. Carenini, C. Conati, in *UMAP Workshops, CEUR Workshop Proceedings*, ed. by E. Herder, K. Yacef, L. Chen, S. Weibelzahl, vol. 872 (CEUR-WS.org, 2012), CEUR Workshop Proceedings, vol. 872. http://dblp.uni-trier.de/db/conf/um/umap2012w.html

53. B. Steichen, O. Schmid, C. Conati, G. Carenini, in *UMAP 2013 Extended Proceedings. First International Workshop on User-Adaptive Visualizations (WUAV 2013)* (2013)
54. G. Marchionini, Commun. ACM **49**(4), 41 (2006). doi: 10.1145/1121949.1121979. http://doi.acm.org/10.1145/1121949.1121979
55. R.W. White, R.A. Roth, Exploratory Search: Beyond the Query-Response Paradigm, Synthesis Lectures on Information Concepts, Retrieval, and Services. G. Marchionini (Ed), vol. 1 (Morgan & Claypool Publishers, 2009). doi:10.2200/s00174ed1v01y200901icr003
56. D. Simkin, R. Hastie, J. Am. Stat. Assoc. **82**(398), 454 (1987). doi:10.1080/01621459.1987.10478448
57. S. Few, Information Management Magazine (2005)
58. S. Koshman, A. Spink, B.J. Jansen, C. Blakely, J. Weber, in *Canadian Association for Information Science Conference* (2006)
59. J. Gemmell, G. Bell, R. Lueder, S. Drucker, C. Wong, in *Proceedings of the tenth ACM international conference on Multimedia*, pp. 235–238 (2002)
60. P. Zhang, in *Proceedings of The Second Americas Conference on Information Systems (AIS'96)* (1996), pp. 16–18
61. G. Kiczales, ACM Comput. Surv. (CSUR) 28(4es), **154** (1996)
62. L. Chittaro, C. Combi, in *Proceedings of the Eighth International Symposium on Temporal Representation and Reasoning, 2001. TIME 2001* (2001), pp. 13–20
63. T. Lavie, N. Tractinsky, Int. J. Hum.-Comput. Stud. **60**(3), 269 (2004). doi:10.1016/j.ijhcs.2003.09.002
64. N. Cawthon, A.V. Moere, in *Information Visualization, 2007. IV'07. 11th International Conference* (2007), pp. 637–648
65. E. Michailidou, S. Harper, S. Bechhofer, in *Proceedings of the 26th Annual ACM International Conference on Design of Communication* (2008), pp. 215–224
66. K. Nazemi, C. Stab, A. Kuijper, in *Human-Computer Interaction. Design and Development Approaches, Lecture Notes in Computer Science*, ed. by J. Jacko, vol. 6761 (Springer, 2011), pp. 480–489. doi:10.1007/978-3-642-21602-2_52
67. K. Nazemi, R. Retz, J. Bernard, J. Kohlhammer, D. Fellner, in *Advances in Visual Computing, Lecture Notes in Computer Science*, ed. by G. Bebis, R. Boyle, B. Parvin, D. Koracin, B. Li, F. Porikli, V. Zordan, J. Klosowski, S. Coquillart, X. Luo, M. Chen, D. Gotz, vol. 8034 (Springer, Berlin, 2013), pp. 13–24. doi:10.1007/978-3-642-41939-3_2
68. Freebase consortium. Freebase—a community-curated database of well-known people, places, and things (2013). http://www.freebase.com/. Accessed Aug 2013
69. J. Waniek, H. Langner, F. Schmidsberger, in *Special Interest Tracks and Posters of the 14th international Conference on World Wide Web* (2005), pp. 900–901
70. T. Richter, J. Naumann, N. Groeben, Comput. Human Behav. **16**(5), 473 (2000)
71. R.M. Hogarth, *Educating Intuition* (University of Chicago Press, Chicago, 2001)
72. C. Chen, Y. Yu, Int. J. Human-Comput. Stud. **53**(5), 851 (2000). doi:10.1006/ijhc.2000.0422
73. T. Richter, J. Naumann, N. Groeben, Das Inventar zur Computerbildung (INCOBI): Ein Instrument zur Erfassung von Computer Literacy und computerbezogenen Einstellungen bei Studierenden der Geistes-und Sozialwissenschaften (Reinhardt, 2001)
74. B.S. Bloom, *Taxonomy of Educational Objectives* (David McKay Co., Inc., New York, 1956)
75. Eurographics Consortium. Eurographics digital library (2013). https://diglib.eg.org
76. F.G. Giesbrecht, M.L. Gumpertz, Latin Squares (John Wiley & Sons Inc, 2005), pp. 118–157. doi:10.1002/0471476471.ch6
77. F. Faul, E. Erdfelder, A.G. Lang, A. Buchner, Behav. Res. Methods **39**, 1149 (2007)
78. F. Faul, E. Erdfelder, A. Buchner, A.G. Lang, Behav. Res. Methods **41**, 1149 (2009)
79. C. Summerfield, T. Egner, Trends Cogn. Sci. **13**(9), 403 (2009). doi:10.1016/j.tics.2009.06.003
80. M. Hassenzahl, M. Burmester, F. Koller, in *Mensch & Computer 2003* (Springer, 2003), pp. 187–196
81. P. Baghaei, N. Amrahi, Psychol. Test Assess. Model. **53** (2011)
82. T.M. Haladyna, S.M. Downing, M.C. Rodriguez, Appl Measur. Educ. **15**(3), 309 (2002)

83. T.E. Wilens, M.H. Verlinden, L.A. Adler, P.J. Wozniak, S.A. West, Bio. Psychiatry **59**(11), 1065 (2006). doi:10.1016/j.biopsych.2005.10.029. http://www.sciencedirect.com/science/article/pii/S0006322305014423
84. T. Kirk, J.D. Roache, R.R. Griffiths, J. Clin. Psychopharmacol. **10**(3), 160 (1990). http://search.ebscohost.com/login.aspx?direct=true&db=psyh&AN=1991-00697-001&site=ehost-live
85. D.R. Jasinski, J. Psychopharmacol. **14**(1), 53 (2000). http://search.ebscohost.com/login.aspx?direct=true&db=psyh&AN=2000-08187-007&site=ehost-live
86. H.W. Reese, J. Exp. Child Psychol. **64**(1), 137 (1997). doi:10.1006/jecp.1996.2333. http://www.sciencedirect.com/science/article/pii/S0022096596923334
87. J. Gaito, Psychometrika **23**(4), 369 (1958). doi:10.1007/BF02289785
88. R.P. Hamlin, Eur. J. Mark. **39**(3–4), 328 (2005). http://search.ebscohost.com/login.aspx?direct=true&db=psyh&AN=2005-08599-006&site=ehost-live
89. D.J. Steele, J.D. Medder, P. Turner, Med. Educ. **34**(1), 23 (2000). http://search.ebscohost.com/login.aspx?direct=true&db=psyh&AN=2000-13244-002&site=ehost-live
90. K.A.D. Stahl, J. Literacy Res. **40**(3), 359 (2008). http://search.ebscohost.com/login.aspx?direct=true&db=psyh&AN=2009-09317-004&site=ehost-live
91. J. Meijer, R. Oostdam, Anxiety, Stress Coping Int. J. **20**(1), 77 (2007). http://search.ebscohost.com/login.aspx?direct=true&db=psyh&AN=2007-04652-007&site=ehost-live
92. J. Giles, D.A.J. Ryan, G. Belliveau, E. De Freitas, R. Casey, Act. Learn. High. Educ. **7**(3), 213 (2006). http://search.ebscohost.com/login.aspx?direct=true&db=psyh&AN=2006-13309-002&site=ehost-live
93. H.N.J. Schifferstein, K.S.S. Talke, D.J. Oudshoorn, Chemosens. Percept. **4**(1–2), 55 (2011). http://search.ebscohost.com/login.aspx?direct=true&db=psyh&AN=2011-11665-007&site=ehost-live
94. G. Öquist, M. Goldstein, in *Human Computer Interaction with Mobile Devices, Lecture Notes in Computer Science*, ed. by F. Paternó (Springer, Berlin, 2002), pp. 225–240. doi:10.1007/3-540-45756-9_18
95. S. Buisine, J.C. Martin, Interact. Comput. **19**(4), 484 (2007). http://search.ebscohost.com/login.aspx?direct=true&db=psyh&AN=2007-10838-004&site=ehost-live
96. M. Breyer, J. Birkenbusch, D. Burkhardt, C. Schwarz, C. Stab, K. Nazemi, O. Christ, in *Advances in Affective and Pleasurable Design* (Taylor & Francis, CRC Press, Boca Raton, 2012), Advances in Human Factors and Ergonomics Series. Proceedings of the 4th AHFE Conference; 14, pp. 8066–8074. http://bibcd.igd.fraunhofer.de/bibcd/INI_Science/papers/2012/12p057.pdf
97. M. Hassenzahl, M. Burmester, F. Koller, in *Usability Professionals 2008*, ed. by H. Brau, S. Diefenbach, M. Hassenzahl, F. Koller, M. Peissner, K. Röse (Stuttgart: German Chapter der Usability Professionals Association, 2008), pp. 78–82
98. I. Torre, User Model. User-Adap. Inter. **19**(5), 433 (2009). doi:10.1007/s11257-009-9067-3
99. G. Fiotakis, D. Raptis, N. Avouris, in *Human-Computer Interaction at INTERACT 2009, Lecture Notes in Computer Science*, ed. by T. Gross, J. Gulliksen, P. Kotz, L. Oestreicher, P. Palanque, R. Prates, M. Winckler, vol. 5726 (Springer, Berlin, 2009), pp. 231–234. doi: 10.1007/978-3-642-03655-2_27
100. M. Tory, T. Möller, A model-based visualization taxonomy. Technical. report. CMPT-TR2002-06, Computing Science Department, Simon Fraser University (2002)
101. J.W. Ahn, P. Brusilovsky, Inf. Vis. **88** (3), 180 (2009). http://ivi.sagepub.com/cgi/content/short/8/3/167
102. J.W. Ahn, P. Brusilovsky, Inf. Process. Manage. **49**(5), 1139 (2013). doi:10.1016/j.ipm.2013.01.007. urlhttp://www.sciencedirect.com/science/article/pii/S0306457313000137

Chapter 9
Conclusions and Future Work

The increasing amount and complexity of data in digital repositories emerge new challenges in particular for accessing the vast amount of data by human. The challenge of human access to data is investigated by different and partially complementary research areas. Information visualization focuses on the human visual information processing to provide an interactive *picture* of the data and to amplify human's cognition and provide insights and knowledge. Semantic technologies aim at formalizing data in a machine-readable way and providing "meanings" based on the formalization to make data accessible for human. The relatively young research field of adaptive information visualization combines approaches from information visualization with methods of adaptive systems to provide personalized and enhanced visual representations of data. Recent research in adaptive visualizations showed significant advances in human information processing. The adaptation techniques were applied to search and exploration tasks, whereas user studies evidenced promising and beneficial results. Although, this research area showed significant advances in human access to vast data, existing methods and approaches show shortcomings and limitations. For information visualization the transformation of data to a human-centered visual representation plays a key-role. Adaptation approaches for information visualization should therefore consider human and data as influence factors that affect the visualizations. Today's system either adapts based on data or on user. An investigation of both was not performed yet. Further existing adaptive visualizations that adapts based on human interaction patterns have to be trained by experts. With each new visual layout the entire system have to be trained with commonly static behavioral patterns as repeated interaction sequences. To our best of knowledge there existed no method that learned from users' behavior initially and adapted to the average behavior, instead of being trained by experts. But in our opinion the main limitation of today's approaches is that the transformation pipeline of data to visual representation is not considered. Although, there are many studies of visual perception, reference models for information visualization, and a great and useful number of methods, applications and their effects to human perception, all these fundamental works are not applied in today's adaptive visualization approaches. This research area did not investigate the human interface adaptation. The focus relies

© Springer International Publishing Switzerland 2016
K. Nazemi, *Adaptive Semantics Visualization*, Studies in Computational
Intelligence 646, DOI 10.1007/978-3-319-30816-6_9

more *to what* can or should be adapted rather than *what can be adapted*. A coherent model that investigates the potentials of information visualization with its various variables that influence our perception and consequently the information acquisition was missing.

In this thesis we faced the identified limitations of adaptive visualizations and presented a novel and coherent model for adaptive visualization for information acquisition. In contrast to existing systems and approaches, we investigated in particular the potentials of information visualization for adaptation. Therefore, our reference model for visual adaptation considers not only the entire transformation pipeline from data to visual representation, it enhances far more the reference model to meet the requirements for individual and canonical adaptive visualizations. Our model provides an adaptation on different visual layers and enhances the state of research. Each of the identified layers can be adapted automatically by various influencing factors. The transformation steps from data to visualization are enhanced to provide a fine granular adaptation of visual parameters. To identify the visual layers that affect the human information processing, we investigated various models and studies on visual perception. Our proposed model considers both human and data as influencing factor and does not require an initial training by experts. For this, we introduced the canonical user model, an approach that investigates the usage behavior of all users with the system and enables an initial adaptation based on the average usage behavior.

9.1   Summary

To achieve our goal and provide a coherent and comprehensive model for adaptive visualization in particular for semantics, we started with an extensive literature review. Therefore, we started with an overview of the various disciplines, techniques, goals, and approaches that are coupled to interactive information visualization. In particular the investigation of human visual perception, visualization tasks, and data models were applied in our conceptual design. We further introduced the reference model for information visualization and the differentiation of visual layers, and models of visual perception that built the foundations of our conceptual design. This was followed by an extensive review existing systems and approaches for semantics visualizations. For obtaining a clear picture, we defined first semantics visualization. Further a classification of semantics visualization for providing a comprehensible picture of the existing systems was introduced. Our state of art review covered the last decade, whereas the existing systems were introduced based on our classification. Our review outlined that none of the existing systems support the exploratory search process, although semantics is predestined to support this kind of search. In addition, we investigated the state of the art in adaptive visualizations. To provide a comprehensible way for conveying different adaptation processes, we introduced three main aspects: influencing factors by means of *to what* can visualizations be adapted, knowledge modeling that refers to the way how the influencing factors can be formalized, and human interface adaptation that refers to visualization and their capabilities for adaptation. The main goal of this chapter was to give a comprehensive and comprehensible state of art analysis for adaptive visualizations. For this purpose we first defined adaptive visualization based on the definition of adaptive systems and the definition of information visualization. Our review on the existing systems covered the last decade. The goal was to find systems or approaches that make use of all the defined adaptation criteria, but at least combine some of them to provide a real benefit out of the visual structures. Our review clearly outlined that the emerging area of adaptive visualizations did not investigate the human interface adaptation in depth, yet. The most systems are replacing visualization types and layouts respectively based on some users' implicit or explicit demands. The focus of today's systems is more *to what* should be adapted rather than *what can be adapted*. With this chapter, we concluded the review on existing systems and approaches.

The second part of our thesis introduced our conceptual model. Thereby first the main outcomes of our literature review were summarized to deduce scientific requirements for our model. Therefore, we first identified the requirements that built the foundation on the conceptual work. Thereafter a high-level design of our conceptual model was presented as overall conceptual model. Based on the conceptual model, a detailed and replicable illustration of each layer and component was given. We described the knowledge model with its three main components of data model, data feature model, and user model. Data model described the way how semantic information are gathered from Web-sources and from non-semantic metadata and enriched with semantic information based on our iterative querying approach. This led to a formal representation of data in form of the proposed data model. Our data feature model illustrated the retrieving of quantitative measured of the underlying data with the same iterative querying approach. In this context, two weighting-algorithms that measure the relevance of semantic neighbors of selected instances were introduced in a detailed way. Thereafter, the user model and the related concepts were described that combine users' interaction behavior with data and visual layouts. Based on a formal specification of users' interactions the approach for determining and weighting user behavior and predicting users' action was introduced. In this context we introduced the formal description of the canonical user model and the group definition followed by user similarity and deviation analysis. Thereafter the general adaption process was described that guides through the entire process of adaptation and illustrates when and how the measured values and models are applied. Further, we introduced our layer based reference model of adaptation that enables a fine granular adaptation and investigates the entire transformation steps of information visualization for adaptation. We concluded our conceptual model with the description of our visualization cockpit model and illustrated how this model can be applied to support the exploratory search with juxtaposed visual layouts.

The last part of our thesis introduced the proof of our conceptual model. Therefore first the architectural design of our SemaVis technology was described based on a design pattern. The general architecture aimed at providing the technical interplay of the introduced approaches, algorithms, and models of our conceptual model. It gave an overview of the implementation strategy and enabled a mapping to the already introduced overall conceptual model. Aside from the general architecture of SemaVis three exemplary application scenarios were introduced that enabled to comprehend the adaptation behavior of our system and evidenced the feasibility of our model. Beside the proof of feasibility, we introduced an empirical study on one of the described application scenarios. The main goal of the empirical user study was to prove the added values of our model and find limitations. We conducted first a preliminary user study to evaluate only the effects of the visual variables and find an appropriate design for our main study. For the main evaluation, we implemented an evaluation-software that collected data, guided the user through the evaluation scenario, and reduced thereby human errors. The evaluation was conducted as a within-subject Latin-Square design with four conditions and two interventions. The conditions were applied to our reference model and enhanced with a textual baseline. The interventions were task-types deduced from exploratory search models and were classified into 40 simple and exploratory tasks. Overall a total number of 53 persons participated in the evaluation. Overall four main hypotheses with nine sub-hypotheses were deduced to measure performance, acceptance, and hedonic values. The results of the evaluation illustrated that all our assumptions are confirmed. The full-adaptive semantics visualization is more efficient, more effective, leads to less cognitive load and effort and to higher satisfaction and user experience. In all our hypotheses, the full-adaptive SemaVis outperformed the non-adaptive visualization, the partially adaptive visualization, and the textual baseline.

9.2   Benefits of the Visual Adaptation Model

We focused in this thesis on a more human-centered view on visual adaptation and investigated in particular the way how information visualization or in our specific case semantics visualization can be adapted to data and user characteristics, e.g. user interests or prior knowledge. Our research focused on the different visual variables and layers that affect the information acquisition process. Consequently, the major benefits of our models regard the visual adaptation for searching and acquiring infor-mation. With our *Overall Conceptual Model*, we identified four layers that lead to a more appropriate adaptation of visualizations. The conceptual model addresses the identified limitations in existing systems and provides a novel model for adapting semantics visualizations based on user and data characteristics. Thereby the sur-pluses of existing models are used and combined with new approaches to provide a more reliable adaptation model. The conceptual model contains four main layers, whereas each of these layers contains components that lead to advanced adaptation of visualizations. The main advances in particular on the interface adaptation were provided by our *Reference Model for Adaptive Visualization* that investigated and enhanced the transformation steps from data to visual representation. The reference model contains four adaptation layers, *Semantics*, *Visual Layout*, *Visual Variable*, and *Visual Interface* and includes the transformation steps of *data transformation*, *visual mapping*, *retinal variable mapping*, and *visual layout orchestration*. Beside the transformation steps, the four layers can be adapted by the conceptual model and the integrated adaptation processes. The main benefit here is the advanced reference model that can be applied to any kind of visual adaptation and enhances the state of the art with the various levels of adaptation based on human visual perception.

Our *User Model* approach comprises both data and user for the adaptation process in contrast to existing systems. The user model includes thereby the combined interaction behavior with data and visual layouts. With the subsumption on concept-level, we further enhance existing approaches for user modeling to a domain-independent model. The introduced user model further makes use of the semantic hierarchy of data. Within a certain knowledge domain, the model provides conceptual information that leads to recommend data from the same semantic concepts. Beside the behavioral analysis of users, an enhanced prediction algorithm that enables the guidance of users' attention to data or load not selected data on demand was proposed. The appliance of a canonical user model that represents the average behavior of all users and leads to a general adaptation of the visualization environment supersedes the necessity of an expert to train the system. Further similarity and deviation measures allow providing a more individualized adaptation to certain users and controlling the level of adaptation. With our *visualization cockpit model*, we proposed a novel approach that enables the composition of different visual layouts on screen. Thereby the visual layouts are either linked with each other, linked with certain data-bases or sub-set of data, or not linked with other visual layouts. The model enables an enhanced exploratory search with visualizations, whereas the placement and arrangement of the visual layouts can be either performed by users or by the system.

We have further provided a proof of our concept with the SemaVis technology. SemaVis is an adaptive semantics visualization that implements our conceptual model for providing a more efficient and effective way of visual information acquisition. The benefits of our adaptation model were evidenced with one application scenario of SemaVis. We therefore conducted a comprehensive empirical user study with four conditions. Beside the textual baseline, a non-adaptive version of SemaVis, a partially adaptive version of SemaVis, and the full-adaptive SemaVis were evaluated. The conditions were applied from our reference model to illustrate the main effects of the adaptation of different visual layers. The evaluation revealed that the full-adaptive SemaVis that is based on our conceptual model outperforms all other conditions in terms of *Efficiency, Effectiveness, Cognitive Load and Perceived Effort*, and *Satisfaction and User Experience*.

Efficiency   We therefore used in our evaluation the task completion time for measuring the efficiency of the different conditions. Task completion time was measured automatically by the implemented evaluation system and stored in the log-files. For a single task it constitutes the time elapsed between the first appearance of the task on the screen and the participant's confirmation of an answer. Thereby two different task-types and their completion time were measured: simple and exploratory tasks. The results revealed that all differences between the different conditions were significant. This was the case for both question types. The full-adaptive SemaVis outperformed all other conditions in terms of task completion time. The effect is present for both types of tasks, exploratory and simple tasks.

Effectiveness   Effectiveness describes the accuracy of goal achievement. This was measured by two dimensions: percentage of correctly solved tasks and percentage of omitted tasks. In both measured dimension the full-adaptive SemaVis outperformed all other conditions. It led to fewer incorrect answers than compared to all other conditions. Further fewer tasks were omitted while using the full-adaptive SemaVis compared to all other conditions. The analysis of measures regarding effectiveness revealed that the appliance of our conceptual model leads to significant improvements in terms of effectiveness.

Cognitive Load and Perceived Effort   The amount of cognitive resources needed for solving the tasks was assessed by the use of auxiliaries (pen and paper), due to the assumption that users write notes down, if their working memory is not capable to remember the intermediate results for complex tasks. The use of paper and pen is therefore an indicator for a higher cognitive load. Further a more formal way for measuring the perceived effort was performed with the sub-scale *Effortlessness* of the INTUI questionnaire. Effortlessness refers similar to the use of further auxiliaries the perceived ease to solve a task. The full-adaptive SemaVis led to significantly less usage of further auxiliaries (pen and paper) during the task solving process. Further the perceived effort measured with the questionnaire was significantly lower while working with the full-adaptive SemaVis in contrast to all other conditions. The full-adaptive SemaVis consumes less cognitive effort and the work is perceived as more effortless.

Satisfaction and User Experience   To measure the satisfaction and user experience, we used three dimensions: the sub-scale *Intuitiveness* of the INTUI questionnaire and the sub-scales *Hedonic Quality* and *Attractiveness* of the AttrakDiff questionnaire. The full-adaptive SemaVis outperformed in all three sub-scales all the other conditions. Consequently, our conceptual model leads to a visual adaptation that increases the users' satisfaction and user experience. This leads to a higher acceptance by users.

In summary, the results of the evaluation illustrate that the full-adaptive SemaVis and thereby our proposed conceptual model is more efficient, more effective, leads to less cognitive load and effort and to higher satisfaction and user experience. The full-adaptive SemaVis outperformed the non-adaptive SemaVis, the partially adaptive SemaVis, and the textual baseline in all our assumptions.

9.3   Prospects for Future Work

Our conceptual model, the implementation as SemaVis, and the evaluation of our model contributed to the research on adaptive visualizations. This section outlines futures directions that were in particular emerged during our research in the course of this thesis.

Data Mining for Knowledge Modeling   Our conceptual model and the related approaches focused on the visualization of semantics. Both, the user model approaches and the data model were determined through the interaction of semantically annotated data or at least metadata. Due to the rapidly increasing semantic repositories as Linked-Data, our model provides a sufficient way for a visual adaptation of these kinds of data. An interesting enhancement would be to apply our model to non-semantic data and use instead of the formalized conceptualizations data mining approaches to generate topics, keywords, or key-features of non-structured document corpora. Text mining methods could be used for instance to extract key-terms from text documents users are interested in and build based on the extracted terms models that represent users behavior. To generate semantics, the extracted terms could be queried in semantic data-sources. With our iterative querying our conceptual model could be then applied without the need of major changes.

Process-oriented Adaptation   As we focused on user and data as influencing factors, we found out during our research that another main influencing factor for adaptive visualization could be the process of task solving that can be divided and defined as activities. There are application scenarios such as the introduced policy modeling scenario, in which the tasks of the different users vary. Not only the goals or different but also the way to achieve the goal in visual manner. An enhancement of our model would be the investigation of the process as repeated sequences of activities and activities as repeated sequences of interactions. Thereby machine learning approaches could be used to detect certain activity patterns automatically and support users in their process of problem solving. The entire process can be visualized for the user and prediction algorithms may highlight the next possible user activity or in best case the entire process that should be passed until the main goal is reached. Users of the system can intervene and correct the process or they may be guided through the entire process. Thereby the activities may be coupled to certain sets of visual layouts that fit best for the current task.

Editable User Model Visualization   SemaVis adapts the visual appearance based on the underlying user model, whereas the user himself is not aware about the weightings, measurements, and in particular topics and visual layouts that lead to the adaptation effect. Another fruitful advancement would be the visualization of the individual user model for the particular user with the ability to change measures and correct the observed and trained system behavior. The advantages are two-fold: (1) the user experiences more transparency, thus she can clearly see and recognize which data and which visual layouts were identified as relevant for her and (2) occurred errors in the user model can be changed or adapted by the user herself. Thus human learns in course of the time and the interests may change, the editable visualization of the user model would be helpful to give users more control on the adaptation of the different layers.

Search Intention Analysis   Our main user analysis approaches start with users' interactions with the visualization. Commonly the first step of a search process is querying a search-term. Although, we investigated in one of the application scenarios the search-term as input for our user model, an analysis of the search term and thereby the search intention would be an appropriate enhancement of our model. Thereby the frequency distribution of words in a certain language can be used as an indicator for determining how specific a search was defined. The more specific a search was expressed, the more precise can the search results can be visualized. In turn, if the search was verbalized with words that are common in a certain language, a more exploratory view on the results can be given to support the knowledge acquisition process of the user.

Long-term User Study   Although, we conducted a comprehensive user study, the underlying user model of our study was the canonical user model. An investigation of how our adaptive visualization affects the individualized view, information acquisition, and learning process would enlighten more facets of our conceptual model in particular for personalized visualization. We suggest conducting a between-subject long-term evaluation with a period of at least six months. During this time counter-balanced groups should work with either with SemaVis that adapts based on the canonical user model or a version that adapts based on the individual user model. This procedure would help to find out the added values, if given at all, on individual level compared to the canonical level.

Appendix A
Publications and Further Readings

The thesis is partially based on the following publications:

1. Nazemi, Kawa; Retz, Reimond; Burkhardt, Dirk; Kuijper, Arjan; Kohlhammer, Jörn; Fellner, Dieter W.: Visual Trend Analysis with Digital Libraries In: i-KNOW 2015: Proceedings of the 15th International Conference on Knowledge Technologies and Data-driven Business. New York: ACM, 2015, pp. 14:1–14:8. (ACM International Conference Proceedings Series).- DOI 10.1145/2809563.2809569
2. Nazemi, Kawa; Burkhardt, Dirk; Hoppe, David; Nazemi, Mariam: Web-based Evaluation of Information Visualization. In: Procedia Manufacturing. 3 (2015), pp. 5527–5534. DOI 10.1016/j.promfg.2015.07.718
3. Nazemi, Kawa; Burkhardt, Dirk; Retz, Reimond; Kuijper, Arjan & Kohlhammer, Jörn: Adaptive Visualization of Linked-Data. In Advances in Visual Computing, Bebis, George et al., (Ed.), Lecture Notes in Computer Science. Proceedings of 10th Internation Symposium for Visual Computing (ISVC) 2014. Springer International Publishing, 2014, pp. 872–883.
4. Nazemi, Kawa; Burkhardt, Dirk; Retz, Wilhelm & Kohlhammer, Jörn: Adaptive Visualization of Social Media Data for Policy Modeling. In Advances in Visual Computing, Bebis, George et al., (Ed.), Lecture Notes in Computer Science. Proceedings of 10th Internation Symposium for Visual Computing (ISVC) 2014. Springer International Publishing, 2014, pp. 333–344.
5. Nazemi, Kawa; Kuijper, Arjan; Hutter, Marco; Kohlhammer, Jörn & Fellner, Dieter. W.: Measuring Context Relevance for Adaptive Semantics Visualizations. In I-KNOW 2014: Proceedings of the 14th International Conference on Knowledge Management and Business-driven Technologies, 2014, ACM Press.
6. Becker, Tilman; Burghart, Catherina; Nazemi, Kawa; Ndjiki-Nya, Patrick; Riegel, Thomas; Schäfer, Ralf; Sporer Thomas; Tresp, Volker & Wissmann Jens: Core Technologies for the Internet of Services. In Towards the Internet of Services: The Theseus Program, Wahlster, W. et. al., (Eds.). Springer, 2014, pp. 59–88.

© Springer International Publishing Switzerland 2016
K. Nazemi, *Adaptive Semantics Visualization*, Studies in Computational Intelligence 646, DOI 10.1007/978-3-319-30816-6

7. Nazemi, Kawa, Retz, Wilhelm; Kohlhammer, Jörn & Kuijper, Arjan: User Similarity and Deviation Analysis for Adaptive Visualizations. In Human Interface and the Management of Information. Proceedings of the 16th International Conference on Human-Computer Interaction. Springer Berlin, Heidelberg, New York: LNCS 8521 Volume I, 2014, pp. 64–75.
8. Nazemi, Kawa; Breyer, Matthias; Burkhardt, Dirk; Stab, Christian & Kohlhammer, Jörn: SemaVis—A New Approach for Visualizing Semantic Information. In Wahlster, W. et al. (Eds.) Towards the Internet of Services: The Theseus Program, Springer, 2014, pp. 191–202.
9. May, Thorsten; Nazemi, Kawa & Kohlhammer, Jörn: From Raw Data to Rich Visualization—Combining Visual Search with Data Analysis. In Wahlster, W. et al. (Eds.) Towards the Internet of Services: The Theseus Program, Springer, 2014.
10. Burkhardt, Dirk; Nazemi, Kawa; Encarnacao, Jose Daniel; Retz, Wilhelm & Kohlhammer, Jörn: Visualization Adaptation Based on Environmental Influencing Factors. In M. Kurosu (Ed.): Human-Computer Interaction, Part I. Proceedings of the 16th International Conference on Human-Computer Interaction. LNCS 8510, Springer International Publishing Switzerland, 2014, pp. 411–422.
11. Nazemi, Kawa; Steiger, Martin; Burkhardt, Dirk and Kohlhammer, Jörn: Information Visualization and Policy Modeling. In Sonntagbauer, P.; Nazemi, K.; Sonntagbauer, S.; Prister, G. & Burkhardt, D. (Eds.) Advanced ICT Integration for Governance and Policy Modeling, IGI Global, 2014, (to appear).
12. Burkhardt, Dirk; Nazemi, Kawa; Zilke, Jan R.; Kohlhammer, Jörn and Kuijper, Arjan: Policy Modeling Methodologies. In Sonntagbauer, P.; Nazemi, K.; Sonntagbauer, S.; Prister, G. & Burkhardt, D. (Eds.) Advanced ICT Integration for Governance and Policy Modeling, IGI Global, 2014, (to appear).
13. Burkhardt, Dirk; Nazemi, Kawa and Kohlhammer, Jörn: Fundamental Aspects for E-Government. In Sonntagbauer, P.; Nazemi, K.; Sonntagbauer, S.; Prister, G. & Burkhardt, D. (Eds.) Advanced ICT Integration for Governance and Policy Modeling, IGI Global, 2014, (to appear).
14. Sonntagbauer, Susanne; Sonntagbauer, Peter; Nazemi, Kawa and Burkhardt, Dirk: The FUPOL Policy Lifecycle. In Sonntagbauer, P.; Nazemi, K.; Sonntagbauer, S.; Prister, G. & Burkhardt, D. (Eds.) Advanced ICT Integration for Governance and Policy Modeling, IGI Global, 2014, (to appear).
15. Burkhardt, Dirk; Nazemi, Kawa and Kohlhammer, Jörn: Visual Process Support to Assist Users in Policy Making. In Sonntagbauer, P.; Nazemi, K.; Sonntagbauer, S.; Prister, G. & Burkhardt, D. (Eds.) Advanced ICT Integration for Governance and Policy Modeling, IGI Global, 2014, (to appear).
16. Burkhardt, Dirk; Nazemi, Kawa; Klamm, Christopher, Kohlhammer, Jörn & Kuijper, Arjan: Comparison of E-Participation Roadmap in Industrial and Developing Countries based on Germany and Kenya. In Proceedings of the 8th International Conference on Theory and Practice of Electronic Governance. ACM Press, 2014 (to appear).
17. Nazemi, Kawa; Retz, Reimond; Bernard, Jürgen; Kohlhammer, Jörn; Fellner, Dieter W.: Adaptive Semantic Visualization for Bibliographic Entries. In: Bebis,

George (Ed.) et al.: Advances in Visual Computing. 9th International Symposium, ISVC 2013: Proceedings, Part II. Berlin, Heidelberg, New York: Springer, 2013, pp. 13–24. (Lecture Notes in Computer Science (LNCS) 8034).

18. Burkhardt, Dirk; Nazemi, Kawa; Sonntagbauer, Peter; Kohlhammer, Jörn: Interactive Visualizations in the Process of Policy Modeling. In: Wimmer, Maria A. (Ed.) et al.:Gesellschaft für Informatik (GI): Electronic Government and Electronic Participation: Joint Proceedings of Ongoing Research of IFIP EGOV and IFIP ePart 2013. Bonn: Köln, 2013, pp. 104–115. (GI-Edition—Lecture Notes in Informatics (LNI) P-221).

19. Burkhardt, Dirk; Nazemi, Kawa; Stab, Christian; Steiger, Martin; Kuijper, Arjan; Kohlhammer, Jörn: Visual Statistics Cockpits for Information Gathering in the Policy-Making Process. In: Bebis, George (Ed.) et al.: Advances in Visual Computing. 9th International Symposium, ISVC 2013: Proceedings, Part II. Berlin, Heidelberg, New York: Springer, 2013, pp. 86–97. (Lecture Notes in Computer Science (LNCS) 8034).

20. Nazemi, Kawa; Kohlhammer, Jörn: Visual Variables in Adaptive Visualizations. In 1st International Workshop on User-Adaptive Visualization (WUAV 2013). UMAP 2013 Extended Proceedings of the 21st Conference on User Modeling, Adaptation, and Personalization. ISSN 1613-0073, 2013, pp. 32–36. (CEUR Workshop Proceedings Vol-997).

21. Stab, Christian; Burkhardt, Dirk; Nazemi, Kawa: Visualizing Search Results of Linked Open Data. In: Hussein, Tim (Ed.) et al.: Semantic Models for Adaptive Interactive Systems. Berlin, Heidelberg, New York: Springer, 2013, pp. 133–149. (Human-Computer Interaction Series).

22. Sonntagbauer, Peter; Rumm, Nikolaus; Kagitcioglu, Hakan; Nazemi, Kawa; Burkhardt, Dirk: GIS, Social Media and Simulation In Integrated ICT Solutions for Urban Futures. In: Network—Association of European Researchers on Urbanisation in the South (N-AERUS) u.a.: proceedings of the 14th N-AERUS Conference 2013: Urban Futures. Multiple visions, paths and constructions? pp.1–12

23. Stab, Christian; Breyer, Matthias; Burkhardt, Dirk; Nazemi, Kawa; Kohlhammer, Jörn: Analytical Semantics Visualization for Discovering Latent Signals in Large Text Collections. In: Kerren, Andreas (Ed.); Seipel, Stefan (Ed.); SIGRAD: Proceedings of SIGRAD 2012: Interactive Visual Analysis of Data. Linköping: Linköping University Electronic Press, 2012, pp. 83–86. (Linköping Electronic Conference Proceedings 81).

24. Burkhardt, Dirk; Nazemi, Kawa: Dynamic Process Support Based on Users' Behavior. In: IEEE Education Society: ICL 2012: 15th International Conference on Interactive Collaborative Learning and 41st International Conference on Engineering Pedagogy. New York: IEEE, Inc., 2012, 6 p.

25. Burkhardt, Dirk; Stab, Christian; Steiger, Martin; Breyer, Matthias; Nazemi, Kawa: Interactive Exploration System: A User-Centered Interaction Approach in Semantics Visualizations. In: Kuijper, Arjan (Ed.); Sourin, Alexei (Ed.); Fellner, Dieter W. (General Chair); Fraunhofer-Institut für Graphische Datenverarbeitung (IGD): 2012 International Conference on Cyberworlds. Proceedings:

Cyberworlds 2012. Los Alamitos, Calif.: IEEE Computer Society Conference Publishing Services (CPS), 2012, pp. 261–267.

26. Stab, Christian; Nazemi, Kawa; Breyer, Matthias; Burkhardt, Dirk; Kohlhammer, Jörn: Semantics Visualization for Fostering Search Result Comprehension. In: Simperl, Elena (Ed.) et al.: The Semantic Web: Research and Applications. Proceedings: ESWC 2012. Berlin, Heidelberg, New York: Springer, 2012, pp. 633–646. (Lecture Notes in Computer Science (LNCS) 7295).

27. Kohlhammer, Jörn; Nazemi, Kawa; Ruppert, Tobias; Burkhardt, Dirk: Toward Visualization in Policy Modeling. In: IEEE Computer Graphics and Applications. 32 (2012), 5, pp. 84–89.

28. Burkhardt, Dirk; Ruppert, Tobias; Nazemi, Kawa: Towards Process-Oriented Information Visualization for Supporting Users. In: IEEE Education Society: ICL 2012: 15th International Conference on Interactive Collaborative Learning and 41st International Conference on Engineering Pedagogy. New York: IEEE, Inc., 2012, 8 p.

29. Nazemi, Kawa; Christ, Oliver: Verbalization in Search: Implication for the Need of Adaptive Visualizations. In: Ji, Yong Gu (Ed.): Advances in Affective and Pleasurable Design. Boca Raton: Taylor & Francis, CRC Press, 2012, pp. 8047–8056. (Advances in Human Factors and Ergonomics Series. Proceedings of the 4th AHFE Conference 14).

30. Breyer, Matthias; Birkenbusch , Jana; Burkhardt, Dirk; Schwarz, Christopher; Stab, Christian; Nazemi, Kawa; Christ, Oliver: Visualizations Encourage Uncertain Users to High Effectiveness. In: Ji, Yong Gu (Ed.): Advances in Affective and Pleasurable Design. Boca Raton: Taylor & Francis, CRC Press, 2012, pp. 8066–8074. (Advances in Human Factors and Ergonomics Series. Proceedings of the 4th AHFE Conference 14).

31. Burkhardt, Dirk; Breyer, Matthias; Stab, Christian; Nazemi, Kawa: Facilitate Access to E-Knowledge for Adult People in Rural Areas. In: Candel Torres, Ignacio (Ed.); Gómez Chova, Luis (Ed.); Martínez, A. Lopéz (Ed.); International Association of Technology, Education and Development (IATED): ICERI 2011. Proceedings: 4th International Conference of Education, Research and Innovation. Valencia: IATED, 2011, pp. 002050–002057.

32. Nazemi, Kawa; Breyer, Matthias; Stab, Christian; Burkhardt, Dirk; Fellner, Dieter W.: Intelligent Exploration System—an Approach for User-centered Exploratory Learning. In: Tzikopoulos, Argiris (Ed.); Zoakou, Anna (Ed.): RURALeNTER: Lifelong Learning in Rural and Remote Areas. Pallini: Ellinogermaniki Agogi, 2011, pp. 71–83. (reprint)

33. Nazemi, Kawa; Burkhardt, Dirk; Praetorius, Alexander; Breyer, Matthias; Kuijper, Arjan: Adapting User Interfaces by Analyzing Data Characteristics for Determining Adequate Visualizations. In: Kurosu, Masaaki (Ed.): Human Centered Design: Second International Conference: HCD 2011. Berlin, Heidelberg, New York: Springer, 2011, LNCS 6776, pp. 566–575. (Lecture Notes in Computer Science (LNCS) 6776).

34. Schäfer, Ralf; Becker, Tilman; Burghart, Catherina; Nazemi, Kawa; Ndjiki, Patrick; Riegel, Thomas: Basistechnologien für das Internet der Dienste. In:

Heuser, Lutz (Hrsg.); Wahlster, Wolfgang (Hrsg.); acatech–Deutsche Akademie der Technikwissenschaften e.V.: Internet der Dienste. Berlin, Heidelberg, New York: Springer, 2011, S. 19-40. (acatech DISKUTIERT).

35. Burkhardt, Dirk; Breyer, Matthias; Glaser, Christian; Nazemi, Kawa; Kuijper, Arjan: Classifying Interaction Methods to Support Intuitive Interaction Devices for Creating User-Centered-Systems. In: Jacko, Julie A. (Ed.): Human-Computer Interaction: Part I: Design and Development Approaches. Berlin, Heidelberg, New York: Springer, 2011, LNCS 6765, pp. 20–29. (Lecture Notes in Computer Science (LNCS) 6761).

36. Breyer, Matthias; Nazemi, Kawa; Stab, Christian; Burkhardt, Dirk; Kuijper, Arjan: A Comprehensive Reference Model for Personalized Recommender Systems. In: Jacko, Julie A. (Ed.): Human-Computer Interaction: Part I: Design and Development Approaches. Berlin, Heidelberg, New York: Springer, 2011, LNCS 6771, pp. 528–537. (Lecture Notes in Computer Science (LNCS) 6761).

37. Stab, Christian; Nazemi, Kawa; Breyer, Matthias; Burkhardt, Dirk; Kuijper, Arjan: Interacting with Semantics and Time. In: Jacko, Julie A. (Ed.): Human-Computer Interaction: Part IV: Users and Applications. Berlin, Heidelberg, New York: Springer, 2011, LNCS 6764, pp. 520–529. (Lecture Notes in Computer Science (LNCS) 6764).

38. Nazemi, Kawa; Breyer, Matthias; Forster, Jeanette; Burkhardt, Dirk; Kuijper, Arjan: Interacting with Semantics: A User-Centered Visualization Adaptation Based on Semantics Data. In: Jacko, Julie A. (Ed.): Human-Computer Interaction: Part I: Design and Development Approaches. Berlin, Heidelberg, New York: Springer, 2011, LNCS 6771, pp. 239–248. (Lecture Notes in Computer Science (LNCS) 6761).

39. Burkhardt, Dirk; Frossard, Frédérique; Barajas, Mario; Obermüller, Marion; Moises, Monika; Nazemi, Kawa: RURALeNTER: Capacity Building through ICT in Rural Areas. In: Candel Torres, Ignacio (Ed.) ; Gómez Chova, Luis (Ed.); Martínez, A. Lopéz (Ed.); International Association of Technology, Education and Development (IATED): ICERI 2011. Proceedings: 4th International Conference of Education, Research and Innovation. Valencia: IATED, 2011, pp. 001348–001354.

40. Nazemi, Kawa; Burkhardt, Dirk; Breyer, Matthias; Kuijper, Arjan: Modeling Users for Adaptive Semantics Visualizations. In: Jacko, Julie A. (Ed.): Human-Computer Interaction: Part II: Interaction Techniques and Environments. Berlin, Heidelberg, New York: Springer, 2011, LNCS 6766, pp. 88–97. (Lecture Notes in Computer Science (LNCS) 6762).

41. Nazemi, Kawa; Burkhardt, Dirk; Stab, Christian; Breyer, Matthias; Wichert, Reiner; Fellner, Dieter W.: Natural Gesture Interaction with Accelerometer-based Devices in Ambient Assisted Environments. In: Wichert, Reiner (Ed.); Eberhardt, Birgid (Ed.); Verband der Elektrotechnik Elektronik Informationstechnik (VDE): Ambient Assisted Living: 4. AAL-Kongress 2011. Berlin: Springer Science+Business Media, 2011, pp. 75–90. (Advanced Technologies and Societal Change 1).

42. Burkhardt, Dirk; Nazemi, Kawa; Stab, Christian; Breyer, Matthias; Wichert, Reiner; Fellner, Dieter W.: Natürliche Gesteninteraktion mit Beschleunigungssensorbasierten Eingabegeräten in unterstützenden Umgebungen. In: Verband der Elektrotechnik Elektronik Informationstechnik (VDE): Ambient Assisted Living: 4. Deutscher AAL-Kongress mit Ausstellung. Demographischer Wandel-Assistenzsysteme aus der Forschung in den Markt. Berlin u.a.: VDE-Verl., 2011, 10 S.; Paper 5.3.

43. Nazemi, Kawa; Stab, Christian; Kuijper, Arjan: A Reference Model for Adaptive Visualization Systems. In: Jacko, Julie A. (Ed.): Human-Computer Interaction: Part I: Design and Development Approaches. Berlin, Heidelberg, New York: Springer, 2011, LNCS 6761, pp. 480–489. (Lecture Notes in Computer Science (LNCS) 6761).

44. Burkhardt, Dirk; Breyer, Matthias; Nazemi, Kawa; Kuijper, Arjan: Search Intention Analysis for User-Centered Adaptive Visualizations. In: Jacko, Julie A. (Ed.): Human-Computer Interaction: Part I: Design and Development Approaches. Berlin, Heidelberg, New York: Springer, 2011, LNCS 6765, pp. 317–326. (Lecture Notes in Computer Science (LNCS) 6761).

45. Burkhardt, Dirk; Nazemi, Kawa; Breyer, Matthias; Stab, Christian; Kuijper, Arjan: SemaZoom: Semantics Exploration by Using a Layer-Based Focus and Context Metaphor. In: Kurosu, Masaaki (Ed.): Human Centered Design: Second International Conference: HCD 2011. Berlin, Heidelberg, New York: Springer, 2011, LNCS 6776, pp. 491–499. (Lecture Notes in Computer Science (LNCS) 6776).

46. Nazemi, Kawa; Breyer, Matthias; Kuijper, Arjan: User-Oriented Graph Visualization Taxonomy: A Data-Oriented Examination of Visual Features. In: Kurosu, Masaaki (Ed.): Human Centered Design: Second International Conference: HCD 2011. Berlin, Heidelberg, New York: Springer, 2011, LNCS 6776, pp. 576–585. (Lecture Notes in Computer Science (LNCS) 6776).

47. Burkhardt, Dirk; Stab, Christian; Nazemi, Kawa; Breyer, Matthias; Fellner, Dieter W.: Approaches for 3D-Visualizations and Knowledge Worlds for Exploratory Learning. In: GÓmez Chova, Luis (Ed.); Mart Belenguer, David (Ed.); Candel Torres, Ignacio (Ed.); International Association of Technology, Education and Development (IATED): International Conference on Education and New Learning Technologies. Proceedings: EDULEARN10. Valencia: IATED, 2010, pp. 006427–006437.

48. Nazemi, Kawa; Breyer, Matthias; Stab, Christian; Burkhardt, Dirk; Fellner, Dieter W.: Intelligent Exploration System—an Approach for User-Centered Exploratory Learning. In: Gómez Chova, Luis (Ed.); Martí Belenguer, David (Ed.); Candel Torres, Ignacio (Ed.); International Association of Technology, Education and Development (IATED): International Conference on Education and New Learning Technologies. Proceedings: EDULEARN10. Valencia: IATED, 2010, pp.006476–006484.

49. Nazemi, Kawa; Stab, Christian; Fellner, Dieter W.: Interaction Analysis for Adaptive User Interfaces. In: Huang, De-Shuang (Ed.) et al.: Advanced Intelligent Computing Theories and Applications: 6th International Conference on

Intelligent Computing. Berlin, Heidelberg, New York: Springer, 2010, pp. 362–371. (Lecture Notes in Computer Science (LNCS) 6215).

50. Nazemi, Kawa; Stab, Christian; Fellner, Dieter W.: Interaction Analysis: An Algorithm for Interaction Prediction and Activity Recognition in Adaptive Systems. In: Chen, Wen (Ed.); Li, Shaozi (Ed.): IEEE International Conference on Intelligent Computing and Intelligent Systems. Proceedings: ICIS 2010. New York: IEEE Press, 2010, pp. 607–612.

51. Burkhardt, Dirk; Hofmann, Cristian Erik; Nazemi, Kawa; Stab, Christian; Breyer, Matthias; Fellner, Dieter W.: Intuitive Semantic-Editing for Regarding Needs of Domain-Experts. In: Herrington, Jan (Ed.); Hunter, Bill (Ed.); Association for the Advancement of Computing in Education (AACE): ED-Media 2010: World Conference on Educational Multimedia, Hypermedia & Telecommunications. Chesapeake: AACE, 2010, pp. 860–869.

52. Nazemi, Kawa; Burkhardt, Dirk; Breyer, Matthias; Stab, Christian; Fellner, Dieter W.: Semantic Visualization Cockpit: Adaptable Composition of Semantics-Visualization Techniques for Knowledge-Exploration. In: Auer, Michael (Ed.); International Association of Online Engineering (IAOE): ICL 2010 Proceedings: International Conference Interactive Computer Aided Learning. Kassel: University Press, 2010, pp. 163–173.

53. Stab, Christian; Breyer, Matthias; Nazemi, Kawa; Burkhardt, Dirk; Hofmann, Cristian Erik; Fellner, Dieter W.: SemaSun: Visualization of Semantic Knowledge Based on an Improved Sunburst Visualization Metaphor. In: Herrington, Jan (Ed.); Hunter, Bill (Ed.); Association for the Advancement of Computing in Education (AACE): ED-Media 2010: World Conference on Educational Multimedia, Hypermedia & Telecommunications. Chesapeake: AACE, 2010, pp. 911–919.

54. Stab, Christian; Nazemi, Kawa; Fellner, Dieter W.: SemaTime–Timeline Visualization of Time-Dependent Relations and Semantics. In: Bebis, George (Ed.) et al.: Advances in Visual Computing. 6th International Symposium, ISVC 2010: Proceedings, Part III. Berlin, Heidelberg, New York: Springer, 2010, pp. 514–523. (Lecture Notes in Computer Science (LNCS) 6455).

55. Hofmann, Cristian Erik; Burkhardt, Dirk; Breyer, Matthias; Nazemi, Kawa; Stab, Christian; Fellner, Dieter W.: Towards a Workflow-Based Design of Multimedia Annotation Systems. In: Herrington, Jan (Ed.); Hunter, Bill (Ed.); Association for the Advancement of Computing in Education (AACE): ED-Media 2010: World Conference on Educational Multimedia, Hypermedia & Telecommunications. Chesapeake: AACE, 2010, pp. 1224–1233.

56. Nazemi, Kawa; Breyer, Matthias; Burkhardt, Dirk; Fellner, Dieter W.: Visualization Cockpit: Orchestration of Multiple Visualizations for Knowledge-Exploration. In: International Journal of Advanced Corporate Learning. 3 (2010), 4, pp. 26–34.

57. Nazemi, Kawa; Cukusic, Maja; Granic, Andrina: eLearning 2.0—Technologies for Knowledge Transfer in European-Wide Network of Schools. In: Journal of Software. 4 (2009), 2, pp. 108–115.

58. Nazemi, Kawa; Ullmann, Thomas Daniel; Hornung, Christoph: Engineering User Centered Interaction Systems for Semantic Visualizations. In: HCI International 2009. Proceedings and Posters: With 10 further Associated Conferences. Berlin, Heidelberg, New York: Springer, 2009, LNCS 5614, pp. 126–134. (Lecture Notes in Computer Science (LNCS)).

59. Nazemi, Kawa; Breyer, Matthias; Hornung, Christoph: SeMap: A Concept for the Visualization of Semantics as Maps. In: HCI International 2009. Proceedings and Posters: With 10 further Associated Conferences. Berlin, Heidelberg, New York: Springer, 2009, LNCS 5616, pp. 83–91. (Lecture Notes in Computer Science (LNCS)).

60. Burkhardt, Dirk; Nazemi, Kawa; Bhatti, Nadeem; Hornung, Christoph: Technology Support for Analyzing User Interactions to Create User-Centered Interactions. In: HCI International 2009. Proceedings and Posters: With 10 further Associated Conferences. Berlin, Heidelberg, New York: Springer, 2009, LNCS 5614, pp. 3–12. (Lecture Notes in Computer Science (LNCS)).

61. Hornung, Christoph; Granic, Andrina; Cukusic, Maja; Nazemi, Kawa: eKnowledge Repositories in eLearning 2.0: UNITE—A European-Wide Network of Schools. In: Li, Frederick (Ed.); Zhao, Jianmin (Ed.); Shih, Timothy K. (Ed.); Lau, Rynson (Ed.); Li, Qing (Ed.); McLeod, Dennis (Ed.): Advances in Web Based Learning—ICWL 2008. Berlin, Heidelberg, New York: Springer, 2008, pp. 99–110. (Lecture Notes in Computer Science (LNCS) 5145).

62. Nazemi, Kawa; Hornung, Christoph: Intuitive Authoring on Web: A User-Centered Software Design Approach. In: Luca, Joseph (Ed.); Weippl, Edgar R. (Ed.); Association for the Advancement of Computing in Education (AACE): Proceedings of ED-Media 2008: World Conference on Educational Multimedia, Hypermedia & Telecommunications. Chesapeake, 2008, pp. 1440–1448.

63. Nazemi, Kawa; Bhatti, Nadeem; Godehardt, Eicke; Hornung, Christoph: Adaptive Tutoring in Virtual Learning Worlds. In: Montgomerie, Craig (Ed.); Seale, Jane (Ed.); Association for the Advancement of Computing in Education (AACE): Proceedings of ED-Media 2007: World Conference on Educational Multimedia, Hypermedia & Telecommunications. Chesapeake, 2007, pp. 2951–2959.

Edited Book

Sonntagbauer, Peter; Nazemi, Kawa; Sonntagbauer, Susanne; Prister, Giorgio, Burkhardt, Dirk (eds): Handbook of Research on Advanced ICT Integration for Governance and Policy Modeling. Business Science Reference: IGI Global. ISBN: 978-1-4666-6236-0, DOI: 10.4018/978-1-4666-6236-0. 2014. Hershey PA, USA. 542 pages.

Appendix B
Supervising Activities

The following list summarizes the student bachelor, diploma and master thesis supervised by the author. The results of these works were partially used in the thesis.

B.1 Diploma and Master Theses

1. Retz, Wilhelm, Nazemi, Kawa (Betreuer): Integration heterogener Methoden der Empfehlungssysteme zur Visualisierungsadaption. Darmstadt, 2013. (Darmstadt, TU, Diplomarbeit, 2013).
2. Borger, Renate; Nazemi, Kawa (Betreuer): Benutzermodellierung für adaptive Informationsvisualisierungen. Braunschweig, 2013. (Braunschweig, TU, Master Thesis, 2013).
3. Breyer, Matthias; Nazemi, Kawa (Betreuer): Benutzerzentrierte Visualisierung von Informationsempfehlungen basierend auf Recommender Systemen. Darmstadt, 2010. (Darmstadt, TU, Master Thesis, 2010).
4. Stab, Christian; Nazemi, Kawa (Betreuer): Interaktionsanalyse for adaptive Benutzerschnittstellen. Darmstadt, 2009. (Darmstadt, TU, Diplomarbeit, 2009).
5. Burkhardt, Dirk; Nazemi, Kawa (Betreuer): Gestenerkennung zur Unterstützung intuitiver Interaktion an computerbasierten Systemen. Görlitz, 2008. (Zittau/ Görlitz, FH, Diplomarbeit, 2008).

B.2 Bachelor Theses

1. Forster, Jeanette; Nazemi, Kawa (Betreuer): Semantische Visualisierung von Suchergebnissen auf der Basis von Linked Open Data. Darmstadt, 2011. (Darmstadt, TU, Bachelor Thesis, 2011).

© Springer International Publishing Switzerland 2016
K. Nazemi, *Adaptive Semantics Visualization*, Studies in Computational
Intelligence 646, DOI 10.1007/978-3-319-30816-6

2. Wibowo, Ferry Darmawan; Nazemi, Kawa (Betreuer): Visualization of Time-Dependent Semantics Data. Darmstadt, 2010. (Darmstadt, TU, Bachelor Thesis, 2010).

Appendix C
Questionnaires of the Evaluation

© Springer International Publishing Switzerland 2016

K. Nazemi, *Adaptive Semantics Visualization*, Studies in Computational
Intelligence 646, DOI 10.1007/978-3-319-30816-6

Table C.1 Offline part I: sociodemographic questionnaire (German)

Fragebogen

vielen Dank für Ihre Teilnahme an dieser wissenschaftlichen Untersuchung.
Bitte nehmen Sie sich nun kurz Zeit für die folgenden Angaben:

VP-Nummer:

Allgemeine Angaben

Geschlecht: ☐ weiblich   ☐ männlich

Alter:			
☐ unter 20	☐ 20−24	☐ 25−29	☐ 30−34
☐ 35−39	☐ 40−44	☐ 45−49	☐ über 50

Bildungsabschluss:
☐ Allgemeine oder fachgebundene Hochschulreife
☐ Fachhochschul − oder Universitätsabschluss
☐ Anderer Schulabschluss

Beruf	
(Markieren Sie bitte einen oder mehrere Berufe)	
☐ Bau, Architektur, Vermessung	☐ Dienstleistung
☐ Elektro	☐ Gesellschafts−, Geisteswissenschaften
☐ Gesundheit	☐ IT, DV, Computer
☐ Kunst, Kultur, Gestaltung	☐ Landwirtschaft, Natur, Umwelt
☐ Medien	☐ Metall, Maschinenbau
☐ Naturwissenschaften	☐ Produkt, Fertigung
☐ Soziales, Pädagogik	☐ Technik, Technologiefelder
☐ Verkehr, Logistik	☐ Wirtschaft, Verwaltung
☐ Schüler, Student	☐ Forschung
Sonstiges:	

Fragen zur Sehfähigkeit

Sehfähigkeit
(Markieren Sie bitte einen oder mehrere Felder)
Sehfähigkeit: ☐ normal ☐ nach normal korrigiert (Dioptrin) ___Dioptrin
Farbsehen: ☐ normal ☐ Rot−Grün−Schwäche
Sonstiges:

Table C.2 Offline part II: agreement with photographic protocol (German)

Einverständniserklärung zur Fotoaufnahmen und Durchführung der Studie

- Ich wurde über den Ablauf und Zweck der Studie informiert und meine Fragen wurden zufriedenstellend beantwortet.
- Ich nehme freiwillig an dieser Studie teil und stimme zu, dass während der Durchführung des Interviews Foto-Aufnahmen gemacht werden, die veröffentlicht werden dürfen.
- Mir ist bewusst, dass meine Teilnahme an dieser Studie vertraulich ist. Alle gesammelten persönlichen Daten werden nicht ohne mein schriftliches Einverständnis an Dritte weitergegeben. Die gesammelten Informationen dienen ausschließlich Lehr- und Forschungszwecken.
- Mir ist bewusst, dass ich die Teilnahme an dieser Studie jederzeit abbrechen kann.

Name, Vorname

Datum, Ort   **Unterschrift**

Table C.3 Online sociodemographic questionnaire

Fragen zur Person

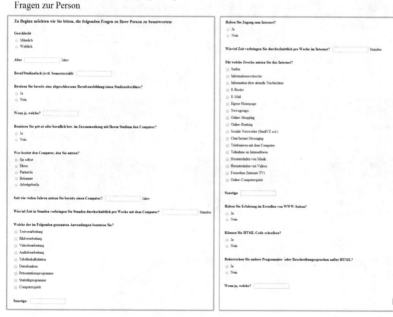

Table C.4 The INTUI questionnaire (from [1])

Please recall the use of the product and **describe your experience using the following pairs of expressions**. The pairs represent extreme opposites, with possible graduations between them.

Perhaps some of the expressions are not quite suitable to the product. Nevertheless, please checkmark one box in each row, indicating which term you deem applicable. Please consider that there are no "correct" or "incorrect" answers – only your own personal opinion counts!

	1	2	3	4	5	6	7			
While using the product...										
...I acted deliberately	☐	☐	☐	☐	☐	☐	☐	...I acted on impulse		G_01
...it took me a lot of effort to reach my goal	☐	☐	☐	☐	☐	☐	☐	...I reached my goal effortlessly		E_01
...I performed unconsciously, without reflecting on the individual steps	☐	☐	☐	☐	☐	☐	☐	...I consciously performed one step after another	P	G_02
...I was guided by reason	☐	☐	☐	☐	☐	☐	☐	...I was guided by feelings		G_03
...I felt lost	☐	☐	☐	☐	☐	☐	☐	...I easily knew what to do		E_02
...I acted without thinking	☐	☐	☐	☐	☐	☐	☐	...I was able to explain each individual step	P	G_04
Using the product...										
...required my close attention	☐	☐	☐	☐	☐	☐	☐	...ran smoothly		E_03
...was inspiring	☐	☐	☐	☐	☐	☐	☐	...was insignificant	P	X_01
...was easy	☐	☐	☐	☐	☐	☐	☐	...was difficult	P	E_04
...was nothing special	☐	☐	☐	☐	☐	☐	☐	...was a magical experience		X_02
...was very intuitive	☐	☐	☐	☐	☐	☐	☐	...wasn't intuitive at all	P	INT_01
...was trivial	☐	☐	☐	☐	☐	☐	☐	...carried me away		X_03
...came naturally	☐	☐	☐	☐	☐	☐	☐	...was hard	P	E_05
...was fascinating	☐	☐	☐	☐	☐	☐	☐	...was dull	P	X_04
In retrospect...										
...it is hard for me to describe the individual operating steps	☐	☐	☐	☐	☐	☐	☐	...I have no problem describing the individual operating steps		V_01
...I can easily recall the operating steps	☐	☐	☐	☐	☐	☐	☐	...it is difficult for me to remember how the product is operated	P	V_02
...I'm not able to express in which way I used the product	☐	☐	☐	☐	☐	☐	☐	...I can say exactly in which way I used the product		V_03

© INTUI (English), http://intuitiveinteraction.net/, Ullrich, D., Diefenbach, S. (2010).

Table C.5 The INTUI questionnaire in German language (from [1])

Bitte vergegenwärtigen Sie sich jetzt noch ein Mal die Nutzung des Produkts und **beschreiben Sie Ihr Erleben der Nutzung mit Hilfe der folgenden Aussagenpaare.** Die Paare stellen jeweils extreme Gegensätze dar, zwischen denen eine Abstufung möglich ist.

Vielleicht passen einige Aussagen nicht so gut, kreuzen Sie aber trotzdem bitte immer an, welcher Begriff Ihrer Meinung nach eher zutrifft. Denken Sie daran, dass es keine "richtigen" oder "falschen" Antworten gibt - nur Ihre persönliche Meinung zählt!

	1	2	3	4	5	6	7		
Bei der Nutzung (des Produkts)...									
...handelte ich überlegt	☐	☐	☐	☐	☐	☐	☐	...handelte ich spontan	G_01
...erreichte ich mein Ziel nur mit Anstrengung	☐	☐	☐	☐	☐	☐	☐	...erreichte ich mein Ziel mit Leichtigkeit	M_01
...handelte ich unbewusst, ohne lange über die einzelnen Schritte nachzudenken	☐	☐	☐	☐	☐	☐	☐	...führte ich bewusst einen Schritt nach dem anderen aus	P G_02
...ließ ich mich von meinem Verstand leiten	☐	☐	☐	☐	☐	☐	☐	...ließ ich mich von meinem Gefühl leiten	G_03
...war ich orientierungslos	☐	☐	☐	☐	☐	☐	☐	...konnte ich mich gut zurechtfinden	M_02
...handelte ich ohne dabei nachzudenken	☐	☐	☐	☐	☐	☐	☐	...konnte ich jeden Schritt genau begründen	P G_04
Die Nutzung (des Produkts)...									
...erforderte viel Aufmerksamkeit	☐	☐	☐	☐	☐	☐	☐	...ging wie von selbst	M_03
...war begeisternd	☐	☐	☐	☐	☐	☐	☐	...war unbedeutend	P X_01
...war einfach	☐	☐	☐	☐	☐	☐	☐	...war schwierig	P M_04
...war nichts Besonderes	☐	☐	☐	☐	☐	☐	☐	...war ein magisches Erlebnis	X_02
...war sehr intuitiv	☐	☐	☐	☐	☐	☐	☐	...war gar nicht intuitiv	P INT_01
...war belanglos	☐	☐	☐	☐	☐	☐	☐	...war mitreißend	X_03
...fiel mir leicht	☐	☐	☐	☐	☐	☐	☐	...fiel mir schwer	P M_05
...war faszinierend	☐	☐	☐	☐	☐	☐	☐	...war trist	P X_04
Im Nachhinein...									
...fällt es mir schwer, die einzelnen Bedienschritte zu beschreiben	☐	☐	☐	☐	☐	☐	☐	...ist es für mich kein Problem, die einzelnen Bedienschritte zu beschreiben	V_01
...kann ich mich gut an die Bedienung erinnern	☐	☐	☐	☐	☐	☐	☐	...fällt es mir schwer, mich zu erinnern, wie das Produkt bedient wird	P V_02
...kann ich nicht sagen, auf welche Art und Weise ich das Produkt bedient habe	☐	☐	☐	☐	☐	☐	☐	...kann ich genau sagen, auf welche Art und Weise ich das Produkt bedient habe	V_03

Table C.6 The AttrakDiff questionnaire (adapted from [2])
Die Visualisierung war...

menschlich	○	○	○	○	○	○	○	technisch
isolierend	○	○	○	○	○	○	○	verbindend
angenehm	○	○	○	○	○	○	○	unangenehm
originell	○	○	○	○	○	○	○	konventionell
einfach	○	○	○	○	○	○	○	kompliziert
fachmännisch	○	○	○	○	○	○	○	laienhaft
hässlich	○	○	○	○	○	○	○	schön
praktisch	○	○	○	○	○	○	○	unpraktisch
sympathisch	○	○	○	○	○	○	○	unsympathisch
umständlich	○	○	○	○	○	○	○	direkt
stilvoll	○	○	○	○	○	○	○	stillos
voraussagbar	○	○	○	○	○	○	○	unberechenbar
minderwertig	○	○	○	○	○	○	○	wertvoll
ausgrenzend	○	○	○	○	○	○	○	einbeziehend
bringt mich den Leuten näher	○	○	○	○	○	○	○	trennt mich von Leuten
nicht vorzeigbar	○	○	○	○	○	○	○	vorzeigbar
zurückweisend	○	○	○	○	○	○	○	einladend
phantasielos	○	○	○	○	○	○	○	kreativ
gut	○	○	○	○	○	○	○	schlecht
verwirrend	○	○	○	○	○	○	○	übersichtlich
abstoßend	○	○	○	○	○	○	○	anziehend
mutig	○	○	○	○	○	○	○	vorsichtig
innovativ	○	○	○	○	○	○	○	konservativ
lahm	○	○	○	○	○	○	○	fesselnd
harmlos	○	○	○	○	○	○	○	herausfordernd
motivierend	○	○	○	○	○	○	○	entmutigend
neuartig	○	○	○	○	○	○	○	herkömmlich
widerspenstig	○	○	○	○	○	○	○	handhabbar

Appendix D
Tasks of the Evaluation

Table D.1 Evaluation Tasks (continued on next page)

No	Task	Answers	Type
1	How many papers were published by "Fellner" in 2008 and 2009?	7, 5, **6**, 4	**Simple**
2	How many papers were published by "Helwig Hauser" in 2010 and 2011?	5, **8**, 7, 2	**Simple**
3	How many papers were published by "Weiskopf" in 2010 and 2011?	**5**, 3, 7, 2	**Simple**
4	How many papers were published by "Sadlo" in 2011 and 2012?	7, 4, 3, **5**	**Simple**
5	In which category had "Keim" the most publications?	Picture/image generation, applications, graphics systems, database applications	**Exploratory**
6	In which category had "Weiskopf" the most publications?	Picture/image generation, applications, graphics systems, hardware architecture	**Exploratory**
7	In which category had "Goesele" the most publications?	Applications, graphics systems, computational geometry, hardware architecture	**Exploratory**
8	In which category had "Schreck" the most publications?	Hardware architecture, applications, information search and retrieval, graphics systems	**Exploratory**
9	In which year had "Sadlo" the first publication?	**2005**, 2006, 2010, 2013	**Simple**
10	In which year had "Dachsbacher" the first publication?	2006, **2003**, 2007, 2011	**Simple**

(continued)

© Springer International Publishing Switzerland 2016
K. Nazemi, *Adaptive Semantics Visualization*, Studies in Computational
Intelligence 646, DOI 10.1007/978-3-319-30816-6

Table D.1 (continued)

No	Task	Answers	Type
11	In which year had "Keim" the first publication?	2008, 2000, 2009, **2004**	**Simple**
12	In which year had "Weiskopf" the first publication?	2006, 2007, **2000**, 2010	**Simple**
13	Which author has the most publications in the area of "GPU Visualization"?	Guthe, **Ertl**, Weiskopf, Strengert	**Exploratory**
14	Which author has the most publications in the area of "Visual Analytics"?	**Ebert**, Miksch, Keim, van Wijk	**Exploratory**
15	Which author has the most publications in the area of "Visual Rendering"?	Walker, Keim, **Chalmers**, Kohlhammer	**Exploratory**
16	Which author has the most publications in the area of "Interactive Visual Analysis"?	Keim, **Hauser**, Ertl, Kehrer	**Exploratory**
17	How many papers about "Semantic Visualization" were published?	**3**, 4, 7, 1	**Simple**
18	How many papers about "Semantics" were published?	3, 7, 1, **4**	**Simple**
19	How many papers about "flow rendering" were published?	7, 4, **3**, 1	**Simple**
20	How many papers about "data cube" were published?	3, 4, **2**, 1	**Simple**
21	Which of the named topics gained recently more interest in research?	Visual Analytics, **Information Visualization**, scientific visualization, realtime visualization	**Exploratory**
22	Which of the named topics gained recently more interest in research?	Geometry rendering, 3D Rendering, GPU Rendering, **Detail Rendering**	**Exploratory**
23	Which of the named topics gained recently more interest in research?	Multivariate, **Perceptual**, information data, surface Meshes	**Exploratory**
24	Which of the named topics gained recently more interest in research?	**Interactive design**, user design, graphics design, rendering design	**Exploratory**

Table D.2 Evaluation Tasks (continued from previous page)

No	Task	Answers	Type
25	With whom had "Kohlhammer" in 2013 no publications?	Davey, Steiger, May, **Ertl**	simple
26	With whom had "Kuijper" in 2013 no publications?	Schmitt, Knuth, **Kohlhammer**, Bender	**Simple**
27	With whom had "Aigner" in 2012 no publications?	Miksch, **Bertone**, Hoffmann, Rind	**Simple**
28	With whom had "Miksch" in 2013 no publications?	**Kainz**, Lammarsch, Bertone, Rind	**Simple**
29	With how many different co-authors had "Schwenk" publications?	**6**, 4, 7, 5	**Exploratory**
30	With how many different co-authors had "Bremm" publications?	6, 4, **7**, 3	**Exploratory**
31	With how many different co-authors had "Neubauer" publications?	6, **8**, 7, 12	**Exploratory**
32	With how many different co-authors had "Chris North" publications?	6, 4, 5, **7**	**Exploratory**
33	In which year had "Ritschel" the most publications?	2004, **2012**, 2009, 2010	**Simple**
34	In which year had "Theisel" the most publications?	**2012**, 2011, 2009, 2010	**Simple**
35	In which year had "Michael Wand" the most publications?	2004, **2013**, 2009, 2010	**Simple**
36	In which year had "Hans-Christian Hege" the most publications?	2008, 2002, **2012**, 2010	**Simple**
37	Which of these authors had after 2010 fewer publications than before 2010?	**Fellner**, Miksch, Kohlhammer, Kuijper	**Exploratory**
38	Which of these authors had after 2010 fewer publications than before 2010?	Schreck, **Van Wijk**, Weiskopf, Elmqist	**Exploratory**
39	Which of these authors had after 2008 fewer publications than before 2008?	Preim, Daniel Keim, **Thomas Ertl**, Mathias Neugebauer	**Exploratory**
40	Which of these authors had after 2009 fewer publications than before 2009?	**Deussen**, Goesele, Michael Wand, Niloy Mitra	**Exploratory**

Appendix E
Complementary and Detailed Results of the Evaluation

Gut Feeling

Table E.1 illustrates the results of the pairwise *t*-Tests for the INTUI sub-scale Gut Feeling. Condition D was given the highest ratings. The differences between the full adaptive visualization and condition A and B were reached significance. However, the difference to condition C was not significant. Further, condition B and condition C did not differ significantly. All remaining effects were significant.

Magical Experience

The results of the pairwise *t*-Tests for the INTUI sub-scale Magical Experience are given in Table E.2, thereby the sub-scale were significant between all conditions. The full adaptive semantics visualizations gained the highest ratings.

Verbalizability

Pairwise *t*-Tests were conducted subsequent to the significant results of the analysis of variance for the INTUI sub-scale Verbalizability (see Table E.3). The results show, that only one of the effects reached significance. Verbalizability was significantly higher in Condition A than in Condition B.

Hedonic Quality–Identity

The pairwise *t*-Tests of the AttrakDiff sub-scale Hedonic Quality–Identity (see Table E.4) illustrates significant effects between all conditions, whereas the full adaptive visualization outperforms all other conditions.

© Springer International Publishing Switzerland 2016
K. Nazemi, *Adaptive Semantics Visualization*, Studies in Computational
Intelligence 646, DOI 10.1007/978-3-319-30816-6

Table E.1 Pairwise *t*-Tests for the INTUI-sub-scale *Gut Feeling*

	Paired differences			t	df	Sig. (2-tailed)
	Mean	Std. deviation	Std. error mean			
C vs. D	−0.25333	1.38146	0.19537	−1.297	49	.201
B vs. D	−0.52083	1.37900	0.19123	−2.724	51	.009
A vs. D	−1.18396	1.75726	0.24138	−4.905	52	.000
B vs. C	−0.23333	1.15261	0.16300	−1.431	49	.159
A vs. C	−0.94167	1.59215	0.22516	−4.182	49	.000
A vs. B	−0.68590	1.49414	0.20720	−3.310	51	.002

Table E.2 Pairwise *t*-Tests for the INTUI-sub-scale *Magical Experience*

	Paired differences			t	df	Sig. (2-tailed)
	Mean	Std. deviation	Std. error mean			
C vs. D	−0.48000	1.16296	0.16447	−2.919	49	.005
B vs. D	−0.91667	1.49263	0.20699	−4.429	51	.000
A vs. D	−2.55503	1.40616	0.19315	−13.228	52	.000
B vs. C	−0.46833	0.89197	0.12614	−3.713	49	.001
A vs. C	−2.13833	1.36034	0.19238	−11.115	49	.000
A vs. B	−1.66346	1.37169	0.19022	−8.745	51	.000

Table E.3 Pairwise *t*-Tests for the INTUI-sub-scale *Verbalizability*

	Paired differences			t	df	Sig. (2-tailed)
	Mean	Std. deviation	Std. error mean			
C vs. D	−0.03333	0.85516	0.12094	−0.276	49	.784
B vs. D	−0.20261	0.81125	0.11360	−1.784	50	.081
A vs. D	0.21384	1.05428	0.14482	1.477	52	.146
B vs. C	−0.18367	0.77298	0.11043	−1.663	48	.103
A vs. C	0.23333	1.14929	0.16253	1.436	49	.157
A vs. B	0.40523	0.98953	0.13856	2.925	50	.005

Hedonic Quality–Stimulation

The results of the pairwise *t*-Tests are given in Table E.5. Significant effects could be observed between all visualizations (conditions B, C, and D) and the textual baseline (condition A). No significant results are given between the visualizations.

Table E.4 Pairwise *t*-Tests for the AttrakDiff-sub-scale *Hedonic Quality–Identity*

	Paired differences			t	df	Sig. (2-tailed)
	Mean	Std. deviation	Std. error mean			
C vs. D	−0.35762	0.67807	0.09589	−3.729	49	.000
B vs. D	−0.56181	0.79752	0.11060	−5.080	51	.000
A vs. D	−1.66914	1.16656	0.16024	−10.417	52	.000
B vs. C	−0.20381	0.44916	0.06352	−3.209	49	.002
A vs. C	−1.30595	0.98811	0.13974	−9.346	49	.000
A vs. B	−1.08173	0.98435	0.13650	−7.925	51	.000

Table E.5 Pairwise *t*-Tests for the AttrakDiff-sub-scale *Hedonic Quality–Stimulation*

	Paired differences			t	df	Sig. (2-tailed)
	Mean	Std. deviation	Std. error mean			
C vs. D	−0.17095	1.04456	0.14772	−1.157	49	.253
B vs. D	−0.24634	1.22368	0.16969	−1.452	51	.153
A vs. D	−2.85337	1.35950	0.18674	−15.280	52	.000
B vs. C	−0.07095	0.81912	0.11584	−0.612	49	.543
A vs. C	−2.76505	1.24455	0.17601	−15.710	49	.000
A vs. B	−2.65916	1.42608	0.19776	−13.446	51	.000

Table E.6 Pairwise t-Tests for the AttrakDiff-sub-scale *Pragmatic Quality*

	Paired differences			t	df	Sig. (2-tailed)
	Mean	Std. deviation	Std. error mean			
C vs. D	−0.33095	0.80408	0.11371	−2.910	49	.005
B vs. D	−0.82418	0.73587	0.10205	−8.076	51	.000
A vs. D	0.04942	1.16632	0.16021	0.308	52	.759
B vs. C	−0.51762	0.67540	0.09552	−5.419	49	.000
A vs. C	0.37429	1.08717	0.15375	2.434	49	.019
A vs. B	0.89377	1.10137	0.15273	5.852	51	.000

AttrakDiff Sub-scale Pragmatic Quality

The results pairwise *t*-Tests of the AttrakDiff sub-scale *Pragmatic Quality* illustrates no significant differences in the ratings between the full-adaptive visualization and the textual baseline. However, the full-adaptive visualization outperformed all other visual representations, in contrast to the textual baseline no significant differences could be found. Table E.6 illustrates the results of the pairwise *t*-Tests of the sub-scale *Pragmatic Quality* of the AttrakDiff questionnaire.

References

1. D. Ullrich, S. Diefenbach, in Mensch & Computer 2010: 10. Fachübergreifende Konferenz für interaktive und kooperative Medien. Interaktive Kulturen (Oldenbourg Wissenschaftsverlag, Stuttgart, 2010), p. 251
2. M. Hassenzahl, M. Burmester, F. Koller, in Mensch & Computer 2003 (Springer, Heidelberg, 2003), pp. 187–196

Printed in the United States
By Bookmasters